Reactionary Mathematics

Reactionary Mathematics

A Genealogy of Purity

MASSIMO MAZZOTTI

The University of Chicago Press
Chicago and London

The University of Chicago Press, Chicago 60637
The University of Chicago Press, Ltd., London
© 2023 by The University of Chicago

Published 2023
Printed in the United States of America

32 31 30 29 28 27 26 25 24 23   1 2 3 4 5

ISBN-13: 978-0-226-82672-1 (cloth)
ISBN-13: 978-0-226-82674-5 (paper)
ISBN-13: 978-0-226-82673-8 (e-book)
DOI: https://doi.org/10.7208/chicago/9780226826738.001.0001

Library of Congress Cataloging-in-Publication Data

Names: Mazzotti, Massimo, author.
Title: Reactionary mathematics : a genealogy of purity / Massimo Mazzotti.
Description: Chicago : The University of Chicago Press, 2023. |
    Includes bibliographical references and index.
Identifiers: LCCN 2022052916 | ISBN 9780226826721 (cloth) |
    ISBN 9780226826745 (paperback) | ISBN 9780226826738 (ebook)
Subjects: LCSH: Mathematics—Study and teaching—Italy—Naples. |
    Mathematics—Political aspects.
Classification: LCC QA27.18 M39 2023 | DDC 510.945/7310903—dc23/eng20230113
LC record available at https://lccn.loc.gov/2022052916

♾ This paper meets the requirements of ANSI/NISO Z39.48-1992
(Permanence of Paper).

Contents

INTRODUCTION

Mathematics as Social Order

The steps which are not brought into question are logical inferences.[1]

In 1806, following years of war, a revolution, and a short-lived Jacobin Republic, the Kingdom of Naples was occupied by a French army and integrated in
the imperial system. The French and their local supporters began to implement an ambitious project of modernization that aimed to transform that
ancien régime society into a centralized administrative monarchy and foster
a liberal economy. The legitimacy of the new sociopolitical order, however,
remained weak and constantly challenged. French reform plans soon ran into
muddy roads, endemic brigandage, popular insurgencies, and the resistance
of powerful local elites. But these were not the only obstacles modernization
faced. One need only open a Neapolitan textbook of the time to realize that
the French were battling in a land where their mathematics was wrong.[2]

This book offers a reconstruction and interpretation of this little-known
episode of mathematical resistance.[3] My immediate goal is to make sense of a
dramatic local shift in mathematical practice and the emergence of priorities
and methods that were alternative to those of then-hegemonic French mathematics. In particular, I'm interested in the success of the belief that "purity"
is an essential trait of veritable mathematical knowledge, a trait that emphasizes the unbridgeable distance between mathematical formalism and the
messiness of human life. The arguments that I develop to make sense of this
shift, however, aim to contribute to broader conversations. This book offers
a contribution to the literature on the situatedness of logico-mathematical
knowledge and its intrinsic historicity; in fact, it argues for understanding
mathematics as culture, and it deploys a notion of mathematical culture that
can be used to this purpose. In turn, as we shall see, a mathematical culture
presupposes and reinforces a specific image of reason, of order, and therefore of social and political life. An image of reason and the mathematical

techniques that best express it open certain social possibilities while clos-
ing others; or, to put it differently, they are functional to a specific distribution of power and agency. Through the prism of this controversy, I recount
the dawn of mathematical modernity as a process of purification, a process
whose outcome was the distinctively modernist perception that mathematical knowledge—and the technologies it legitimates—are neutral tools that
can be used unthinkingly in the manipulation of both natural and social realities. This is a book about the genealogy of this image of purity-neutrality
and its political meaning.[4]

*

By saying that mathematical knowledge is historical, I do not refer to the
fact that its concepts and practices emerged at specific points in time and
space. Rather, I argue that mathematical knowledge and its credibility should
be studied as proper historical objects. Theorems and problems, as well as
the practices of proving and solving them, must be situated historically to be
properly understood. The historicity I refer to is not chronology; it permeates
the very contents of mathematical knowledge and the way in which specific
collectives understand and deploy that knowledge. The relation of history to
mathematics is not that of a colorful background to a core of timeless truths:
history is a constitutive dimension of mathematical knowledge. Therefore,
mathematical objects and practices should be understood as essentially situated in space and time, shaped by the circumstances of their creation and use.
To argue for the historicity of mathematics, in other words, is to argue for
mathematics as culture, and indeed for the existence of a plurality of *mathematical cultures*.[5]

   I ground my argument in a detailed reconstruction and interpretation of
mathematical life in Naples around 1800. Mathematical practice was being
transformed across Europe in this period, when key features that we attri-
bute to modern mathematics, such as the emergence of pure mathematics
as the foundational core of the discipline, originated. I look at this process
of mathematical modernization from a *marginal* point of view.[6] Naples was
the capital of a kingdom known as a destination of the grand tour and for
natural marvels such as Vesuvius, but it hardly figures in the narratives of
eighteenth-century mathematics. While Paris, the uncontested mathematical powerhouse in Europe, remains constantly visible in the background of
my story, I focus my attention southward, though not because I consider the
Neapolitan case typical; nor am I primarily interested in telling stories that
have been long forgotten, although I must admit that I have a certain gusto
for the unfamiliar. I focus on Naples because there, between the 1780s and the

1830s, the enlightened regime of European mathematics was challenged radically and authoritatively.[7] Key Neapolitan scientific institutions, including the leading university of the kingdom, rejected as untrustworthy what they called the "very modern mathematics" incarnated by exemplary French works. In its place, an alternative mathematical culture was consolidated, legitimated, and put to work, and new mathematical futures were imagined.

One of the main challenges for the historical study of mathematics is the difficulty of breaking through the pervasive image of mathematical knowledge as produced in an essentially cumulative fashion and characterized by a distinctive permanence and internal unity of meaning.[8] In this image, the universal nature of mathematical knowledge is taken as given rather than as a fascinating problem. Historical research has made it clear that this conception is a historiographical illusion, the product of a posteriori reconstructions.[9] Granted, it's notoriously difficult to individuate ruptures and meaning shifts in this scientific domain—difficult, but not impossible. Detailed microhistorical analysis of past mathematical practices, for example, can reveal discontinuities and shifts.[10] Such analysis emphasizes how certain techniques and concepts were shaped by assumptions and priorities distant from our current concerns. It's a move inspired by a genealogical method that aims to reveal the historical novelty of concepts and practices often taken for granted and identify transitions between different social configurations and knowledge regimes.[11] Microhistorical analysis intends to find cracks in what might otherwise look like solid, all-encompassing, and universally accepted regimes of knowledge. It is arguably on the edges of such regimes, in settings where training is less intensive and standard practices less routinized and pervasive, that one can expect to find discontinuities, meaning shifts, resistance, and other traces of alternative mathematical cultures. Hence the methodological choice to focus on the margin—on sites where a regime of knowledge is more likely *to lose its grip*.

The present study focuses on a controversy between two groups of mathematicians active on the margin of late-eighteenth-century mathematical Europe. As shown by exemplary work in science studies, focusing on controversies is another method to maximize the possibility of encountering discontinuities. This move can help bring to the fore the flexibility of mathematical concepts, conflicts between alternative mathematical cultures, and processes of consolidation or dissolution of mathematical regimes. In what follows, I reconstruct the mathematical cultures of these two groups, emphasizing their situatedness and the resources deployed to make their mathematical knowledge credible. As noted by Andy Warwick, mathematics does not travel more easily than, say, experimental knowledge or technology; its reproduction and deployment in different settings is no less demanding.[12] But the efforts required to construct sameness and universality in mathematics have remained largely undetected. By looking at European mathematics from the margin, it might be easier to see this universality as a historical object that can be collectively constructed, sustained, and dismantled.

Mathematics Has a History

Reconstructing this episode of mathematical resistance means to focus on mathematical cultures, and on their institutions, resources, practices, skills, and broader orientations. The mathematical cultures under scrutiny manifested themselves primarily through a controversy that reoriented mathematics in Naples between the 1790s and the 1830s. Reacting against what they perceived as the invasion of French mathematics, leading local practitioners argued that mathematical procedures urgently required a much higher degree of rigor and precision, and that "pure mathematics" needed to provide solid foundations to the rest of the discipline.

Other mathematical communities across Europe shared similar concerns.[13] Many began to perceive logical gaps that required urgent attention. Significant parts of mathematics, some believed, needed to be reconstructed on new, more secure foundations. How should we understand the emergence and immense fortune of these beliefs? By and large, responses have been of the teleological kind: there *were* flaws, as we can see today "from a logical point of view." This view is behind the bad reputation of the mathematical eighteenth century. In a landmark study of the history of mathematics, Carl Boyer referred to the elaboration of eighteenth-century calculus as "the period of indecision," while Morris Kline famously labeled it the "illogical development." In this perspective, the foundational concerns and the research programs spurred by those concerns are, quite simply, self-explanatory.[14]

By contrast, this study problematizes the emergence of what has been effectively described as the foundational *anxiety* that characterizes much of nineteenth-century mathematical life.[15] Naples, in this respect, offers an excellent vantage point, due to the early manifestation of this anxiety and its striking radicalism. In Naples, foundational concerns emerged in a controversy over geometrical problem-solving techniques in the late 1780s and early 1790s. The main actors were members of two competing schools, the synthetic and the analytic. Under the banner of the synthetic school, a group of mathematicians led by Nicola Fergola (1753–1824) championed the use of sophisticated geometrical methods as the only legitimate and trustworthy methods to establish mathematical truth. Their methods were inspired by the classical tradition, and the language and style of their problem-solving procedures aimed to revive what they regarded as the Greek golden age. The synthetics saw the purity and elegance of their methods as threatened by what they described as a recent aberration in mathematics, an analytic trend that saw the essence of mathematics in an algebraic and algorithmic approach best incarnated in the work of Joseph-Louis Lagrange (1736–1813). To the members of this analytic school, as they called themselves, what mattered was not the solution of a single geometrical problem via inspection of the figure, no matter how ingenious, but rather the capacity to interpret that problem as a case of some family of problems whose solutions could be obtained mechanically through the solution of systems of algebraic equations. To these Neapolitan mathematicians, choosing one set of concepts and techniques over the other had serious practical, epistemological, and even moral implications. In the eyes of leading synthetic mathematicians, training young students in the unrestricted application of algebraic methods was tantamount to depraving their minds—and the consequences of that training were catastrophic for both mathematics and social order. To understand this anxiety and the perceived need for rigorization it caused, we must begin by reconstructing the meaning of these apparently extravagant allegations.

That these questions are amenable to historical analysis is not an obvious claim. Traditionally, historians of mathematics have been more concerned with tracing lineages of ideas rather than with the people who carried them.[16] "For in truth, not everyone agrees that mathematics has a history," wrote Reviel Netz, referring to a divide among historians of mathematics that was effectively captured by David Rowe's distinction between "cultural" historians and "mathematical" historians, "those who study the history of mathematics primarily from the standpoint of modern mathematicians."[17] For mathematical historians, the history of mathematics reconstructs the reasoning behind past mathematical practices, thus clarifying their rational nature and, if necessary, detecting mistakes. In this perspective, one makes sense of past mathematics by showing how it fares in relation to an alleged universal mathematical rationality, which tends to coincide with our own current practices; history might be curious and entertaining but is ultimately just noise. History is the layers of contingency that need to be peeled off to reveal the rational (or irrational) core of a piece of mathematics: what is truly important *cannot* be historical. In this view, questions like those addressed in this book would be addressed through rational reconstructions; standard labels like "the rigorization of calculus" were coined to refer, at once, to the intentions of the historical actors and to the judgment of the historian.

The cultural history encouraged by Rowe, by contrast, is an attempt to articulate or problematize the notion of a taken-for-granted universal mathematical rationality. To varying extents, historians have pointed at the historicity of mathematical reasoning and its outcomes, including allegedly timeless notions such as rigor, exactness, proof, and logical deduction.[18] By the mid-1990s, Rowe noted, the "monolithic image of mathematics" had visibly shifted toward "investigations informed by a variety of methodologies and set firmly in a socio-cultural context."[19] Numerous sophisticated reconstructions of the conditions and contexts of past mathematical practices have been written since then, exploring the intricacies of what Michael Barany aptly named "the unruly past of ruly knowledge." In the current historiography, this heightened attention to context coexists with sophisticated attempts to trace diachronic connections and a renewed interest in the heuristic potential of rational reconstructions. The tension between these different axes of research has indeed been the focus of many recent productions.[20] To paraphrase a classic passage by Steven Shapin: one can either debate the possibility of doing the history of mathematics, or one can do it.[21]

Interpreting the History of Mathematics

In this book, I outline a sociohistorical interpretation of the emergence and legitimization of certain trends that characterized mathematical life in Europe around 1800. I'm especially interested in offering an interpretation of my actors' claims that certain eighteenth-century ways of proving theorems and solving problems were inadequate and dangerous. Offering such an interpretation does not mean ignoring their technical concerns but rather understanding why they saw certain technical choices as the only viable ones.

I construct my interpretation by situating mathematical perceptions, resources, and practices within specific cultures, which are sustained by collectives (in this case "mathematical schools") that orient themselves in specific ways in their social worlds.[22] This process of contextualization and reconstruction allows me to see possibilities and constraints and therefore retrieve relevant meanings and motivations. To use an expression by historian of art Michael Baxandall, I try to map the "patterns of intention" that innervate specific mathematical practices. Or, in anthropological parlance, I treat the mathematical discontinuity under study as a deep rather than a shallow phenomenon.[23] Deep enough, at least, to be legible as a sign that the disagreement invested the very idea of *reason* that grounded those mathematical practices.[24]

In his notes for those who survived the eclipse of reason, Martin Jay identifies "a reciprocal feedback loop" connecting the theoretical idea of reason, however it might be understood, and "the multiple practices of rationalization."

Among those practices of rationalization, mathematical practices hold a special place, as they are often portrayed as quintessential expressions of reason in action. This book contributes to the exploration of this entanglement, focusing on the way mathematical practices embody and sustain specific images of reason. The mathematical controversy we'll scrutinize was *also* a controversy between two incompatible ideas of reason. Hence it was a most salient but essentially unsolvable controversy; one that could only, as it did, fade away into insignificance.[25]

Through this controversy we can glimpse a particular moment in the history of reason, one—I argue—characterized by the dawn of a *reactionary reason*. It was a reactionary turn that shaped mathematical, scientific, and political modernity. To emphasize this point, I link mathematical practices to ideas of reason through a series of other cultural phenomena, including science, engineering, political economy, politics, philosophy, theology, and landscape painting. While keeping mathematical problem-solving techniques at— literally and figuratively—the center of the book, we shall recognize in each story, no matter how apparently distant and unrelated, the blueprint of the same basic cognitive and political conflict. To many of the protagonists of this book the elaboration of a new and reactionary mathematics was indeed the first and necessary step toward the overall restoration of social order.[26]

The choice to focus on a controversy is also an homage to the early days of the Edinburgh school in the sociology of scientific knowledge, and to its pioneering case studies that searched for the interests that shape even the most technical of knowledges.[27] The initially outrageous tenets of those studies are still key to the historical study of mathematics and formal knowledges: that we should write the history from the point of view of the actors, take a symmetrical approach in the analysis of what we now consider true or false knowledge, or look at what actors did rather than rely solely on their methodological claims.[28] This book builds upon the literature in the sociohistorical study of mathematics that ensued, especially those works that pioneered the study of mathematical knowledge as situated knowledge, focused on the broader significance of processes of abstraction and mechanization, and emphasized the social shaping of formal techniques such as statistics.[29]

Many of the important lessons to be drawn from this literature are embedded in the notion of "mathematical culture" deployed in this book. There is a long tradition of identifying styles in mathematics, a notion that has been variously understood as having to do with individual temperament, national character, the Zeitgeist, or the direct expression of a culture in a Spenglerian sense. The term "mathematical culture" has been given a more familiar anthropological and sociological meaning by David Rowe, who uses it to refer to the set of resources available to a specific group of mathematicians.[30] The ethnomethodological and sociological study of formal deduction has offered some other relevant usages of similar terms, such as Eric Livingstone's "culture of proving," which refers to the shared culture that supports the mathematician's strategic moves, and Donald MacKenzie's deployment of the same term in the plural to emphasize instead social conflict and difference.[31] These authors' usages are typically deflationary, in the sense that they do not gesture at grand, all-encompassing worldviews but rather refer to the contingent heterogeneity of resources and practices involved. I embrace this notion of heterogeneous mathematical cultures, which echoes Ludwig Wittgenstein's rendering of the mathematician-as-bricoleur—and his effective demolition of the illusion that the mathematician's moves are dictated by an underlying universal rationality conceived as an invisible, all-powerful, and perfectly rigid logical machine. Wittgenstein's disarmingly elementary examples show that no formal deduction can ever be self-evident. There is no rigid logical machine, only people who act rigidly when they practice logic. Those examples also show that deductive procedures have meaning only if immersed in forms of life that would tacitly and contingently define their criteria for application. In David Bloor's sociological reinterpretation of those famous aphorisms, mathematical concepts and rules become self-evident only if and when they turn into "institutions"—when they are inserted in a network of concepts and practices supported by the collective interests of a social group.[32]

In this book, a mathematical culture cannot be properly understood in separation from the specific forms of interaction and social organization that create the conditions for its emergence. Technical choices and instrumental action will be described as always and necessarily sustained by social priorities and objectives. This notion of mathematical culture is inclusive of practice, and can be defined as a set of mathematical resources and practices shared across an expert group whose social organization and functioning are best understood in terms of a goal-oriented status group.[33] Rather than pursuing misleading demarcations between mathematical contents and social context, this notion emphasizes the collective and goal-oriented nature of mathematical knowledge as well as the performative dimension of mathematical cultures, as they incessantly contribute to the representation, constitution, and transformation of social worlds.[34]

Mathematics and Power

Around 1800, a mathematical culture emerged in Naples that was alternative to hegemonic French mathematics. That such a significant cognitive discontinuity appeared at a time of major social and political upheaval is hardly coincidental. The social transformations in the Kingdom of Naples were not essentially different from those in other parts of Europe. What was distinctive, though, was the intensity with which Neapolitans experienced these transformations and the extreme historical conditions that structured their experience.[35] Due to these conditions, the Neapolitan critique of French mathematics reveals an urgent and unmediated political dimension. The rejection of certain practices and the redefinition of the relation between mathematical knowledge and empirical reality were, in the eyes of the relevant actors, politicized acts. They were a *reaction* against the aberration of a *revolutionary* mathematics that had dissolved the "natural" social order, undermined the principle of authority, and afforded a redistribution of agency that empowered the traditionally subordinate, bringing on the stage new political subjectivities. Recognizing the political dimension of certain mathematical choices does not mean that these choices were not truly mathematical: they were mathematical and political at once. After all, in Naples, French analysis had shaped the political imagination of local Jacobins and had been defended by French guns, deployed by civil engineers to reshape the landscape of the country, resisted by local communities, and mathematically refuted by reactionary elites. How could one split the two stories apart without missing the point of it all?

The standard reference for the relationship of science and polity in revolutionary and Napoleonic France has long been Charles Gillispie's monumental two-volume assessment.[36] For all its scholarly prowess, however, his inquiry into the founding institutions of French modern science has surprisingly little to say about power. He conceives science and polity as two separate, autonomous spheres of action that at times interact—as when scientists search for financial support or when politicians try to use ready-made scientific knowledge and technological artifacts to achieve their political goals. Gillispie's account is dichotomic and normative, presenting science and politics as two distinct spheres of action that are, and should remain, separate. Interaction—contamination, one might say—is dangerous, and will inevitably produce bad science as well as bad politics. In other words, Gillispie portrays the relation of science and polity as *external*, not essential to the nature of either element. Reviewing Gillispie's first volume, Keith Baker notes the implausibility of this position given the multitude of historical evidence that indicates the reciprocal shaping of science and political action in late-eighteenth-century and revolutionary France. Consider, for example, the collaboration between Turgot and Condorcet on the problem of the standardization of weight and measures. Facing a crisis of political legitimation, Turgot's strategy was to use the growing authority of modern science to achieve political consensus by transforming a political problem into a cognitive one. In his view, "natural reason, rather than political will, was to be the source of order and authority in political affairs."[37] What Baker describes is a situation in which the relation of science and polity is, in fact, *internal*.[38]

Moving away from Cold War historiographical categories, historians of science have demonstrated how, far from being a neutral analytical tool, the separation of polity and science should be seen as historically situated: in fact, largely, as an outcome of the historical processes studied in this book.[39] In Naples, this theme was visible primarily in the emphasis with which some Neapolitan mathematicians insisted on the necessity of defending the alleged purity of mathematical knowledge—as opposed to the impurity of knowledge derived from, and therefore polluted by, empirical considerations. How to interpret this anxiety-ridden defense of mathematical purity? In what follows, I turn this process of purification and its outcomes into historical objects whose emergence needs to be reconstructed and interpreted through a genealogical approach—a genealogy of purity.[40] Consistent with the definition of mathematical culture introduced above, this genealogy tries to reconstruct the concrete ways in which socioeconomic formations created the conditions of possibility within which certain mathematical objects—including the modern concept of "pure mathematics"—were constituted and legitimated.[41] At the same time, I explore the performative role of these mathematical objects and techniques, the way in which they were deployed as organizing principles to structure social and political life.

These considerations have further sharpened my use of the notion of mathematical culture. As mentioned above, the term refers to the resources available to a group of actors *and* to the ways those resources are deployed to address the goals and priorities of that group—culture and practice. Also, I do not limit my use of the term to refer to the heterogeneity of the mathematical resources and practices involved but emphasize the patterns of social interaction that are proper within that group and its processes of socialization.[42] These patterns direct our gaze toward larger social configurations. In fact, the group's goals and priorities can only be produced and sustained through situated collective action; they are shaped by the group's broader orientations to the natural and social world.[43]

In early-nineteenth-century Naples, debates in mathematics and politics centered on the relation between the practice of geometrical problem solving and that of political problem solving. One could not intervene effectively in the political discourse without tackling the question of the nature of mathematics; nor could one do mathematics without ipso facto taking a political side. The notion of mathematical culture and the genealogical method are especially apt to capture this historical moment, as they are porous and invite contamination. They emphasize the *proximity* of polity and science rather than their distance and mutual exclusion. We'll see that this proximity will at times look like an identity: in my reconstruction mathematics is not "influenced" by politics; neither am I arguing that social conditions shape mathematics in some generic sense. Rather, mathematical reasoning and social power enter the narrative as different aspects of the same process, as if they were the same thing under different forms of awareness, and a change in one would not cause a change in the other but would rather entail it.[44] In this way I articulate, within the history of mathematics, a fundamental science studies insight, one often expressed through the idiom of coproduction: to produce knowledge means always and necessarily to produce a certain form of social order.[45]

Mathematical cultures and the ideas of reason they embody orient practitioners in their attempts to order the world. These acts of ordering and re-ordering are ideal sites for the historian to study how technical knowledge can be constitutive of specific distributions of power. In fact, this book is as much about mathematical practices as it is about distributions of power and agency.[46] As we shall see throughout the Neapolitan story, new mathematical possibilities entail new political possibilities, while mathematical restrictions entail a restriction of viable political options. The first half of this book reconstructs the emergence of a *revolutionary mathematics* that embodied a radical social vision centered on a massive redistribution of agency across society and on the empowerment of subordinate groups. The second half focuses on the *reactionary mathematics* that challenged these new political subjectivities in the name of tradition, the principle of authority, and the "natural" social order. Reactionary mathematics, as we shall see, was a mathematics forged in battle, no less militant than its revolutionary counterpart. The modernist image of techno-mathematical purity and neutrality arose from a battlefield: the same battlefield upon which Europe's sociopolitical future was being decided.

The Structure of the Book

The book is structured as a scaling exercise articulated in two movements. At its center is the Neapolitan controversy over mathematical methods, which functions as the main organizing principle. The first movement consists in zooming in on the controversy. Chapter 1 focuses on analysis and the power of abstraction. We move from the eighteenth-century notion of analysis as a set of heterogeneous practices and explore briefly its significance. I emphasize how analysis came to be seen by many as a savoir faire that should inform political and administrative decision-making, according to the principle that "to administrate is to calculate." This analytic reason coexisted with a culture of sensibility that permeated eighteenth-century philosophy and science, and the combination of analysis and sensibility provided powerful tools for the critique of traditional social institutions. I then reconstruct the reception of analysis in Naples and describe how its practices were integrated into the local enlightened and reformist discourses. I explore some of analytic reason's concrete embodiments and examine how they were deployed for the material reorganization of production, thus shifting from analytic reason's critical capacity to its extractive capacity. The mechanization of olive oil mills is exemplary of the analytic rationalization of a traditional production process that created a new kind of commodity and forced a transformation of social relations in the producing regions. The partial failure of this sociotechnical project reveals the full extent of the resistance against modernization, its heterogeneity, and the impossibility of neatly separating "tradition" and "innovation." Finally, I argue that the late Neapolitan Enlightenment, characterized by more pessimistic and palingenetic visions of the future, was not a rejection of previous analytic and sensationalist priorities but rather a form of radicalization, one that created the conditions for a full-fledged revolutionary discourse in the 1790s.

Chapter 2 shows how some Neapolitan mathematicians and political activists mobilized analysis and sensibility in their attempt to revolutionize Neapolitan culture and society. I follow closely the scientific and conspiratorial activity of Carlo Lauberg (1762–1834) and Annibale Giordano (1769–1835), who ran a popular private school of chemistry and mathematics, became leading figures in the local Jacobin movement, and, in 1799, led the revolution and the short-lived Jacobin Republic. The Republic itself was conceived as a political system founded on the logic of analysis. To these Jacobins, mathematics was a revolutionary knowledge and through its practices, they believed, a more equal and just society could be constituted. There could be no meaningful separation of mathematics and politics in the Jacobin worldview, as the analytic reason embodied in their mathematics was the same that shaped their political vision and guided their revolutionary action. In other words, for these Jacobins analysis was a tool of liberation—mathematical and political; its practices were necessarily impure, mixed up with empirical reality, and that is why they were effective. Based on the Neapolitan evidence, I suggest an overall revision of the notion of "Jacobin science" and of its historiographical significance. What earlier critiques of Jacobin science were targeting was the very possibility of a *revolutionary* science—a science that understands itself as agent of political change. The joint action of local counterrevolutionary forces and international anti-French contingents defeated the newborn Republic, opening a season of political trials that wiped out the local mathematical culture associated with analysis, social mathematics, and revolutionary action.

In chapter 3, the empire of analysis strikes back. I recount the return of the French to Naples and the success of analysis during the so-called French decade (1806–1815). I focus on the ambitious project of modernization implemented by the French-Neapolitan governments and on their attempts to reorganize the state and the administration through the deployment of a new kind of expertise: the analytic knowledge of technical elites such as civil engineers, statisticians, and cartographers. Analysis was celebrated as a universal language for representing and transforming the natural and social worlds. But the structural conditions of this renewed analytic enthusiasm were different from those of the 1790s. The Napoleonic normalization of political life and the imperial system to which Naples was annexed created the conditions for a clear-cut separation of analytical techniques and political revolution. It was a process of purification that aimed at isolating an alleged technical core of analytic formalisms from the Jacobin political agenda. Intriguingly, the impossibility of critiquing the political status quo had mathematical implications, which resulted in a methodological compromise between synthesis and analysis. The philosophical framework for this political and mathematical compromise was initially ideology (the analytic study of the origins and connections between ideas), followed by other approaches that emphasized the *relative* autonomy of individual reason and expressed the interests of the postrevolutionary agrarian elites, a social bloc that supported constitutional liberalism and modernization up to 1848. Envisioned as a tool of emancipation and liberation by the Jacobins, analysis was deployed primarily—though not exclusively, as we shall see—as a tool of social control and extraction during the Napoleonic age and the Restoration.

Chapter 4 considers the *shape* of the southern kingdom: the different ways in which its physical and human landscape could be represented and transformed. I focus on the most prominent group of technical experts established during the French period: the corps of civil engineers. These engineers, trained and selected in an internal school, were perceived as embodying analytic savoir faire at its best. Their action aimed literally to reshape the Neapolitan landscape and the forms of life that it supported. Administrative centralization and techno-scientific standardization were the leitmotifs of their campaigns. This new generation of analytic mathematicians founded and legitimated their practice on the ideal of a powerful but not all-pervasive algorithmic rationality. Their mathematical tools' critical capacity was not universal, but it could be legitimately exercised within well-defined areas of intervention. Through a brief survey of projects supported by the director of the corps of engineers, Carlo Afan de Rivera (1779–1852), we see how analysis was being deployed to overcome the resistance of local authorities and local knowledges. The goal was to rationalize the administration, the economy, and the very physical landscape. I then shift attention to the *resistance* met by this normalized version of the analytic reason. Who were the enemies of the engineers? The intense sociotechnical controversy over modernization was a battlefield for competing social groups, and the legitimacy of the engineers' analytic rationality was often the key matter of concern. I examine the views of a first-rank politician as a means to enter the world of reactionary politics, and to reconstruct how reactionaries imagined the shape of the kingdom. This reactionary vision found expression in the production of one successful group of painters who portrayed Naples and its surroundings in the early nineteenth century. Their productions originated a series of romantic visual tropes that became popular across Europe. I argue that their way of seeing embodied reactionary assumptions and priorities that radically questioned the modernization project and its mathematical and moderate liberal foundations.

At the center of the book, in the intermezzo, I offer a concise description of the methods used by the members of the two Neapolitan schools, the synthetic and the analytic, which competed for mathematical hegemony between the 1790s and the 1830s. The narrative is not organized chronologically; nor does it aim to offer a complete survey of the controversy. Rather, I reconstruct the mathematical methods of the two schools through a couple of elementary examples, and refer to their main resources, priorities, and pedagogical styles. I consider the epistemological assumptions that informed the two methods, but hew close to the mathematical world under scrutiny, as is customary in this kind of reconstruction. I briefly outline the historiographical trajectory of this controversy, noting how "backwardness" has been the key interpretive category: the synthetics were defending a backward mathematics against the innovative mathematics of the analytics. The implausibility of such an interpretation is made clear by the relation of the intermezzo to the other chapters: the meaning of the controversy is indeed refracted through each one of them.

With chapter 5 we begin to zoom out of the technical controversy, along a trajectory that mirrors the movement of the first four chapters. We thus shift to some broader considerations on what I call the "geometry of reaction"— mathematics as practiced by the synthetic school. I survey the social features of this group and examine their research and educational goals, as well as their self-understanding within the history of mathematics. I argue that their distinctive, Greek-like mathematics was not the expression of an outmoded local tradition but rather a *new* phenomenon, the outcome of precise choices. Central to this mathematics was a distinctive and very modern anxiety to defend the alleged purity of mathematical knowledge from the contaminations that had corrupted it during the eighteenth century and the revolutionary period. But a pure mathematics, disconnected from empirical and social reality, required reliable mathematical-logical foundations to be credible. The political move of isolating mathematical knowledge from social reality created a major mathematical problem, the problem of foundations. I reconstruct synthetic mathematics as a mathematical culture that offered a possible solution to that problem.

Chapter 6 continues to zoom out, opening up a vista on what I describe as a scientific counterrevolution in Naples. Mathematics was perceived as key to a broader reaction against the alleged degeneration of the modern world, but all sciences were supposed to join the fight. I explore various forms of reactionary science that emerged in Naples between the 1790s and the 1830s, moving from the exemplary work of Nicola Fergola, leader of the synthetics but also a prominent natural philosopher. I consider university textbooks and the research of famed physicians and experimental natural philosophers to reconstruct their "apologetic empiricism," an approach grounded on the redefinition of the role and scope of human reason and therefore of legitimate scientific practice. Mapping apologetic empiricism means mapping the web of individual actors and institutions that sustained the political and religious reaction in Naples.

Chapter 7 zooms out further, offering an overview of the broader political and cultural reaction within which this scientific counterrevolution was embedded. This world is that of a newly defined reactionary reason, which was supposed to fight the revolutionary excesses of the analytic reason—and indeed replace it. I explore, in particular, how the culture of reactionary Catholicism redefined the relationship between polity, religion, and knowledge. I trace this form of anti-Enlightenment back to its eighteenth-century roots, showing that Naples was the site of an early and radical elaboration of anti-encyclopedic themes and practices. I sharpen the contours of reactionary reason by considering the productions of French traditionalism, the literature of the aristocratic reaction, and Neo-Scholastic philosophy. As diverse as they were, these movements reacted against a common enemy: the challenges to an order in which agency was considered the prerogative of traditional elites—be they in culture, religion, or politics. In each of these domains, the tide of secularization needed to be reversed and the principle of authority reinstated. At the root of all religious and political upheavals was, ultimately, the rebellion of individual reason, emblematized in the persona of the hubristic godless mathematician. Reinstating the principle of authority implied rejecting the analytic reason—individual, a priori, universal, and ahistorical—while saluting the dawn of a reason that was collective, a posteriori, contingent, and eminently suited to historical thinking—a reactionary reason.

Chapter 8 articulates my historical interpretation for the emergence of pure mathematics as a foundational field at this particular juncture of the nineteenth century. I argue that this emergence can be seen as part of a broader project to "return to order" that characterized much of social life starting from the revolutionary experiences of the 1790s. The foundational concerns and the emphasis on the need to keep mathematics pure were mathematical responses to urgent political problems. The same is true for the faith in the power of algebraic algorithms that characterized the competing analytic practice (boundless in the Jacobins, limited in the later Napoleonic and Restoration technicians). I then reflect on the locality of my own interpretation and on the possibility of generalizing it to the broader European context. What can be generalized, I argue, is my interpretation of foundational anxiety as a reaction to the use of analysis to legitimize and guide action in social and political matters, hence seeing the preoccupation with purity as an attempt to restrict such action mathematically. The emphasis on purity and rigor are the cornerstones of a new mathematics for the age of the return to order. That one should call for a return to synthetic geometry rather than for a rigorization of analysis is not essential in this respect but depended on local conditions. I consider the mathematical work of Paolo Ruffini (1765–1822) and Augustin Cauchy (1789–1857) to illustrate this point. The deep opposition in my story, therefore, is not between geometry and algebra. Rather, it is between mathematics as a pure, rigorously defined and self-contained body of knowledge and mathematics as a set of highly general and universally applicable algorithmic procedures, seen as expressions of all-encompassing universal rationality. In other words, it is about two different kinds of reason and their political significance. I conclude by comparing the results of my study to those of other assessments of the relation between mathematics and modernity. For all its apparent backwardness, I conclude, the anxiety that pervaded the synthetics' reactionary mathematics takes us straight to the core of our mathematical and political modernity.

Adventures of the Analytic Reason

Analysis loomed large in eighteenth-century European culture. Some philosophers, politicians, and administrators were fascinated by its perceived transformative power; others were repelled. Few remained indifferent. We begin our exploration in well-trodden territory: how did the French philosophes understand analysis? The analytic way of thinking, in their view, was combinatorial, calculative, and algorithmic. But it was not a conventional game: its power derived from its connection to the profound structures of reality. Nor was it mindless; on the contrary, analysis was emblematic of the workings of human reason. The critical potential of these claims cannot be missed: analysis could be deployed to question traditional cognitive and social hierarchies. The reason of these philosophers was an analytic reason—the organizing principle of new ways of life.

The thread of eighteenth-century analysis takes us swiftly to the world of sensations and emotions. A wealth of scholarship has dispelled the illusion that one could neatly separate an "analytic" from a "sentimental" Enlightenment. The eighteenth century was, at once, an age of reason and an age of sentiment. The champions of analysis were indeed fostering an epistemology that grounded all human knowledge and morality in sensations. Sensory impressions, understood as micro-collisions, were considered the building blocks of both ideas and emotions, bridging the gap between the rational and the sentimental, the natural and the social. Both reason and natural sentiments were to be mobilized to transform moral and social life.

The next step takes us southward. Through a concise reconstruction of the Neapolitan Enlightenment, we follow the fortunes of analysis at those lower latitudes in the second half of the eighteenth century. For many Neapolitan philosophers and reformers, planning the future of their country was an exercise of analytical imagination. Reformist plans typically moved from recognizing the backwardness of the kingdom's socioeconomic conditions to promoting an analytic reorganization of institutions, production processes, and markets. When these transformations met resistance or failed altogether, reformers denounced the baffling ignorance of relevant actors, as well as their illicit interests. In particular, enlightened producers of olive oil—a lucrative business in the kingdom—found that certain modern machines imported from northern Italy did not seem to work on their estates. These machines, I suggest, embodied the logic of analysis and could not function properly without a profound transformation of the socioeconomic landscape. Simon Schaffer, explorer of the concrete embodiments of machine philosophy, noted that there is "an intimate connection between the machinery of natural philosophers' concerns and that of the new entrepreneurs." The mechanistic and analytic ambitions of eighteenth-century philosophers cannot be understood apart from the mechanization of production processes or the rise of a rationalized labor regime. The Neapolitan case is revealing because it illustrates the joint *failure* to consolidate both such a regime and the culture of analysis. The shape of milling stones and the design of presses talk to us about resistances to the plans of rational management of the Neapolitan enlightened elites.[1]

Disposing of the misleading opposition between a rationalistic and a proto-romantic Enlightenment is also useful because it allows us to reconstruct the metamorphosis of analytic reason in Naples during the 1780s and 1790s. That the reformist discourse was developed largely in the sites and languages of Freemasonry and framed within historicist perspectives should not be seen as a rejection of earlier analytic ideals. Reformers continued to believe it possible to transform social reality through the understanding and manipulation of its underlying rational structure. Clear to many, however, was that the gradual reformist approach had failed. Instead, the revival of Giambattista Vico's cyclical philosophy of history, alongside a renewed fascination with natural catastrophes, seemed to point toward a more radical process of palingenesis.

Reason as Analysis and Sensibility

To its eighteenth-century supporters, the term "analysis" did not refer to a theory or a particular mathematical technique. Rather, it was a way of thinking and a way of doing: a savoir-faire. Authors used analysis to refer to a variety of cultural resources and practices that could be deployed to frame and solve all kinds of problems; the unifying element was its underlying logic. The scope of analysis expanded rapidly from mathematics to natural philosophy, engineering, and medicine, reaching even the moral and political sciences. Essentially, its adherents perceived analysis as a novel way to represent and classify things, understand all their possible relationships, and intervene to modify them. Supporters of this savoir-faire pointed at the impressive developments of the mathematical sciences: from Descartes onwards, they argued, analysis had brought to the previously divided disciplines of geometry and algebra uniformity, procedural clarity, and empirical effectiveness. The contrast between this development and the persistence of traditional cognitive styles in the moral sciences was increasingly the source of an intolerable scandal.[2]

One of the eighteenth-century meanings of analysis was the method of solving mathematical problems by reducing them to equations. Problems in geometry or mechanics were translated into linear sequences of symbols and reduced to their simplest components. Each component could then be considered and solved in isolation from the others as an autonomous problem, which made for a streamlined and effective problem-solving process. D'Alembert was convinced that the analytic reason was best incarnated in mathematics and referred to the methods of integral calculus as the epitome of analysis in action. But the significance of analysis extended well beyond the domain of mathematics: analysis, he argued, "gives us the most perfect examples of the manner in which one should use the art of reasoning." This analytic art of thinking came to be regarded as an exemplary incarnation of reason itself. For Voltaire, "the only way man can reason on the objects [of experience] is analysis," while Condillac insisted that analysis was not only the best method for discovering all kinds of truth but also the best pedagogical method. To him, the ancients' predilection for synthesis was their greatest cognitive limitation. Overall, the philosophes saw no obvious limitations to the application of analysis, attributing its heuristic power to the use of a symbolic language that permits the decomposition of complex ideas into simple ones. Thus all possible relationships can be represented in the clearest and most immediate way.[3]

To understand the development of eighteenth-century science, we ought to take into account that many practitioners were striving to appropriate and deploy into their own fields what they perceived as a unitary analytic paradigm. This perception of analysis as a universal procedure was built upon John Locke's epistemology, especially his sensationalism (the theory that all knowledge originates in physical sensations) and his call for a well-defined language to communicate them. The articulation of the relation between senses and thought remained problematic though, and most scholars agreed that sensationalism had to be somehow mitigated, for example through the recognition of certain active faculties of the mind, or by positing signs as the intermediaries between perception and reflection. To Condillac, a system of signs is what makes thinking possible; hence, the art of reasoning is coextensive with a well-made language: "The creation of a science," he wrote, "is nothing else than the establishment of a language." Scientific languages are incarnations of analytic systems and therefore of correct reasoning. In this sense, in the *Encyclopédie* they are described as "a kind of calculus."[4] By the mid-eighteenth century, this constellation of beliefs had become so pervasive that, as noted by Ernst Cassirer, the very notion of science was seen as coextensive with that of a calculus. Michel Foucault described this striking feature of eighteenth-century scientific culture as a fundamental trait of a new episteme characterized by classification, decomposition and composition, and the use of artificial signs.[5]

Condorcet was a prominent champion of the belief that the conjunction of observation with an artificial symbolic language would produce major advancements in both the natural and social sciences. Keith Baker noted that, for Condorcet, "there could be no true science without the model of mathematics, nor was there a subject that could be regarded by definition as not potentially susceptible of mathematical treatment."[6] With Condorcet, the combinatorial analysis of mathematics—not just analysis as decomposition in the Lockean sense—becomes the essential model for reasoning and the discovery of new truths. Algebraic analysis is a combinatorial practice in which a calculus of signs applied to the elementary constituents of a field of experience reveals their essential relations—hence his vision of the sciences as "mechanical arts." One should not mistake Condorcet's combinatorial view of mathematics with the modern notion of a formal system. For Condorcet, mathematics is not a game governed solely by its internal logic. On the contrary, his boundless faith in the universal validity of analytic methods rested on an ontological and epistemological assumption: that of the fundamental uniformity of reality. This uniformity was a necessary condition for the validity of procedures that aimed to decompose and compose all concepts and representations—"ideas"—to discover new truths. As a condition, it ultimately warranted the precision and certainty of mathematics itself, which Condorcet understood as an enormous exercise of analysis applied to an ontological unity.[7]

In this perspective, talking of "pure mathematics" in the modern sense would be meaningless. Condorcet was part of a tradition that saw the development of mathematics as essentially interwoven with that of the physical sciences. It was a "mixed" and, by later standards, impure mathematics, driven by the application of infinitesimal calculus to problems of mechanics, where new mathematical concepts were deemed interesting because they provided a conceptual framework to make sense of aspects of the material world, essentially by making them measurable. These practitioners considered the problems of mixed mathematics as always, in principle, reducible to problems of pure mathematics, where "purity" meant maximum abstraction from empirical reality rather than a form of knowledge essentially disconnected from it. Indeed, physical problems were reduced via a progressive elimination of physical circumstances—a series of abstractions operated by human reason that revealed nature's basic architecture. When considered at the appropriate level of abstraction, every material phenomenon could become the object of a process of analysis.[8]

Convinced that analysis could lead to the discovery of truths relative not only to physics but also to "a great number of questions in metaphysics, morals, and politics," Condorcet gave his social and political arguments a strikingly mathematical form.[9] Alexandre Koyré perceptively noticed how Condorcet treated the problem of giving a political constitution to France as, *literally*, a problem of mathematical integration.[10] In this as in other matters, analysis would replace traditional synthetic procedures, which, Condorcet lamented, derive unverifiable consequences from arbitrary definitions. Condorcet certainly had a taste for abstraction and calculation. His mathematical writings reveal an almost complete absence of visual imagination. However, he never associated the algorithmic procedures of algebra with a routinized, unreflective, or demeaning kind of labor but with intelligence and reason actively engaged in a fight against prejudice and tyranny. Such calculations were, in the words of Lorraine Daston, "the concrete practices of abstract reason in the Enlightenment": an essential nexus existed between highly abstract reasoning and concrete administrative and political action. Mathematics was, first of all, "a concrete instrument of civilization."[11] Condorcet's ambitious project of a "social mathematics"—which Baker describes evocatively as "a mathematical assault upon the political sciences"—meant that philosophers could break down complex political, economic, and moral ideas into their elementary components, explore their relationships and, if necessary, intervene in these relationships. Analysis became an instrument to remedy human weaknesses and errors.[12]

The deployment of analysis and rational thinking, however, required a structural reordering of knowledge. Objects needed to be properly identified and classified before they could be analyzed and manipulated. Éric Brian has shown how, for Condorcet, there existed an essential relationship between the procedure of reduction by analysis and that other exemplary eighteenth-century practice: classifying by means of analytic tables. To analyze a problem meant to present systematically all the possible cases that could originate with the conditions of that problem. Similarly, an analytic demonstration must consider every possible case falling under the theorem. Condorcet believed that a proof is complete when all the possible cases have been taken into consideration. The validity of such a proof was founded not only on the formal coherence of its passages—as in the modern conception of proofs—but also in the systematic coherence of the set of possible cases.[13] It follows that, in proving a theorem or in solving a problem, the analyst was bringing out something like a subjacent classificatory table. Or, to put it differently, that the composition of classificatory tables and the solution of problems through analysis were founded on one and the same analytic savoir-faire. The reorganization of the structure of knowledge, based on new and reliable classifications, was thus a necessary condition for the expansion of the empire of analysis.

Condorcet's body of work shows well how, concretely, analytic savoir-faire, sensationalism, and sentimentalism (the theory that all sentiments originate in physical sensations) came together in eighteenth-century reformist and radical culture. It was based on this conjunction that he could call for the unification of the natural and moral sciences, which included the study of the mind itself and of the relations between individuals. The study of politics was therefore declared methodologically contiguous with the study of psychology and physics. Taken up by reformers, Jacobins, and idéologues, this contiguity would frame innumerable scientific, pedagogical, social, and political projects. Edward Jones-Imhotep has shown the convergence of analytic rationality and culture of sensibility in the making of that quintessential instrument of Jacobin governance: the guillotine. This machine, he argues, was a "sentimental technology" because its efficiency was understood, at once, in analytic terms and in terms of sentimentalism, the public psychology endorsed by the revolutionaries.[14]

We have reached here a key feature of eighteen-century culture. It was the age of Voltaire, for sure, but it was the age of Rousseau too; the age of reason and the age of sensibility; the age of rationalist mechanism and the age of organicism and vitalism. Looking back at this moment with the eyes of the mid-twentieth century, scholars have long interpreted the presence of these threads as a radical opposition between Enlightenment and Romanticism—or, more subtly, between a scientific and a sentimental Enlightenment. But projecting a neopositivist dichotomy onto this distant world does not enhance historical understanding. Instead, recent scholarship has convincingly argued for the complexity of the relation between these components of eighteenth-century culture, and indeed for their profound connections. It's not only Rousseau who argued that the senses shape the way the mind reasons and makes moral judgments. Around the mid-eighteenth century, Locke's epistemology is appropriated by Condillac, Diderot, Helvétius, Holbach and many others across Europe to argue that sensations and sensibility were the veritable building blocks of all rational and moral thinking.[15]

Research on sentimentalism and the culture of sensibility in the Enlightenment has shown how, like analysis, they run through a variety of modes of inquiry and traditions, from philosophy to religion, from economics to literature. They shaped new ideas of the human body and of the human sciences that, in turn, redefined notions of corporeality, citizenship, and rights. Sentimentalism was key to the articulation of progressive social and political projects, as it emphasized everyday moral and emotional meaning over transcendental values, thus providing new rationales for individual and collective behavior. It was also a powerful resource for egalitarian and democratic discourses, and for the constitution of individuals—citizens—who invariably embodied "the general, the abstract, the universal" within "a uniform, national space." Through the language of sensibility, authors constituted individuals as autonomous and, ultimately, interchangeable—in literature as in economics or philosophy. The radical critique of the orders and the corporative institutions that made up the ancien régime was constructed upon new, sentimental and sensationalist, conceptions of the self. These same conceptions would be used by Jacobins and *idéologues* to address the urgent problem of collective behavior and social order as they tried to dismantle pre-revolutionary structures.[16]

Like any other part of culture, science was invested by the sentimental wave. To Jessica Riskin, "sentimentalism was integral to the method of Enlightenment science as a whole." She identifies a core group of French philosophers who placed sensibility at center stage and explored the relations between the moral and intellectual faculties, preparing the terrain for new theories and practices of socialization. She labels them "sentimental empiricists," in a convincing rebuttal of simplistic representations of Enlightenment science as "cold rationality." Anne Vila shows how the monistic paradigm of sensibility was central to a revolution in medical theory as well. It was a paradigm that could unify the physical and the moral nature of humankind. Leaving behind seventeenth-century models and problems, in the work of scholars like Albrecht von Haller it was the physical body that more fundamentally constituted the individual, and therefore became the key site for understanding and guiding individual and social behavior. The discourse on sensibility is also key to the rise of eighteenth-century vitalism, a notoriously slippery phenomenon that we can think of, for our present purposes, as a holistic approach to the study of living things that emphasized the autonomy of life and the self-organizing properties of matter. The new preeminence of the senses thus had multiple and profound effects on scientific life, including a questioning of the privileged nexus between vision and cognition. In an essay that focuses on blind practitioners, Lissa Roberts criticizes a historiography of science that has systematically obfuscated the role of sensuous engagement in the making of eighteenth-century science. The Kantian normalization of Enlightenment as a culture of sight—characterized by the control of understanding over "blind senses"—has indeed been key to shaping retrospectively our perception of eighteenth-century reason.[17]

The Neapolitan Enlightenment

The philosophes had launched their assault against traditional cognitive authorities through an effective combination of sensationalist epistemology, a new classification of knowledge, and a newly defined, universally applicable analytic method.[18] But Paris was not the only place in which the heuristic and critical potential of the conjunction of the culture of sensibility and analysis were being explored. Similar resources and concerns can be found across Europe, often as part of reinterpretations of seventeenth-century rationalist philosophies in the light of the new science's stunning achievements. The discussion about the power of analysis was usually framed as part of broader debates on philosophical, economic, and political change. In Naples, the analytic savoir-faire as a cognitive and administrative tool found supporters within the local Enlightenment, an articulated movement that spanned from the 1740s to the 1790s.

While analysis was far from being the only important motif in the Neapolitan Enlightenment and in the reformist culture associated with it, it was a persistent one, and one that shaped the philosophical imagination of more than one generation. The remainder of this chapter does not offer a survey of this movement but rather follows the thread of analysis that runs through it, taking for granted an understanding of Enlightenment in terms of its concrete socio-material conditions and diverse local manifestations, one that extends its scope to spaces, actors, processes, and artifacts deemed irrelevant in earlier and more rarefied assessments.[19] In this perspective the term "Enlightenment" has two main meanings. On the one hand, it refers to the emergence or acceleration of long-term processes of social change that were not necessarily univocal, and that cannot be explained away through the notion of a rising transhistorical rationality, as scientific reason itself turns into an object of historical investigation. On the other, "Enlightenment" is an actors' category, and refers to the narrative of a fundamental watershed, characterized by the rejection of the old ways of doing things in favor of a new rational method (or set of methods) that is far superior to any method devised by the ancients.[20]

As argued by Franco Venturi, one would understand little of the fortune and forms of enlightened culture across Italy without mapping them onto local governments' reform campaigns.[21] The case of Naples shows this correlation clearly. The kingdom had recently regained its independence during the war of Polish succession, after being for two centuries the eastern frontier of the Spanish empire and its first line of defense against the Turks. The crowning of King Charles in 1735 ushered in a period of cultural and social transformation, as the new monarch considered the existing feudal-communal structure of the state inadequate to sustain his dynastic ambitions. While one should not simply identify the Neapolitan Enlightenment with the politics of reform within the monarchy, its representatives were all to some degree involved in the reformist effort. Their relentless campaign for social progress, often framed in the language of political economy, was inseparable from their philosophical and scientific experience.[22]

The degree to which power was decentralized in the southern kingdom was unparalleled in Italy. The monarchy had no bureaucracy of its own and relied on the royal courts and tribunals—which combined juridical and administrative roles—to fight feudal pretensions. Much of the kingdom, however, was ruled by feudal courts and tribunals, which defended the interests of absentee landowners, the Church, and powerful corporate bodies. This situation placed the monarchy in an extremely weak financial situation. The new king, who belonged to a branch of the Spanish Bourbons, began wrestling with those innumerable self-governing corporations that traditionally shared power with the prince. The battle he was facing was clearly difficult and long, but he could count on support within Neapolitan society, especially among urban professional elites, non-absentee landowners, and the numerous underpaid men of letters who, like Giambattista Vico, scraped by in the overcrowded capital—the third most populous metropolitan area in Europe.[23]

One of the distinctive traits of this reformism was its focus on concrete issues: public education, the administration of the state, the legal system, and political economy.[24] In the words of one of its leading historians, the Neapolitan Enlightenment "was above all an urge to reform, a support, even a direction for transformation."[25] In their fight against "backwardness" and for "civilization," reformers strived to transform what they saw as inefficient and overly complex economic and juridical structures. Among their main polemical targets was the Roman Church's power over Neapolitan affairs.[26] While this debate was hardly new, this time reformers saw the church's privileges as just *one* of the many irrational aspects of southern society. Thanks to reformers' convergence with the crown's absolutist ambitions, the battle to free the state from church control could become part of an anti-feudal program that aimed to implement free trade policies, redistribute wealth, regulate the growth of trade and manufacturing, and redesign legislation.[27]

The emergence of this reform movement in the 1740s was accompanied by a reorientation of Neapolitan philosophical and scientific culture. As in other parts of Europe, those who had gathered under the banner of Newton's natural philosophy challenged academic knowledge. Newtonianism was not a well-defined or univocal set of doctrines and methods: the varieties of continental Newtonianism differed remarkably from each other and from British ones, and—as argued by J. B. Shank for the French case—their selective readings of Newton's texts changed in time, shaped as they were by local concerns and preexisting philosophical traditions, most notably those inspired by the works of Descartes, Malebranche, Leibniz, and Wolff.[28] The Neapolitan government viewed this cultural renewal favorably and supported it by reforming the university and creating the Royal Academy of Sciences. Among the protagonists of this early reformist phase was Antonio Genovesi (1712-1769), one of Vico's students and the most influential philosopher and economist in eighteenth-century Naples.[29]

Genovesi began his career as a theologian with a keen interest in metaphysics and logic, but by the mid-1740s his attention had shifted to economic and administrative matters, which he framed using an empiricist perspective inspired by authors like Locke, Newton, and Condillac. Rather than as a conversion or radical break, this shift appears like a strategic repositioning, as religious authorities had effectively silenced Genovesi's innovative theological work—which downplayed the role of miracles while emphasizing natural theology.[30] In Newton, Genovesi saw a paradigm for the harmonization of modern science and Christianity. His understanding of Newtonianism, built on the apologetic rendering of the Boyle lectures, was shared by influential members of the Roman Curia. And it was in Rome that his friend Celestino Galiani (1681-1753), a monk, had been replicating Newton's optical experiments as early as 1707—probably the first replications ever performed on the Continent. Galiani moved to Naples in 1731 to take up a bishopric; he soon led the university as major chaplain and participated in the creation of the new Academy of Sciences. This strain of southern Newtonianism was characterized by an appreciation of natural theology, and an emphasis on experimental practice rather than on the power of the new mathematics.[31]

In 1754, Bartolomeo Intieri, a wealthy land agent, endowed what has been described as the first chair of political economy in the world and offered it to Genovesi. As a professor of "commerce and mechanics," Genovesi began lecturing on "public happiness" and the rational administration of the state. In his inaugural lecture "on the real goal of letters and sciences," Genovesi argued that all intellectual activity—including the most speculative and abstract studies—should be directed toward the improvement of the people's material conditions of life, and that reason was the "universal art" that could secure the indefinite progress of these conditions. Following reason's guidance, philosophers had devised "a kind of geometry that is not idle, but perfects the arts," and "a kind of physics that promotes our welfare, without being magic." Bringing together sensationalism and the universal language of analysis, Genovesi argued that reason works best when immersed in the material world, when it's embodied and in action. In fact, "one cannot say that reason has achieved its maturity in a country where it is still placed in the abstract intellect, rather than in the heart and in the hands."[32]

Genovesi remained firmly within the cultural horizon of enlightened absolutism and Catholic orthodoxy. While he appreciated the transformative power of Locke's theory of ideas and French sensationalism, he feared the potential materialistic and politically subversive implications of these approaches and loathed those freethinkers who aimed to pervert the "order" and "internal texture of the world." Genovesi built his defending wall in the depths of the human mind, where the intellect reigns supreme, and where certain "internal sentiments" mysteriously emerge that are "the core of our true knowledge." These sentiments are not Cartesian innate ideas, but rather constitutive elements of human reasoning: the sentiments of existing, of being free, of right and wrong, of truth and falsehood. Moving from these sentiments and through pure intellection the mind can reach fundamental logical, metaphysical, and theological truths, from the law of contradiction to the existence of God. The pure truths of mathematics and ontology participate at this level of evidence and certainty. It is only in a logically posterior moment that the intellect engages with sensation and imagination to produce most of human knowledge, where all one can aim for is moral certainty.[33]

Between 1754 and 1769, the year of his death, Genovesi trained a new generation of reformers—administrators, landowners, lawyers, academics—who would campaign for a combination of free-trade policies, public incentives, and the rationalization of agriculture and manufactures. All these issues came together in 1763–64 as famine ravaged the southern kingdom. Reformers argued that the state's irrational administration, its system of regulations and monopolies, was the main culprit in the disaster. The famine also revealed, brutally, the gap between the living conditions of the oligarchy that ran the city, the inner elite of the Neapolitan aristocracy, and those of the rest of its swelling population, further increased by waves of starving immigrants. To

Genovesi, one important lesson was the need to invest in public education, especially training in modern agricultural techniques. He openly accused the Neapolitan church, the great landlords, and lawyers in the capital of putting their corporate interests before any reform. The abuse, he argued, would continue until their nefarious alliance was broken up. These themes are central to Genovesi's *Lectures on Commerce* (1765–67), which incarnated a new civil philosophy: that overcoming all juridical exemptions and privileges must be the fundamental goal of any "civilized state."[34]

Genovesi's civil battle was taken up by a group of his students, which included politicians, like Domenico Caracciolo (1725–1789), and students of law and society, like Ferdinando Galiani (1728–1787), Giuseppe Galanti (1743–1806), Gaetano Filangieri (1752–1788), and Mario Pagano (1748–1799). While embracing Genovesi's empiricism and sensationalism, these reformers paid less attention to the epistemological and metaphysical issues that intrigued their teacher. Rather, they followed his invitation to foster an impure science, one created by philosophers not isolated from the rest of culture and society; one that would focus on the real problems of the country and value practical results over speculations. In order to achieve practical results, reformers detached these problems' technical dimension from traditional metaphysical concerns and reduced their solution to a matter of rational administration.

By the late 1760s, however, reformers were growing frustrated. The structure of southern society, with its entrenched feudal-communal institutions and striking inequalities, had remained untouched. Although deprived of political relevance, the feudal aristocracy had maintained most of its juridical and economic privileges. In the provinces, the authority of the central government was still filtered by intermediate corps that turned the exercise of power into a complex mediation between local interests. Faced with such a dispiriting situation, reformers began to question the legitimacy of the feudal system as a whole, and the complete abolition of feudalism became the sine qua non in most reform plans. As lucidly recognized by Genovesi, this political change required the creation of a robust public sphere. Hence the reformers' growing emphasis on the need to educate the people and engage the "middle order" (*ceto mezzano*; the "public opinion" for Filangieri)—which they saw as a political actor independent from the feudal system that could replace the crown as the main force of social transformation.[35]

Political space for reform opened up in 1776. As a new prime minister took over, leading Neapolitan Enlightenment figures entered the ranks of the administration and the government. Pagano joined the Admiralty Tribunal and Filangieri and Palmieri the Council of Finances, while Domenico Grimaldi and Giuseppe Galanti were nominated inspectors (*visitatori*) of the kingdom. Domenico Caracciolo became viceroy of Sicily in 1781: he questioned the privileges of the barons and abolished the Tribunal of the Inquisition, a branch of the Spanish Inquisition that had been operating on the island for half a millennium; by 1786 he was prime minister.[36] To this period date a reform of the university; the re-establishment of the Royal Academy of Sciences; the creation of the *Cassa sacra*, an institute of credit designed to support small landowners; and a reform of the army, which drastically reduced the aristocracy's right to command. That only the reforms of the army and the navy were, eventually, successful reveals that the long-term priorities of the monarchy had not changed much from the earlier phase, and that reformers' and the crown's objectives had remained, by and large, different.[37]

This later phase of the Neapolitan Enlightenment was characterized by a convergence of reformism and Freemasonry. After Genovesi's death, the university was replaced by city palazzi and country villas as the main sites for elaborating enlightened discourses. The models of sociability offered by the *conversazione* and the Masonic lodge were appealing for their internal egalitarianism and their relative autonomy from religious and political power. Their success also attested to, as suggested by John Robertson, "the difficulty of establishing a genuinely 'public' sphere in the kingdom."[38] By the 1780s, every provincial town had its own lodge, where noblemen, bourgeois landowners, professionals, ecclesiastics, and literati met regularly to discuss the future of their country.

In Naples as elsewhere, Freemasonry was a multifarious phenomenon. The Neapolitan lodges of this period fell roughly into two main groups. One was more mystically oriented, inspired by "strict observance," and emphasized hierarchical elements. The other followed the "Scottish rite" and was concerned with the themes of egalitarianism, constitutionalism, and the "regeneration" of southern society. Most reformers were associated with this second kind of lodges. Provincial reformers would typically host a lodge in their mansions, while those based in the capital gathered at discreet locations such as the villa of the Duke de Gennaro, in the outskirts of the city. In 1786, many reformers joined the newly created Neapolitan branch of the Illuminati, an association originating in Bavaria that aimed at the regeneration of humanity based on moral renewal and the diffusion of scientific knowledge.[39]

The reformism of the late Neapolitan Enlightenment varied. Provincial reformism was more problem-specific, often based on technical assessments and supported by men who were themselves part of those productive processes under scrutiny. Reformers based in Naples were more likely to be public administrators, lawyers, university professors, or part of the precarious intellectual underworld as teachers and booksellers. Their contribution to the reformist discourse tended to be theoretical, dealing with questions like the role of the monarchy in a modern state or the science of legislation.[40] Some historians have interpreted the tension between these two strains of the Neapolitan Enlightenment as a tension between reform and utopia. The predominance of the more abstract and philosophical utopic component would have contributed to the eventual failure of this movement and the repression of the 1799 revolution.[41] This interpretation recalls earlier assessments of the revolution, which explained its failure in terms of excessively "abstract thinking" and the passive importation of foreign ideas and values that didn't meet the real needs of the Neapolitan people.[42] I depart from this interpretation in at least two respects. First, I emphasize the *continuity* rather than the tension between the two strains, which becomes visible when we follow the thread of analytic reason. I also turn the charge of abstraction and passivity into an object for historical investigation rather than accepting it at face value. One of the chief accusations raised against analysis was indeed that it brought a foreign and abstract form of reasoning to bear onto matters—scientific, social, and political—best treated with traditional and local methods.

Theoretical or practical, reformist publications of the late Neapolitan Enlightenment were all permeated by an increasing sense of urgency. Knowledge about the social and economic conditions of the kingdom had been compiled by two generations of reformers. Now implementation was needed. As a provincial reformer wrote in 1787, "a little practical application [of the reformist principles] is what the interest of the people now cries out for."[43] But time had run out. In July 1789, news of the "facts of France" triggered the progressive closure of spaces for reformist intervention, as a new government took a reactionary turn in both foreign and internal policy. The arrest and execution of the French monarchs threw the Neapolitan crown into a state of panic—heightened by the fact that the queen was a sister of Marie Antoinette. In 1793, Naples joined the anti-French coalition, and in 1794 the discovery of a "Jacobin conspiracy" set off a vast police operation to wipe out those who had not realigned. Among the last voices of Neapolitan reformism were Nicola Fiorentino (1765–1799), author of a general reform plan (1794), and Giuseppe Galanti, author of the multivolume *Political and Geographical Description of the Sicilies* from 1786 until 1794, when publication was interrupted by the authorities—large-scale political and economic analysis now looked suspicious. In his volumes, Galanti performed an impressive exercise in the science of statistics, based on years of data collection across the provinces. He aimed to demonstrate through numbers the structural dysfunctions of the kingdom and connect them to their common root: feudalism. Galanti argued that feudalism had lost its original function and was a ghost that haunted the kingdom, an institution that had transformed itself into a tyrannical mechanism of oppression. The elements of reciprocity that defined feudal landownership had disappeared, and feudal estates had turned into large monopolistic and tax-free private properties. The solution, Galanti believed, was in the creative force of the market. But he was losing hope. His narrative was dominated by a sense of impending doom and by his bitter disappointment for the apathy of those social groups, like the provincial bourgeoisie, that he had hoped to mobilize in support of reform. Galanti remained a monarchist who saw himself as the adviser of the enlightened prince. But, as he was writing, many of those who believed in the necessity of demolishing the feudal system were experiencing new forms of political mobilization and preparing more-radical political action.[44]

Reason's Machinations

Understandably, the Enlightenment's ambitions and achievements have attracted much more attention than has the resistance they met, though that resistance was often formidable and even insuperable. To dismiss the composite world of the anti-Enlightenment as one of mere social and cultural inertia would be misleading. More specifically, in the Neapolitan case, to understand resistance to reform as the straightforward expression of a conservative feudal aristocracy opposed to an enlightened bourgeoisie would not do justice to the complexity and fluidity of the social landscape. This simplified dualism would make the eventual failure of the reform movement an incomprehensible outcome that would call for fatalistic explanations based on the alleged innate indolence of the southern people, which in fact abound in the literature on the "southern question."[45]

Instead, let's focus on material culture in order to take a fresh look at these social dynamics of inertia and resistance. Across the southern kingdom, Enlightenment often meant an interest in concrete issues of technological innovation. Reformers claimed that most problems derived from the fact that inhabitants did not know their homeland. Basic infrastructures such as roads, ports, and canals had to be rationally redesigned based on a new, scientific knowledge of the country's landscape, and the economic potential of its natural resources was yet to be fully understood, let alone exploited. These reformers were often landowners who had an interest in rationalizing the productive processes under their control, developing new agricultural techniques, and commercializing new products.[46] The introduction and substantial failure of new olive oil technologies is emblematic in this respect. It signals an active and widespread resistance to analytic practices and materialities as forces of social transformation.[47]

Domenico Grimaldi (1735–1805), from the province of Calabria, introduced on his lands machinery and irrigation techniques he had seen in the
Po Valley. The Neapolitan economy was essentially agrarian, and producers
like Grimaldi needed to move from a subsistence to a commercial framework to intercept foreign demand. Grimaldi is best known for his efforts to
mechanize production of olive oil—the single most important Neapolitan export. He knew that new technologies had been introduced in oil-producing
regions of Provence and the Italian Riviera. Oil produced with the new machinery was faring well on international markets, and its price had begun
rising significantly. Why did Calabrian producers resist innovation? A keen
reader of the *Encyclopédie*, Grimaldi was convinced that the main reason for
their inertia was ignorance. The region needed agricultural schools as well as
producer societies that would finance technical innovation.[48]

Grimaldi replaced traditional Calabrian oil mills (*trappeti*) with so-called
Genoese mills. After some successful experiments, he hired Genoese specialists and processed his entire crop with the new method in 1771. The results
were clear: the Genoese press required less labor and produced more and
better oil. The new mill was also smaller and cheaper, so producers could
distribute more mills over their land, solving the problem of periodic over-
production. Based on these results, Grimaldi forecast the rapid diffusion of
the new machines across the kingdom. In 1783 Calabria was stricken by a
massive earthquake; to Grimaldi this disaster presented an opportunity not
to be missed. Most traditional oil mills had been damaged and could be replaced with Genoese ones. But doing so required opening new lines of credit
for small landowners and entrepreneurs whose businesses had been hit by
the earthquake and the American War. Grimaldi asked for state loans and
for the creation of a society of Calabrian oil producers. The government's
response was lukewarm, and the opportunity provided by the earthquake was
lost. Grimaldi's experience remained an isolated case in Calabria. Eventually,
he met financial failure and political doom.[49]

Why did Calabrian producers resist technical innovation? The new mills
were more powerful than the old ones, maximized production, reduced waste,
and could be operated around the clock. By modifying the design of milling stones, replacing lever presses with screw presses, and powering the mills
with vertical waterwheels or using oxen rather than horses, reformers increased the new mills' total power. Waste was reduced by introducing a series
of interconnected water basins that washed olive remains in order to extract
the last drop of oil. These changes, however, were far from being self-evident
improvements. Traditional machines were designed to crush olives at a relatively late stage of maturation so did not need much power. Grimaldi needed

more power because he aimed to mill younger olives to produce the low-acid olive oil international markets demanded. Low-acid oil was too expensive for local consumption but fetched high prices on the northern European markets for its utility as lubricant for industrial machinery. Grimaldi was not simply interested in producing more oil more cheaply: he wanted to produce a different kind of oil altogether, for an entirely different market. Within this export-oriented framework, traditional machines and labor processes became irrational, inefficient, and wasteful.[50]

Traditional machinery had been designed to capture oil and olive remains at various stages of the manufacturing process, as reclaiming these byproducts was one source of millers' profit. Getting rid of waste was only viable if a single owner controlled the entire production process, enjoying the full property of both land and machinery. In Calabria, the predominant feudal-communal system made it difficult for a legal agent to obtain such full propriety. The province was divided into small autonomous administrative units, called *università*, which included common, private, and ecclesiastical lands. Each land was subject to a number of different rights owned by different legal agents, be they individuals or communities. For instance, peasants enjoyed certain rights on private lands (e.g., grazing), whereas the feudatory held the monopoly over the production or transformation of specific products (e.g., milling olives). The *università* were therefore rarely responsive to pressure for commercialization and specialization.

The stability of the feudal-communal system had been further reinforced by specific kinds of economic and financial activities. For example, the credit system that allowed most of the agricultural producers (landowners or tenants) to survive was in the hands of a restricted group of financiers and merchants of the capital. Grimaldi had denounced the "illegal profits" made by speculators through the *contratto alla voce*, whereby merchants from the capital, exploiting the chronic lack of agricultural investments, could buy crops one year in advance at remarkably low prices. These economic elites had no incentive to radically modify the sociotechnical landscape of regions like Calabria in order to compete on the international markets, as they enjoyed a de facto monopoly on both the sale of credit to landowners and the traditional trading routes—such as the shipping of high-acidity olive oil to Marseille to feed the soap industry. Absentee landowners also had no incentive to replace traditional machinery and labor processes, as the diffusion of new and relatively cheap mills would put an end to the large feudal mills' monopoly over entire olive-producing regions. Furthermore, moving toward the production of low-acidity oil would require a radical change in commercialization strategy, and a new positioning on international markets. Those markets interested in Calabrian oil for making soap would be lost should prices start to rise, while large, risky investments would be required to enter new markets in competition with northern Italian and French producers.[51]

But Grimaldi faced other problems as well. In Calabria, few workers were willing to operate the mechanized mills. The feudal-communal system granted the peasantry relatively secure land access through traditional sharecropping tenures. Traditional oil mills temporarily hired peasants, men and women, who worked for only a few hours a day. But the new mill setup stripped chief millers of their discretion and authority and demanded crews work longer hours with fewer breaks: brigades of men alternated at the new machines to keep them going around the clock. "Nothing is more damaging to a fully equipped oil mill than being at rest," wrote a well-known Apulian entrepreneur.[52] Genovesi had theorized an analytic reason that was not confined to the intellect, but embodied and in action. Entrepreneurs like Grimaldi read Genovesi's arguments as an invitation to redesign the production process in every detail, imposing their rational vision upon the unthinking labor of machines and workers. Eliminating the chief millers was therefore a key step. In Grimaldi's new oil mills labor was divided analytically and organized in a sequence of elementary tasks. The continuous, regular, and synchronized functioning of the machines required a precise choreography and a new, more intense and coordinated kind of work. Technological innovation could succeed only where the local workforce could be effectively disciplined to its new role. In Calabria, resistance to the new work discipline was such that Grimaldi had to relocate workers from Genoa as he began experimenting with the new machinery. When his workers became the targets of local peasant attacks, Grimaldi was puzzled; to him, the traditional production process was obviously irrational and wasteful, while the new machines allowed for a rational—indeed analytical—organization of labor.[53]

The public administration also resisted technological change. In the province of Apulia, the feudal-communal system of land was less pervasive than in Calabria, which allowed for a partial modification of machinery and a gradual reorientation toward the production of oil for the international market. However, the Apulian revenue system seemed to prevent the full transformation of the manufacturing process. According to Giuseppe Palmieri (1721–1794), a local reformer, a well-known critic of duties on olive-oil exports, and, unlike Grimaldi, a civil servant, the excessive duties on oil exports were curtailing growth. He was especially critical of the system of the *arrendamenti*, whereby the government hired private companies to raise taxes on specific products—most notably olive oil, silk, and soap. Palmieri emphasized the perverse effects of this system, which led to tax increases at every stage of production and transportation. Merchants and landowners simply passed on the additional costs to tenants and day laborers. The complexity of the revenue system was, to Palmieri, a clear expression of the central government's inability to control its local branches and challenge local interests. A reform of the tax system, however, would require the standardization of the system of weights and measures. Only a standardized system, Palmieri thought, would allow the government to regain control over public revenues and implement a uniform duty on oil and other exports.[54]

By the 1790s, it was clear that the resistance of shareholders and administrators of the *arrendamento* for oil and soap was too strong to overcome. In his comments on the failure of the reform of weights and measures, Palmieri recognized that the staunchest opposition came from members of the *arrendamento*'s lower provincial administration, who felt immediately threatened. Interestingly, not even the powerful lobby of oil merchants in the Apulian port of Gallipoli—the most open to international trade—supported reform. The Spanish had designed the fiscal system during the seventeenth century to obtain short-term revenue. With the kingdom's independence, those revenues had become vital for the bureaucratic apparatus of the Neapolitan state and benefited a few well-positioned investors. As had been the case with other feudal-communal institutions, even the overly complex traditional system of weights and measures had been stabilized by financial and merchant elites who had carved out lucrative protected positions, turning the "irrational" variety of weights and measures into a source of profit.[55]

Finally, one important reason the crown failed to keep up with the reformers' expectations had to do with the exceptional economic situation of the overcrowded capital. Four hundred thousand people—more than one-tenth of the kingdom's population—lived in Naples, including most absentee landowners. With so many disincentives to invest in manufacture and agriculture, they had a high propensity to consume, almost exclusively in the capital. Thus the city's economic existence depended on the continuous influx of public revenues and baronial incomes from the provinces. The stagnation of provincial agriculture and the overgrowth of the capital were two sides of the same coin. Any serious attempt to direct the landowners' profits toward agricultural investment or dismantle the feudal-communal system would produce, in the short term, a dramatic rise of unemployment and unrest in the capital. And there was nothing the rulers of Naples feared more than the angered crowds of Europe's most rebellious city.[56]

To sum up, what reformers portrayed as the outcome of ignorance and inertia was in fact a complex sociotechnical system sustained by an alliance of old and new economic interests, typical of a region in which the bourgeoisie had grown into the interstitial spaces of the feudal-communal land system. Reformers attacked this system's concrete manifestations: "ignorant" resistance to technological innovation, "excessive" duties, or "confusing" systems of measure. The interests they defended would be best served by a combination of controlled liberalism and the free market. Resistance, however, was strong, coming from large sectors of the public administration, particularly its local branches; shareholders and administrators of companies that monopolized the production and taxation of specific products; great merchants of the capital who controlled exports and transferred the risks of a fluctuating market to agricultural laborers; local traders who exploited the confusion of the measures system; and bankers who sold credit to impoverished tenants at 20 percent interest. Reformers also faced the resistance of agricultural laborers. The failure of technological innovation in Calabria was a troubling symptom of a deeper resistance to the new enlightened order that aimed to replace what remained of the feudal-communal world. Peasants' and tenant farmers' living conditions were deteriorating due to the combined effects of feudal reaction—which was eroding customary-use rights and enclosing land—and commercial speculation. In this context, peasants perceived the new machinery and methods as direct threats to their entire way of life. It's no coincidence that in 1799 Calabria would be the center of a massive peasant insurrection that contemporaries identified as a civil war and called "the great anarchy." Peasants rose in arms under the banners of the Catholic counterrevolution to slaughter landlords and sack cities. However, it was not feudal castles they set on fire this time, but the city houses of the "new men," the entrepreneurs—those who supported the philosophical principles of the Enlightenment, the political ideals of the French Revolution, and the analytic management of mechanized production.[57]

Analysis and Palingenesis

Entrepreneurs and reformers like Grimaldi saw analysis as a set of resources for transforming the sociotechnical landscape of the southern kingdom. Unlike their counterparts in northern Italy or France, however, Neapolitans could not count on a shared culture of analysis to support their projects. Analysis had to be appropriated and legitimated, and doing so meant supporting mathematical training. Historians have long noticed the relatively low social and epistemological status of mathematics in the lively Neapolitan philosophical tradition that links Giovanni Battista della Porta's Renaissance naturalism to Giambattista Vico's critique of Cartesian analysis. Local elites valued mathematical training less than their foreign counterparts did, as shown clearly by the reception of Newtonianism.[58] Compared to northern scientific centers like Bologna and Padua, where a technical understanding of Newtonianism was predominant, the Roman and Neapolitan reception were characterized by a keen interest in the philosophical and apologetic dimension of the *Principia* and in the experimental strategies of the *Opticks*. From the 1710s, northern Italian scholars close to the group of the Bernoullis had contributed to the development of the techniques of infinitesimal calculus. In the south, developing these mathematical techniques was not a priority; integral and differential calculus entered higher education only in the second half of the century.[59]

Despite the reformers' efforts to raise the status of scientific disciplines, academic careers in natural philosophy and mathematics remained precarious, and professors typically supplemented their meager salaries with private tuitions. Scientific life and the circulation of people, texts, and instruments in Naples in the mid-eighteenth century was characterized by a close interaction with northern Italian centers, especially the University of Bologna. The new mathematics and its applications, however, do not seem to have attracted significant attention. Its positioning within the university curriculum reveals that Neapolitan elites perceived the new mathematics as, by and large, irrelevant.[60] Historians have marveled at the "singular lack of comprehension" of eighteenth-century southern Italian scholars and institutions for the potential of modern science. We seem to face yet another case of scientific backwardness here—perhaps the most emblematic. How should we make sense of it? As in the case of technology, considering the concrete conditions that made it difficult for mathematics to emerge as an authoritative discipline and a legitimating discourse in the southern kingdom is instructive.[61] Those northern Italian areas that were more receptive to mathematical innovation had well-established mathematical traditions and housed groups of experts in the Leibnizian-Bernoullian calculus. These experts were all more or less directly involved in the management of networks of rivers, channels, and waterworks that crisscrossed the Po Valley—a technical infrastructure of great economic and political relevance. For local governments to sustain expert groups in what they called the "science of waters" was an unquestionable priority. These expert groups played a role comparable to that of the experts of oceanic navigation for the Atlantic colonial powers and essential to the development of new mathematical tools for the representation and manipulation of empirical reality.[62]

In the southern kingdom, by contrast, central and local administrations had no immediate need for such expert groups. The practice of mathematics remained related to elementary teaching and, in some cases, to philosophical inquiry. Significant exceptions were the reform projects for the navy and the army, which contained some concrete attempts to update and expand the mathematical curriculum. The only institutions that routinely taught infinitesimal calculus were indeed the military academies.[63] The government systematically moved the most talented teachers from the university to the academies, where they taught courses and wrote textbooks in mathematics, mechanics, hydraulics, artillery, and nautical science, with few incentives to engage in original research. This situation was indicative of precise political choices, above all the decision to concentrate the reformist effort on training a new military elite that would defend the kingdom's commercial interests along the Mediterranean trade routes.[64]

By the 1780s, however, the belief that mathematical training—more precisely *analytical* training—was essential to the overall success of reform was on the rise.[65] And it found powerful political supporters, like Prime Minister Domenico Caracciolo, who blended the reformist agenda of his teacher Genovesi with Condorcet's faith in the boundless power of analysis, and who believed that a rational administrative system could only be designed and implemented by an elite of technicians trained in modern mathematics. Caracciolo had been the Neapolitan ambassador in Paris and had liaised between the first generation of philosophes and Neapolitan reformers. A personal friend of d'Alembert, Caracciolo was also an early admirer of Lagrange, who arrived in Paris from Turin under his patronage. Interestingly, in 1781 Caracciolo wrote to his former protégé to invite him to lead the Neapolitan Academy of Sciences. By that time Lagrange was well established in Berlin, working under conditions that the Neapolitans could not match—as he politely pointed out in his refusal.[66] Undeterred, Caracciolo would continue to pursue his technocratic vision until his death in 1789, two days after the storming of the Bastille.

The later phase of the Neapolitan Enlightenment was characterized by a growing tension between the faith in an all-powerful analytic method and a historically grounded pessimism about the possibility of southern society's escape from its destiny of backwardness and inequality. Reformers continued to work to introduce analysis as a privileged tool for reducing administration to calculation, but their efforts were now articulated within Vichian philosophies of history that emphasized the historicity of knowledge and the inescapable cycles of human society. These motifs were thrown into sharp relief by the 1783 Calabrian earthquake. That a civilized society could be established by means of a gradual and painless transition had become hard to believe. Instead, reformers began to equate the overcoming of the feudal system with some sort of great natural catastrophe that could wipe out existing social structures. As argued by Sean Cocco, connections between natural and social change were hardly new in Naples, a city whose very landscape was a constant reminder of the potentially destructive effects of ominous telluric forces. Reformers drew upon Vico's idea of cycles not only for understanding the life of civilizations but also for studying the effects of natural catastrophes, investigating the historical origins of social institutions like feudalism, and evaluating their current political function. Fascinated by the "spectacle of the revolutions of nature," they studied the "anthropological phenomena" relative to the earthquake, emphasizing the connection between nature and society. For some, like Mario Pagano, the earthquake was more than just a metaphor: it showed how the very basis of any civil society could be dissolved by natural events, bringing humanity back to its original state of equality and liberty. The catastrophe was therefore an agent of social change: upon such destruction one could finally build a new, regenerated and just society.[67]

In this context, readings of Vico's *New Science* were integrated with other motifs from the composite Hermetic philosophy that circulated in the Neapolitan lodges, such as pantheism, vitalism, the metaphor of the universe as an animal, and a dislike for facile mechanist reductionism. This constellation of beliefs—which Vincenzo Ferrone suggestively called "neonaturalism"— permeated Neapolitan culture in the last quarter of the century.[68] Neonaturalism and the Vico revival have been correctly interpreted as a pessimistic reaction to the substantial failure of enlightened reformism. This failure brought into question previously held assumptions, like the vision of history as a progressive process. Instead, cyclical philosophies of history, the catastrophic evolution of the earth, and the inevitable decadence of all civilizations became popular among reformists, as did the tendency to subsume the evolution of the human and natural sciences under historical categories.[69]

Neonaturalism, however, should not be seen as a radical rupture with the earlier Neapolitan Enlightenment and its science-driven reformism.[70] Rather than a crisis of reason, it signaled a reassessment of the concrete possibilities of social transformation. This reassessment mobilized cultural resources that had been part of the reformist discourse all along. There has been a historiographical tendency to cast certain traits of the late European Enlightenment as irrational involutions that questioned earlier infatuation with Newtonian science. Robert Darnton's study of mesmerism in France is emblematic in its presentation of mesmerism as the underside of the French Enlightenment. And yet, there are problems with this interpretive line. Jessica Riskin has shown how mesmerism had profound and ambiguous connections with the established science of its time. Above all, it was not a rebuttal but rather an extreme version of that culture of sensibility that had contributed to define the world of the *Encyclopédie*.[71] In the Neapolitan case too, what has been described as the collapse of Enlightenment should be seen in continuity with, and indeed as an acceleration of, trends already visible in the earlier phase. Filangieri and Pagano, the two leading figures in the 1780s, certainly adopted a historicist perspective, but they strived to integrate it with their constitutional, egalitarian, and scientific projects. The language and themes were shifting, but a fundamental faith in analysis as a method and a framework persisted. Consider, for example, how in their writings, pantheism and historical cycles coexist with faith in the fundamental rationality and knowability of both natural and social reality. In this sense, they held true to the end to Genovesi's original call to discover the "true laws of politics and economics," as "politics, like economics, has its own certain and eternal principles, and therefore its own theorems and problems."[72]

It's precisely the faith in a rationally ordered reality that drove Filangieri's *Science of Legislation* (1780–1789), the most influential text of the late Neapolitan Enlightenment.[73] To Filangieri, reason and moral sense speak a universal language and always prescribe "the same laws"—hence legislation has its own fundamental principles, just like any other science. Granted, "the state is a complex machine," but although its components and moving forces vary, the laws that govern them are fixed. Hence the science of legislation can be precise and certain. Only the scientific knowledge of the principles of legislation would allow governments to identify the most appropriate legal framework for social and economic progress.[74]

Filangieri's work is representative of the lively debate on constitutionalism and the rights of man that took place in the Neapolitan lodges during the 1780s, as an anti-despotic campaign inspired by the newborn American democracy was gaining ground. Such was Filangieri's enthusiasm for the American experience that he not only kept correspondence with Benjamin Franklin but also considered moving to the New World to contribute drafting the American constitution.[75] Like other cadets of aristocratic families, Filangieri felt out of place: a pacifist who asked to invest in education rather than in the army, he detested court life and gradually lost faith in the crown as an agent of change. As the American myth drifted in the background—too distant, he now believed, from the concrete conditions of the southern kingdom—he became convinced that the political impasse needed a pedagogical-epistemological solution. The minds and emotions of future "citizens" needed to be shaped by a sensationalist program of education, one that would respond to their material and spiritual needs by fostering wisdom and civic virtues through appropriate "impressions." Only the joint regeneration of citizens and the legal system would produce the necessary transformation. Interpreting Genovesi's notion of public opinion as referring to "men of good will," Filangieri could now appeal directly to the "will of the people."[76]

In 1786 Filangieri participated in the foundation of the egalitarian lodge of the Illuminati, and a few months later he met his "brother" Wolfgang Goethe, with whom he agreed as to the despotic nature of the Austrian and Neapolitan governments. Shortly after, in 1788, Filangieri's life was cut short by tuberculosis. In his eulogy, fellow reformer and Freemason Donato Tommasi (1761–1831) pointed out that Filangieri had envisioned the creation of a whole system of knowledge, a "new science of the sciences" that would explain the history of nations as well as the general history of man. The sciences, Tommasi continued, do not deal with the essences of things but with their relations, and all sciences are linked to each other because all truths are connected. The universal system that connects all truths—a majestic analytic table—is the object of the science of the sciences, of which the science of legislation is one component. The connection between Filangieri's systematizing effort and his battle against the monsters of despotism, militarism, and feudalism was crystal clear to his fellow Freemasons: only by discovering the "natural and eternal laws" of man and society could one renew a corrupt and deeply dissonant world and create social harmony—a society of equal and educated citizens.[77]

After Filangieri's death, his battle for constitutionalism, republicanism, and egalitarianism was taken up by Mario Pagano. A provincial self-made man and a Freemason, Pagano was a professor of criminal law at the University of Naples and had begun exploring the connections between the natural and moral order in an attempt to clarify the notion of "juridical proof." He became convinced that any reliable method for collecting circumstantial evidence should be based on a clear understanding of nature and of its relation to the laws that govern the moral world, "which are as constant and necessary as the physical laws." It is only on this basis that the judge "can be convinced and persuaded by circumstances."[78] Pagano's main work, *Political Essays on the Principles, Progress, and Decadence of Societies* (1783–1785), is emblematic of the combination of historicism, vitalism, and unshakable faith in the universality of reason that came to define the radical strain of the late Neapolitan Enlightenment. The *Essays* built on Vico's *New Science*, a formidable baroque encyclopedia whose labyrinthine construction lends itself to multiple interpretations. What impressed Pagano was Vico the brilliant visionary, his capacity to recognize structures and patterns that run through human history, but without canceling out contingency or complexity. Vico, for Pagano as for other Neapolitan radicals, was mostly a point of departure,

though. Pagano was fascinated by recent scientific developments, from volca-
nism to the discovery of deep time to Lavoisier's new chemistry, and he strove
for a science of history in which the same laws ruled both the physical and
moral worlds. Pagano believed that the "new science" could become an exact
science with precise laws, like mathematics. He proposed a systematic philosophy of history articulated in seven progressive stages of social evolution,
where change was driven not by providence but by nature itself. Pagano's nature was dynamic and pulsating, a force that keeps shaping and reshaping human societies, mostly through catastrophic events. Behind human progress
and the establishment of new forms of cultural and social life lies a primitive,
instinctive vital force that operates *in* humans and *through* humans. In his
narrative, civilizations rise and fall against the background of an immutable
and harmonic order: there is no gap between the history of humankind and
the history of nature, but an essential continuity.[79]

Pagano's monistic perspective can be traced back to his teacher Domenico
Cirillo (1739–1799), a professor of botany and practical medicine whose epistemological and political perspectives loomed large in the late Neapolitan
Enlightenment. A student of Genovesi and a Freemason, Cirillo was an important link between Neapolitan literati and the republic of letters: he cor-
responded with d'Alembert, Diderot, and Voltaire and was affiliated with
numerous European scientific academies. His critical reflections on the Neapolitan sociopolitical situation were published under the deceptively low-key
title of *Academic Speeches* (1789). They ranged from the eulogy of Filangieri
to the nature of sensibility and compassion, to the current state of hospitals
and prisons, to the philosophy of Rousseau—whose egalitarian arguments
Cirillo admired. An early follower of Lavoisier in chemistry, Cirillo had devoted most of his experimental work to the study of plant physiology and
the central role of "vital fluid" in supporting life. These studies provided the
framework for his understanding of the moral as directly related to the physical. For example, he reduced the universal sentiment of compassion to the
physical constitution of human beings, explaining that "nature has the essential attribution of maintaining peace and equality among its creatures, and
it is perturbed by the presence of misery, by the expression of sorrow." To
Cirillo, this connection between the physical and the moral demonstrated
that all "spiritual affections" are strictly linked to the body and to matter, and
that equality among men is a fundamental natural principle.[80]

But how could an equal society be finally achieved? For Pagano, Cirillo
and many others, that objective was still within reach. Despite their growing
frustration, they did not give up believing in the human capacity for self-
understanding, self-improvement, and civil progress. However, they doubted

that this progress would unfold as a continuous and smooth process. In Pa-gano, the passage from one historical stage to the next is invariably troubled, traumatic, and characterized by violent natural and social upheavals. He described it as the outcome of two opposite forces: one producing ruptures and desegregation, the other fostering unification and convergence. In other words, the much-needed social transformation of the kingdom could only be the result of a revolution.[81]

Mathematics at the Barricades

Antonio Genovesi championed a vision of science that would not be contemplative and idle but address "public happiness"—the real needs of the people. His students Filangieri and Pagano were convinced the true laws of social life could be discovered and used to establish a science of society that would legitimate a more just—republican and egalitarian—social order. In the 1780s and 1790s, a generation of science teachers and military officers brought the mathematical tools of analysis to bear on this radical discourse. The political significance of their technical expertise became fully visible during the crisis of the 1790s, especially in the writings and action of a group of revolutionary political activists who called themselves Jacobins.

Neapolitan Jacobinism is a well-established historiographical area, yet the involvement of its leading members in scientific life has been generally considered marginal to their political vision. Instead, exploring the nexus of science and politics is essential to understanding the transformation of Neapolitan life during the revolutionary age. In this chapter, the existential trajectories of long-forgotten mathematicians, men who experienced a profound connection between their scientific and political aspirations, take us to the heart of Neapolitan Jacobinism. These men's lives elucidate the disillusionment of reformist elites and their response to the crown's inaction. The continuities with enlightened reformism are clear, and many Jacobins—Mario Pagano, to name one—came directly from that tradition. Sensationalism was the foundation of a unified conception of knowledge that included the mathematical, natural, and moral sciences. The true laws of nature and society, they argued, can be discovered by deploying a set of basic principles and methods. Religious truths, on the other hand, are grounded not in experience or reason but in sentiment, and they belong to the personal rather than the public sphere. Overall, the cultural and political practices of Neapolitan Jacobinism were informed by what its adherents called the "spirit of analysis," a set of principles and techniques that were believed to express the workings of human reason at its best.

Almost every noteworthy figure in Neapolitan Jacobinism received some mathematical training. A basic understanding of analysis was an essential part of the Jacobin worldview, as were republicanism, egalitarianism, and anticlericalism. Analysis innervated their problem-solving methods in mathematics, their theory of knowledge, their moral and social arguments, and their political action. The very structure of their secret society—a network of Jacobin clubs—was a working model of how analytic savoir-faire could be deployed in matters of social organization. Neapolitan Jacobins aimed to find a universal method for discovering useful truths and handling pressing social and political problems. By the mid-1790s, they had become convinced that their vision of a just and equal society could be realized only through the concrete deployment of a new mathematics—a *revolutionary mathematics*.

A Jacobin Conspiracy

Carlo Lauberg (1753–1834), Jacobin leader, scientist, and teacher, is emblematic of the intersection of Neapolitan Jacobinism and science. Following him through the last decade of the eighteenth century means reconstructing Neapolitan Jacobinism itself. Most of what we know about his activity is derived from the surviving proceedings of political trials that began in 1794. Lauberg had been teaching mathematics privately since around 1790 with a young mathematician named Annibale Giordano (1771–1836).[1] They ran a school in Vico dei Giganti, a dark and narrow street in the heart of the ancient city, where students and friends met regularly to discuss mathematics and modern chemistry, especially the works of Lagrange and Lavoisier. Lauberg was a chemistry lecturer at the military academy, while Giordano, aged nineteen, was already a well-known mathematician and a professor in the same institution. He had studied with Nicola Fergola, and in 1786 he had presented the members of the Royal Academy of Sciences with an original synthetic solution for the so-called Cramer's problem. The solution was hailed as an example of synthetic elegance and found its way into the acts of the Italian Society of Sciences—the first mathematical work from Naples to appear in the prestigious journal in a long time.[2]

In May 1792, the two teachers opened a larger school in piazza Santa Caterina. They lectured twice a week to an audience of young men "of distinct condition," discussing mathematics, chemistry, and "democratic doctrines."

At the end of each lecture, a group of students would stay to discuss the works of Neapolitan reformists and French philosophers. On such occasions, Lauberg—an imposing, charismatic figure—spoke about "politics, the facts of France, and about the success of talented individuals living under popular governments." The school attracted a wide range of students, hundreds of them, many from well-off provincial families who had moved to Naples to study law or medicine. They shared the school's benches with lawyers, physicians, military officers, and highly skilled artisans, especially cabinetmakers, gilders, and clockmakers. A few clergymen could be spotted, as well as a few scions of the capital's great aristocracy, like Gennaro Serra (1772–1799), son of the Duke of Cassano, and Ettore Carafa (1767–1799), son of the Duke of Andria.[3]

The Santa Caterina school was promptly infiltrated by the police, who were trying to map the fast-changing landscape of conspiratorial groups across the kingdom. In 1790, as the government was growing suspicious of the Masonic anti-despotic discourse, the police had begun reporting unusual lodges in Calabria that had taken an interest in the facts of France. Apparently, they were not affiliated with any lodge of the capital but with a lodge in Marseille.[4] Behind them, the police suspected, was Abbé Antonio Jerocades (1738–1803), a former student of Genovesi and erudite man of letters who had authored a successful collection of poems celebrating Masonic values in 1785. His writings echo the palingenetic visions of the late Neapolitan Enlightenment, and he had been among those who had mourned Filangieri's untimely death in 1788. On his return from a trip to Marseille in 1789, Jerocades fostered the transformation of some Calabrian lodges into conspiratorial cells inspired by the French Jacobin clubs.[5]

Between 1791 and 1792, Jerocades was in Naples, where he visited Lauberg's school and was instrumental to the creation of new lodge-clubs "in the style of Marseille." Lauberg and Giordano joined one of these clubs. Participants discussed recent international events, read gazettes and excerpts from Condorcet's *Chronique de Paris* (1791–1793), criticized the crown, and saluted the French revolution as an "attack against tyranny." They argued that violent rebellion was legitimate, as it was based on the "rights of reason" and necessary for "civil progress." Liberty and equality, their main objectives, could be fully achieved only within a republican state. They discussed religion freely and argued that official religions were mere instruments in the hands of despotic governments.[6] Lauberg's closest collaborators were proselytizing well beyond the clubs: among dockworkers, for example, and at the opera theater, where they established connections with members of the nobility sympathetic to their anti-monarchic stance.[7] Meanwhile, select students from the

Santa Caterina school and other scientific circles connected to Lauberg were affiliated with the clubs, as were prominent figures like Troiano Odazi (1741–1794), a student of Genovesi who held the chair of his teacher at the University of Naples. The network was also growing its international connections: by August 1792, Lauberg was receiving the Parisian press directly from the new French ambassador.[8]

In December 1792, an unexpected event triggered a further transformation of these Marseille-style clubs. The Neapolitan government, now fiercely anti-French, had plotted to prevent France from being represented in the Ottoman Empire. The French government retaliated, dispatching a naval squad under the command of Admiral Louis de La Touche-Tréville to the Gulf of Naples, threatening to bomb the city. The French admiral was received at court in a series of diplomatic meetings. Parallel to the official negotiations, between December 1792 and January 1793, La Touche and his officers contacted local Jacobin leaders. The admiral visited the Santa Caterina school and attended some of Lauberg's chemical demonstrations. During one of his visits, La Touche addressed the students, inviting them to emulate the French nation. Giuseppe Cestari, a Jansenist priest and a Freemason, organized a dinner for local Jacobins and French officers. The admiral later reciprocated on board the flagship *Languedoc*; he approved a general reform of the Neapolitan lodges and promised military support in case of an armed insurrection.[9]

As the French fleet left the Gulf, the participants to those dinners were put under strict police surveillance, while religious authorities dispatched Jerocades and Cestari—two priests—away from the capital.[10] Realizing that the police were onto them, Lauberg decided to plan for short-term action. Jacobin propaganda intensified, and support came from unexpected quarters, such as trading and financial elites who saw their interests threatened by the government's new anti-French politics. In May 1793, Lauberg became the head of the Society of the Friends of Liberty and Equality, and thus the "political author of the conspiracy." In a sign of the new confidence that pervaded the movement, the society translated the French Constitution of 1793 and printed it in two thousand copies to distribute across the kingdom.[11]

In July 1793, Lauberg met with his closest collaborators to plan the abolition of the surviving traditional lodges and their absorption in the network of Jacobin clubs. Among the participants was Ferdinando Visconti (1772–1847), a young artillery officer and mathematician and a link between Lauberg and the army.[12] Lauberg reminded everyone of the vital objectives that should guide their actions: the end of feudal abuses, a popular and republican government, liberty and "perfect equality" for every citizen, and the rejection of "every religion, as extraneous to the natural order and invented by sovereigns and supreme authorities to guarantee their stability." The participants also approved a last reorganization of the Jacobin network into the Neapolitan Patriotic Society. Its structure was a veritable incarnation of the principles of analysis. The building blocks were the "elementary clubs," small autonomous units with no more than eleven members, each of them run by a president and a secretary, though without any hierarchy between functionaries, as "in these clubs they hold to the equality of all individuals."[13] Each elementary club elected a deputy, and all deputies met in a deputy club, again with no more than eleven members. Each deputy club elected a representative to be sent periodically to the central club. Through the deputies, and only through them, the different levels were "harmonically" linked to each other. Lauberg was elected "central point" of the society. The central club had executive and legislative power. Clearly, the organization was shaped by the need to shelter the identity of its members. But it was also an exemplary embodiment of analysis, made up by a set of identical and interchangeable elementary units—the members—who all enjoyed the same rights and duties, only differing in function. The Neapolitan Patriotic Society was a social experiment in which analysis legitimated and performed a radical egalitarianism.

The counterrevolutionary discourse would be quick in turning these analytic features into the "Jacobin machine," a deadly device for the control of public opinion, political life, and the state. Discussing the French case, Michel Vovelle convincingly dismantles the historiographical image of a monolithic Jacobin machine, showing the diverse, pragmatic, and contingent character of Jacobin strategies. Yet, the Jacobin machine metaphor is revealing and worth attention. Vovelle traces its trajectory up to François Furet, who describes the Jacobin club as "a machine for producing unanimity." Looking at the use of this metaphor from Naples, it becomes clear that it functions not only to emphasize discipline, organization, and capacity of control, but also the essentially extraneous and polluting character of what is, ultimately, a set of manipulation techniques. In Naples, the Jacobin machine is foreign, disconnected from local political traditions and extraneous to the body politic. But in France too its effect is one of contamination, this time from the inside. In both cases it is the purity of the body politic that must be defended from the mechanical-analytic threat.[14]

By fall 1793, the Neapolitan Patriotic Society had been reorganized, and the "democratization of the spirits" was in full swing. A club met in the workshop of the clockmaker Andrea Vitaliani (or Vitaliano, 1761–1799), discussed the political conditions of the country, and pondered questions such as the existence of God and the immortality of the soul. Members agreed that the people have "the faculty of accepting or rejecting kings."[15] Similar points were discussed at more elegant gatherings in the villa of Marquis Giovanni Letizia, one of the links between the Patriotic Society and those aristocratic Freemasons who looked with interest at the republican experiment in France. Between December 1793 and January 1794, new elementary clubs were opened within the army, at a major Neapolitan hospital, and in the provinces. Jacobin leaders knew they needed the support of the provinces for the revolution to be successful and believed that the social tensions in the rural world might be to their advantage. The new provincial affiliates were mostly "new men" who had been investing in former feudal land and gaining influence in local administrations. Hence, in the provinces the term "Jacobin" began to be associated with wealthy landowners, speculators, and professionals: the new oppressors who pushed for the enclosure of common lands and the abolition of communal rights. The deadly fight between Jacobin villains and the "low people" (*popolo bascio*) would figure prominently in contemporary popular ballads.[16]

Lauberg's students at the Santa Caterina school played a major role in proselytizing in the provinces. In December 1793, twenty-year-old Emanuele de Deo (1772–1794) traveled back from Naples to his native province of Apulia. He carried copies of the translation of the new French constitution and, trying to stir up revolutionary enthusiasm among friends, read aloud satirical poems against the monarchy. Mentioning the possibility of an insurrection in the capital, he argued that the people have the right to dethrone the kings and execute them. His activity did not go unnoticed: the young man was arrested, and the police searched Lauberg's school and house. They could not find anything conclusive, although officers seized many "pages covered in mysterious algebraic signs."[17]

At the ·Patriotic Society's recommendation Lauberg fled the country in January 1794.[18] By March, the magistrates had enough elements to proceed with searches and arrests.[19] A special tribunal was set up to take to trial those involved in what was called "the Jacobin conspiracy." The accusation was grave: "conspiracy against religion, monarchy, and the state." Judgment was reached and sentence passed in just a few hours, without granting the basic rights of the accused.[20] Mario Pagano denounced the improper use of this procedure and offered to defend some of the accused. He himself, however, would be arrested in 1796. Troiano Odazi was arrested and died in prison under unclear circumstances. Annibale Giordano was sentenced to lifelong imprisonment in a fortress; artillery officer Ferdinando Visconti was sentenced to ten years. Arrests continued for a couple of years, both in the capital and in the provinces. Numerous aristocrats and high state functionaries were also arrested, "suspected of Jacobinism." Ettore Carafa was arrested in 1795; he escaped from prison in 1798. Overall, a few hundred were implicated, from shopkeepers to magistrates and nobles.[21] The student de Deo was hanged in October 1794, together with another student from out of town and the brother of the clockmaker Vitaliani. They were small fry, but their execution was a clear message to Jacobin leaders who had fled the country or remained undetected. A large crowd gathered around the gallows on the day of the execution, and tension must have been palpable: the soldiers' reaction to a false alarm left six dead and thirty-five injured.[22]

Neapolitan Jacobinism

While they have been often studied in the isolation of their national contexts, Jacobin networks had connections and looked at each other for strategies, symbols, and organizational models. In 1794 "Jacobins" and "patriots" were on trial for conspiracy and insurrection in Piedmont-Sardinia, Genoa, Rome, Bologna, and the Veneto, as well as in Germany, Austria, Hungary, Poland, England, and Ireland. Similarly transnational was the Jacobinophobia detectable from Naples to Scotland, where Tory professor of natural philosophy John Robison confirmed the existence of a single massive conspiracy "against all the religions and governments," whose main actors were Jacobins and Freemasons; some of his English colleagues went on to argue that Jacobins ate pies filled with human flesh.[23]

In 1796, the arrival of a French army in the Italian peninsula opened what would be remembered as the "Jacobin triennium" (1796–1799), a period characterized by the creation of numerous sister republics that looked at the French republic of 1792–1794 as a symbol of liberation. Whether the term "Jacobinism" should be used to describe these movements and what exactly it would mean in this context are long-debated and controversial issues. Franco Venturi cautioned against using the term for movements active after the crisis of French Jacobinism and emphasized the distinctive features of the Neapolitan movement. Other historians have insisted on the theoretical and programmatic limits of Italian Jacobinism, which they characterized as comprising intellectual groups unaware of the concrete social conditions of their countries. Terms like "republican," "democrat," or "patriot" are currently preferred to "Jacobin" to describe the protagonists and events of the triennium, creating some confusion between the fluid actors' semantics and later analytic categories.[24]

By contrast, I decided to retain the term "Jacobin." I refer to the historical use of this term: it was an actors' category, used by certain political activists and—with gusto—by their enemies. Other terms were used as well, and the Neapolitan protagonists of this chapter would also identify themselves as "republicans" and "patriots," especially after 1795. These terms, however, were no less charged, ambiguous, and strategically deployed. Choosing "Jacobin" over other possibilities means emphasizing one component of the articulate democratic world of the Italian peninsula, which tended to split, when facing key decisions, into moderate and radical—Jacobin—factions. The Neapolitan Jacobins elaborated a doctrine and tried to implement policies that were directly inspired by French Jacobinism. Theirs was a republican, democratic, and egalitarian movement that saluted with enthusiasm the 1793 French constitution and its sweeping plans for democratization and redistribution of wealth. Its doctrine was grounded in citizen virtue, natural law, natural rights, constitutionalism, popular sovereignty, and a strong universalism. The immediate political objective of this movement was an insurrection and the overthrowing of the monarchy.[25] The conquest of the state would be used to transform society on behalf of the will of the people. Even after Thermidor, it has been often noted, they kept referring to *robespierrisme* as "the heroic moment of the French Revolution." They had concrete connections with that moment—such as the collaboration with Filippo Buonarroti, the *agent national* sent by Robespierre to support Italian Jacobins and their early democratic experiments. From about 1795, the Neapolitans also began to see the unification of Italy as a necessary condition for political and social revolution—hence the important overlap with patriotism and Italian nationalism. And during the 1799 Neapolitan Republic—which the French Directory never officially recognized—they aimed to assemble an entirely new state machine and worked on a series of laws that would push egalitarianism well beyond its juridical meaning.[26]

Finally, I use the term "Jacobinism" to turn its semantic ambiguity, its tension between the historical and the conceptual, into a resource. In the economy of this book, Jacobinism stands for a crucial shift from previous reformist experiences, a "qualitative mutation," as Michel Vovelle called it. It's a decisive shift to the political, instantiated in the in-practice experimentation of new forms of egalitarianism and democratic life. I refer to Jacobinism to identify a moment of rupture, in which political action—including the use of force and vanguardist agency—is used to bring about social change *now*, escaping the apparent fatalities of history as well as the logic of reform. To reconstruct Neapolitan Jacobinism and its political and mathematical imagination is the first of the two main steps of this book. The second step reconstructs the *reaction* to this revolutionary rupture. I'm obviously interpreting broad, *longue durée* processes, but focusing on the dyad of revolution-reaction and the crucial passage of the 1790s is a strategy to sharpen my argument about the nexus of politics and mathematics. It is within the Jacobin shift to the political that, most clearly, mathematics became *essentially* revolutionary. It's within the reaction against that shift that the *very possibility* of a revolutionary mathematics was closed down.[27]

✳

Neapolitan Jacobins' social profiles reveal a prevalence of the urban petty bourgeoisie, though they vary significantly. Among them were lawyers, military officers, soldiers, ecclesiastics, students, physicians, academics and literati, artisans, small urban traders, shopkeepers and their workers, servants—and a few nobles, too.[28] Many were young, like Emanuele de Deo, but others had long been arguing for social reform, like Mario Pagano, the professor of law, reformer, and Freemason who joined the Società Patriottica and was imprisoned as a Jacobin in 1796. In the provinces, the majority were entrepreneurs, non-absentee landowners, and members of the minor baronage: emerging local elites that constituted the "public opinion" invoked by reformers. Domenico Grimaldi, the olive-oil entrepreneur, was part of a Calabrian progressive lodge that became a Jacobin club in the early 1790s. Under surveillance from at least 1793, he was arrested in 1798. He was in prison during the revolution of 1799, but his son Francesco Antonio, a military officer, joined the cause of the Republic.[29]

The documents of the anti-Jacobin trials confirm the movement's composite character and its ramifications in various sectors of Neapolitan society.[30] Especially noticeable is the presence of artisans and small traders from the capital. This evidence runs counter to interpretations of the Jacobin conspiracy and the events of 1799 as a revolution led by a small group of utopist intellectuals. Conservative and liberal interpretations have insisted on the Jacobins' misguided attempt to import foreign forms of political organization and their reluctance to attend to the alleged real needs of the people. Consistent with these interpretations, in the nationalistic readings of the nineteenth and twentieth centuries, the Jacobin triennium is an ephemeral aberration, extraneous to the Risorgimento of the Italian nation. In Benedetto Croce's more benevolent interpretation, the Neapolitan Jacobins are martyrs, heroic precursors of the fin de siècle southern liberal bourgeoisie, but they are still dreamy, abstract: "great idealists and bad politicians."[31] On the contrary, Jacobin plans and ambitions were rooted in the rich Neapolitan reformist experience and its detailed assessments of the social, political, and cultural conditions of the southern kingdom. Even republican catechisms had a strikingly pragmatic dimension. During the 1799 Republic, for example, Onofrio Tataranni, a priest and reformer who had been fascinated by Bourbon social experiments that brought together utopian communities and mechanized production, wrote a catechism to defend the values of liberty and equality. Both, he claimed, depend on the "universal law of Nature, which is Reason." And both need to be sustained by education, humanistic and scientific. Especially valuable are "the physical and chemical sciences," because they are essential to agriculture; a lengthy description of agricultural techniques that should be introduced in the kingdom followed.[32]

Antonio Gramsci argued that these Jacobins understood better than anyone else the reality in which they were immersed and conceived the state as "a concrete form of an unfolding Italian economic development." Against the charges of "abstraction," he emphasized the concrete nature of their political action.[33] Thus, for example, they realized the importance of building a popular following through street theater and republican catechisms, and mobilized politically local devotions and rituals: during the 1799 Revolution, they would turn Saint January into a Jacobin. The problem, as acutely noted by John Davis, "was not that the republicans did not try to appeal to popular religious sensibilities, but rather that those sensibilities were manipulated more skillfully by their enemies." Renata de Lorenzo notes the similarity between the patriotic and xenophobic "sentiments" mobilized by the Jacobins in France and those mobilized by the counterrevolution in Naples. And Paolo Viola comments on the revolutionary potential of the counterrevolutionary campaign of 1799, which begins with a popular insurgence against the king's representative, accused of plotting with the French and thus betraying the Neapolitan nation.[34] The Neapolitan Revolution was "passive" not because it was utopic or because the peasant masses were necessarily and invariably hostile to change, but because it ultimately failed to bring together the city and the provinces through a revolutionary agrarian reform. The Neapolitan Jacobins were doomed when—under extreme financial and military circumstances—they failed to bring peasants and dispossessed day laborers into political life. These groups' growing discontent would fuel massive counterrevolutionary insurgencies instead.

But let's return to Lauberg. We left him on the run in January 1794, when he sailed northward on a boat arranged by Jacobin affiliates. He probably landed near Rome, where he must have spent some time in jail. In September of that year, he was in the south of France, where he married a local woman and enrolled in the French Army of Italy as an officer pharmacist.[35] By year's end, most Neapolitan Jacobins who had fled their country had regrouped in France or in the French-occupied town of Oneglia on the Italian Riviera, where the *robespierriste* Filippo Buonarroti (1761–1837) had begun a political experiment of republican self-governance and national liberation.[36] Lauberg

and many other Neapolitan and Italian Jacobins played a role in the administration of Oneglia, a veritable laboratory of democratic and egalitarian practices. The reputation of the Neapolitan Jacobins was on the rise, and Buonarroti was convinced that "if Italy is destined to be free, the real revolution will
begin in the torrid climate of the Vesuvius." In 1796, they were with Bonaparte's army and for the next three years were active across northern Italy
as public speakers, journalists, and political organizers. Their objective was
twofold: national unification and revolution. The French authorities looked
at their mix of democratic radicalism and Italian patriotism with suspicion
and referred to them polemically as *anarchistes*. In 1797, Bonaparte himself
worried about the growing number of Neapolitan expats in Genoa; they, he
wrote, "have always brought to Italy disorder and anarchy."[37]

Lauberg settled in Milan with former students Matteo Galdi (1765–1821)
and Flaminio Massa (1770–1805). They organized patriotic theaters and societies of public education, planted liberty trees, and founded periodicals
intended to persuade local elites of the advantages of republicanism and
Francophile politics. As a town was occupied, these activists moved in to de-
mocratize and regenerate it. Lauberg was active in Bergamo, Brescia, Verona,
and Venice, where he was president of the Society for Education and lectured
on the "politico-moral principles" to be taught—in dialect, if possible—to
the people. He must not have been perceived as a mere French agent if the
citizens of Brescia wanted him as their representative in a difficult negotiation
with the French military command. His activity was supported by generals
Joubert and Championnet, committed republicans who were among the few
who saw the Italian operations as a war of liberation.[38]

The ideal form of government for the yet-to-be liberated Italy was much
discussed. Galdi and other radical patriots close to Lauberg published essays
arguing that an indivisible democratic republic would best suit the unified
nation. They rejected any form of federalism as dangerous to the integrity of
the state: a republic must unite the people in a "single body." Their staunch
opposition to intermediate political institutions echoes both the French debate and the reformist critique of the Neapolitan feudal-communal system.
The rejection of federalism was not simply a matter of institutional principles: only a centralized state could proceed to a more equitable redistribution
of wealth, starting from assigning land to dispossessed peasant families. National unification, in other words, was a necessary condition for the realization of their political and social vision. In the republic to come, the institution
of hereditary nobility and any other source of "particularism" in jurisdiction,
customs, and currency needed to be abolished. On these issues, the opposition between radicals and moderates expressed a fundamental social conflict:

the moderates, who tended to align with the French Directory's normalization, resisted plans for wealth redistribution and argued for the administrative autonomy of towns and regions.[39]

The fundamental unity of natural and moral phenomena was Lauberg's basis for turning scientific knowledge into the most effective guide for political action. Ultimately, he claimed, political and social progress are grounded in the human body and unfold naturally once the impediments of ignorance and oppression are removed. It was for this reason that scientific elites were called to lead the nation: mathematicians, physicists, and chemists needed to join the ranks of politicians and legislators.[40] In a 1797 speech hailing the decisive occupation of Mantua, Lauberg made an interesting analogy: "As sensations, in the limited sphere of our individual existence, shape our heart and our intellect, so political collisions, which are the sensations of mankind, gradually shape human reason." In this way, "reason emerges, majestic, from the chaos of the centuries; its laws reunite all men in a single family." Reason itself emerges historically as a natural consequence of the organic constitution of human beings, and political revolutions are necessary because they "are founded on our own physical constitution, and on our relations to the beings around us." The revolutionary process doesn't depend on a few individuals, and, most importantly, cannot be stopped. It can, however, be accelerated by the conjunction of science and "republican courage."[41]

The battle for reason was at once scientific and political. Superstition and despotism, the "two tyrannies," could not be fought and defeated separately. In 1798, in a speech at the Constitutional Circle in Milan, Lauberg exhorted the audience to find in scientific knowledge their most powerful revolutionary weapon:

> Republicans, thus far the slaves have looked for justice in imaginary archetypes; you should study man—he is sensitive, and his relations [to external reality] will take you to the true ideas of good and evil. Tyrants have imagined a being invested with their own attributes in order to justify their crimes; you are rational, so find in your own reason the model that will inform your actions. Nature is not a vain name, it exists, and its existence confounds itself with the eternity of the centuries. It's active, and its ultimate, imperceptible molecules are forces that, with their necessarily constant action, produce every single phenomenon in the universe. This is what you should study, in order to find the Ariadne's thread that will guide you in your systems. If you depart from its principles, you will fall in the inextricable labyrinth of human opinions.[42]

The idea that a revolutionary political program should be grounded in analysis and sensationalist epistemology informed Jacobin publications, including what is perhaps the most representative text of its kind: an Italian and French bilingual booklet titled *Catechism of the Rights of Man*, dating probably to late 1794. It's a question-and-answer commentary on the first thirty-five articles of the *Déclaration* that introduced the French Constitution of 1793. It contains typical themes from French Jacobinism, like the legitimation of tyrannicide, as well as more distinctive Neapolitan motifs, like the role of scientific education in grounding the state and its laws. The authors were two southerners with a passion for analysis: Ascanio Orsi (1770–?) and Michele de Tommaso (1765–1830). A monk and a priest, they had studied and conspired with Lauberg in Naples, fled the city in 1794, and settled in Oneglia, where they penned this important document, which was quickly deployed as a textbook and for propaganda. Tommaso taught mathematics and sensationalist philosophy in Oneglia, and among the books that Buonarroti ordered for him in Paris were twenty copies of Condillac's *Logic*. For Tommaso, neglecting the analytic method had been "one of the main sources of our errors."[43]

The *Catechism* opens with a concise introduction to sensationalism, presented as the scientific basis of "natural equality," the first and foremost right: all human beings share the same physical constitution and therefore the same needs, passions, and faculties. Equality, "the basis of men's happiness," must be preserved through law. Laws that defend particular interests are therefore unnatural and illegitimate. If a society is unequal, "everything must be changed from the foundations: destroy every inequality." While "perfect equality" is unattainable, the law should ensure that no individual amasses excessive wealth or lives in poverty. Public education is necessary to allow citizens to discover and use their natural reason and thus ground their coordinated action, which in turn sustains national sovereignty. Sovereignty, which is indivisible and inalienable, is of the people, "the one and only nation." In defense of their sovereignty, citizens have the right of resistance and insurrection, and every citizen has the right to assassinate a tyrant. Now is the time, the authors concluded, to educate other men and take up arms. Theirs was a clear and urgent call for revolution in Italy at a time when French military support seemed to falter, and French Jacobinism a thing of the past.[44]

Lauberg's Analytic Program

The liberated reason that had finally emerged from the fogs of history to guide the Neapolitan Jacobins' action was individual, autonomous, analytic, and legislative: let reason free to flourish, they argued, and everything else will follow. In their vision science, epistemology, morals, politics, and history were seamlessly integrated. What was needed was removing oppression, impediments, and distortions: only in this way could the objectives of liberty, democracy, and equality be achieved. But what kind of science were they referring to? Above all it was the universal savoir-faire of analysis that captivated their imagination, together with sensationalist epistemology and the operative, anti-metaphysical stance of much late-eighteenth-century reformist thought. What was new was the context in which they were acting: reformism had been replaced by revolutionary praxis and by an urgent and deadly political-military struggle. Minds and hearts had to be won quickly and en masse.

Jacobin science is not a darling of French Revolution historiography. According to a persistent view, between 1792 and 1794 French Jacobins attacked science in the name of Rousseauian irrationalism, populist rhetoric, and a "vulgarly utilitarian" understanding of knowledge. Cold War historians, in particular, insisted on the Jacobins' devastating attack on science, symbolized by Lavoisier's execution. For Gillispie, Jacobin philosophy of science was directed against the theoretical and highly mathematical Newtonian science, which they considered abstruse and of little practical interest. By contrast, a more intuitive and popular science inspired by the philosophy of Rosseau and Diderot should be promoted. Gillispie unmasks the ideological component in the Jacobin position: theirs is a science that enhances specific social interests through the celebration of the experience and competences of their social basis, while devaluing ancien régime scientific institutions and corporations. The Jacobins, he concludes, distorted and degraded science by polluting it with political considerations: what they did was akin to later totalitarian distortions of science like Lysenkoism, and that's why he cannot force himself to speak of a "Jacobin science."[45]

While Gillispie identifies perceptively the political dimension of Jacobin science, there are problems with his argument. Historiographically, he assumes that one can neatly separate an "analytic" Enlightenment (Newtonian, mathematical) from a "sentimental" Enlightenment (empirical, sensationalist)—with Jacobins mobilizing the latter against the former. But, as we argued in the previous chapter, these dimensions of eighteenth-century culture cannot be disentangled, as most natural philosophers endeavored precisely to strike a balance between the sensory and the abstract, the experimental and the rational. The idea of a pure Newtonian Enlightenment attacked by the partisans of passions and irrationality is a historiographical figment—admittedly an enduring and powerful one. Not surprisingly, Gillispie has to downplay the analytic elements that he encounters describing Jacobin science and technology, such as the very analytic French Republican calendar.[46]

Some historians have also emphasized how French Jacobins attacked the elitism and privileges of royal scientific institutions, their monopoly on technical judgments, rather than science per se. To Roger Hahn, Jacobin scientific vandalism is only a "myth," while Henry Guerlac notes that when the Jacobins executed men of science they were executing "financiers, politicians, and public officials"—the persecution of science is an a posteriori perception.[47] Finally, from a methodological point of view, Gillispie's reconstruction is obviously asymmetrical. On the one hand we have the polluted science of the Jacobins that requires social interpretation, and on the other the pure science of their enemies—a self-explanatory mirror of rationality and nature. And yet, again, Gillispie is perceptive: this *is* a crucial historical juncture for the notion of pure science. But not because pure science, supposedly situated in a fanciful domain of transhistorical objectivity, came under Jacobin attack. Rather, pure science—pure mathematics first of all—emerged *in reaction* to the programmatically politicized science of the Jacobins. François Furet, in his famous exchange with Michel Vovelle on the legacy of the French Revolution, dismissed Jacobinism to dismiss once and for all the question of *revolutionary politics*—using force and vanguardist agency to bring about political change. Gillispie, thirty years earlier, had dismissed Jacobin science to dismiss once and for all the question of *revolutionary science*—a science that understands itself as an agent of political change.[48]

Lauberg's analytic program is revolutionary science in action. It reveals how science and politics merged seamlessly in the experience of the Neapolitan Jacobins. A man with a passion for Lavoisier and many enemies in academia, Lauberg had first encountered analysis as a student at the Military Academy, whose leadership was affiliated with the lodges of the enlightened Freemasonry and shared a Francophile scientific and political orientation. Deciding against a military career, he had pursued mathematics and natural sciences instead, studying with local celebrities like Domenico Cirillo, and eventually graduated in medicine. In 1777, Lauberg became a secular priest in the congregation of the Piarists, which provided free education for the poor; his first publications were booklets for his students: one on chemistry and one expounding the operations of human understanding in sensationalist and analytic terms.[49] Sensationalism and analysis were, in Lauberg's hands, an acid that could dissolve the errors of past philosophies, as in an early philosophical commentary that linked different philosophical systems to the personalities of their authors and to the "circumstances of their lives": they were the source of the impressions used by the active mind to construct theories. As for religions, they were mere instruments of power, and Giordano Bruno an early martyr of reason.[50]

In the early 1790s, Lauberg was absorbed by the study of mathematics and chemistry. His research was oriented toward practical applications, such as the manufacture of colorants and the large-scale production of sulfuric acid. His entrepreneurial ambitions, however, like his academic ones, remained unfulfilled. A sympathetic biographer reports that Lauberg's experiments "were very successful, but they were not encouraged." Later French biographers overlooked Lauberg's career as a conspirator and his major role in the Neapolitan Revolution. Such omission is probably related to Lauberg's own reticence after he settled in Restoration Paris. At that point, his revolutionary past could have only harmed his career and family. Thus, according to these biographies, in 1794 Lauberg went to France simply because he "wanted to participate to the scientific movement." All things considered, this claim is less misguiding than it might seem.[51]

Shunned by the university, Lauberg's teaching was appreciated by the director of the Military Academy. In 1789 he held a temporary position in this institution, which was an entryway for both analytic mathematics and the new chemistry—the first Italian translation of Lavoisier's *Elementary Treaty of Chemistry* (1791) was directed to its students.[52] Lauberg was considered for the chair of mathematics. Thanks to Fergola's pressure, however, the chair went to Annibale Giordano. On that occasion, Lauberg submitted an essay on mechanics that was already a manifesto for analysis. Lauberg presents the science of motion "through new analytic perspectives," to reach its fundamental principles "in an easy and direct way." Doing so means clarifying the conventions at the basis of this science, then deriving from them all its laws. At this point, "the solution of every single problem is reduced to a matter of pure calculation." Following Lagrange's then-recent example, Lauberg argued that all theorems of mechanics are different expressions of the principle of virtual velocities. More generally, every science "is nothing but the combination of all the simple ideas that constitute the complex idea of a phenomenon, and of the conventions that have been established." All these possible combinations can be expressed through a general equation, which can then be used to solve particular problems relative to that phenomenon. Thus, in mechanics, problems will be solved regardless of any "useless" metaphysical consideration about the "essences of bodies" or the "nature of forces."[53]

Lauberg competed for university chairs in 1791 and 1792 but without success. By 1792 he had left his religious congregation and opened the private school of Santa Caterina with Annibale Giordano. The two published their own two-volume textbook: *Analytic Principles of Mathematics* (1792), in which they reduced arithmetic and geometry to their most elementary notions and laws, then reconstructed them "with analytic order." Any science, they held, is built upon a class of sensations and a set of conventions. Scientific knowledge links these constituents to reach general scientific laws. In mathematics, the

constituents are sensations and conventions about magnitudes; physics deals
with "phenomena resulting from the activity of matter." Metaphysics, once
freed from "darkness," is the study of human sensibility and the investigation
of the most general physical laws. Morals deal with the analysis of sensations
caused by human needs, and politics addresses the problem of meeting individual and societal needs. Therefore, "if Physics, Metaphysics, Morals, Politics are merely the analysis of the effects of the activity of matter, of human
sensibility, and of the control of this sensibility relatively to human needs, as
Mathematics is the analysis of quantity; then, Mathematics being an exact
science, so also must we consider the other ones, when we regard them without mystery, and from the correct point of view."[54]

The choice of the analytic method as *the* method in mathematical textbooks was therefore the key step for the whole scientific and political enterprise of Lauberg and Giordano. If one aims "to promote public education,
and to eradicate the old prejudices, the only way is to accomplish this simpli-
fication of the sciences, so that—reduced to the analysis of our sensations—
they no longer constitute that congeries of isolated truths which is presented
by the method of Composition [i.e. the synthetic method]." In mathematics
as elsewhere, synthetic reasoning is confusing and limited because it does
not reveal the underlying analytic order of reality. "These considerations,"
the authors continued, "made us regard the textbooks of mathematics and
philosophy compiled according to the synthetic methods as unworthy of the
education of man." These books "offer a history of single truths, instead of
presenting the method of invention that contributed to the development of
the human spirit." The analytic method had opened a new era for the mathematical and empirical sciences; it was now time to use it for the development
of the science of society.[55]

The Regeneration of Science and Religion

Mona Ouzouf has shown how revolutionary "regeneration" concerned at
once the physical, political, moral, and social spheres: its intended outcome
was not simply a new kind of body politic but, consistent with its religious
roots, a new kind of individual.[56] For the Neapolitan Jacobins, the analytic regeneration of knowledge and politics needed to include the religious dimension as well. The conspirators of 1794 shared a radically critical view of the
Catholic Church and in some cases of religion tout court. Giordano's police
file contained abundant evidence of his "lack of belief in matters of religion."
Lauberg had been heard claiming that "hell is a tale invented to control the
people," that "there is neither hell nor heaven," and that "death is the end

of everything." The existence of God and the immortality of the soul were openly questioned in Vitaliani's workshop, and police informants reported that Lauberg's students scoffed at images of the saints.[57]

These positions became more nuanced during the triennium 1796–1799. Lauberg continued to condemn institutional religion as an instrument of oppression, but he seemed aware of the social relevance of religious sentiments. In his speeches he deployed an assemblage of elements from various cults born in France to replace Catholicism, without abandoning the imagery of the Goddess Reason.[58] On Christmas Day 1797, in Milan, he spoke in favor of Theophilanthropy, a new cult that aimed at salvaging an allegedly pure religious core, separating it from the temporal structures of the Catholic Church, which was held responsible for supporting despotism and feudalism. Lauberg opened with a praise of religious freedom, then explained that the new cult was based on love for the Supreme Being and for fellow humans. According to this regenerated religion, he continued, God created all human beings equal and put in their heart the aspiration to self-preservation and happiness. Individual happiness, however, depends necessarily on public happiness, so all efforts must be directed to the well-being of the entire society. Therefore the new cult reinforced republican virtues. He concluded: "Let us enter the temple of Reason, daughter of God." The Milanese audience agreed that Theophilanthropy was "in conformity with nature and reason" and "worthy of republicans."[59]

Meanwhile, anti-Catholic polemics intensified. Flaminio Massa, a former student of Lauberg and Pagano, portrayed Catholicism as a "monstrous edifice of chimeras and scandalous tales" built by "criminal fanatics"; thus, he saluted a public chemical experiment that, he believed, had debunked once and for all the miracle of Saint January, a pillar of Neapolitan popular devotion.[60] Lauberg and Galdi argued that Roman Church properties had to be returned to the people and planned a military expedition that would head straight to the Vatican. In an animated assembly on this issue, the young daughter of a local chemist jumped on the stage and promised her hand to the one who would bring back to her the head of the pope. When the "theocratic tyrant" was deported from Rome, Lauberg hailed the victory of both republican arms and philosophy: "the monster"—superstitious religion—had been slain in Rome, and now "philosophy hunts it down in the most hidden refuges, and in the heart of man." Attempts to restore the old monstrous principles would be struck down by ridicule: "We will laugh at the Gothic sacerdotal barbarism that is being prepared."[61]

Overall, Theophilanthropy and other new cults did not fare well in Italy. More influential was the role of the so-called Jacobin priests in weaving republicanism and egalitarianism into a renewed form of Christianity, inspired by the alleged simplicity of the primitive church.[62] We have already encountered some of these priests, and indeed they made up 16 percent of those executed after the fall of the Neapolitan Republic.[63] Calls for a renewal of Christianity were hardly new within the Catholic world. In Naples, the eighteenth-century fight of the crown against Rome had been led by Jansenist clergymen who championed moral rigor and asceticism against what they saw as the corrupt baroque devotion of the Roman Church. The expulsion of the Jesuits from the kingdom in 1767 had been one of their great victories, although they had been less successful in eradicating those forms of superstitious religiosity that characterized Neapolitan popular piety. Some of these Jansenist priests joined the Jacobin campaign for a regenerated religiosity, among them Gennaro Cestari (1753–1814), whose sketch of the Jacobin tree of knowledge we'll now briefly consider.

Gennaro and Giuseppe Cestari were members of the local progressive Freemasonry, then joined the Jacobin clubs.[64] They were among Lauberg's closest collaborators, and in 1799 they participated in the establishment of the Neapolitan Republic, for which Gennaro wrote the official catechism. After the fall of the Republic, Gennaro, like many other *émigrés*, settled in Milan. Reflecting on his Jacobin experience, he wrote about the necessity for a regeneration of humanity and its implications for society, morals, science, and religion. In 1803 he published an *Essay on the Regeneration of the Sciences*, a key document for understanding the epistemological perspective of Neapolitan Jacobins and their attempt to—using Robert Darnton's expression—trim the tree of knowledge.[65]

Gennaro Cestari was convinced that the sciences required a radical reorganization and spoke for all those who were "dissatisfied with the present state of human knowledge." Contrary to the belief that the sciences had reached their peak and were doomed to decline, Cestari claimed that the accumulation of scientific knowledge was only beginning. He compared the construction of the system of the sciences to that of a "complex machine": the sciences are all linked, and if one science falls behind, the entire mechanism will function imperfectly. Historical investigation, he held, revealed that the sciences were not yet completely free from the "ancient barbarism" that survived in some of their doctrines and in their very language. Even the *Encyclopédie* contained too much of the old philosophy: rather than limiting their considerations to what is clear, certain, and true, the authors had included topics such as scholastic philosophy, rhetoric, and heraldry. As a result, their encyclopedia was a heterogeneous mix of "natural, supernatural, truth, falsehood, opinions, superstitions, and conjectures." Moreover, their system lacked a single, all-pervasive organizing principle: the connections it establishes are not real but "imaginary," and other systems of knowledge could be imagined that would be equivalent. But, then, what was left of the original ambition of a scientific system of true knowledge? [66]

Cestari also criticized the *encyclopédistes* because they seemed more interested in the speculative faculties of the soul than in its "active, operative faculties" or in the passions—the source of every useful invention. The Neapolitan Jacobins considered any attempt to isolate a core of purely theoretical knowledge as misguided. In reality, Cestari argued, scientific knowledge does not derive exclusively from intellectual faculties but is the outcome of a combination of these faculties with senses and passions. Passions are, at once, the basis of morality and of the natural and social sciences. It follows that the traditional division of human understanding into the three faculties of memory, reason, and imagination, and the corresponding division of human knowledge into disciplines, should be abandoned. Instead, a new system must be devised that reflects the true principles of all knowledge: "sensibility" and the "faculty of abstraction."[67]

Moving from this familiar sensationalist perspective, Cestari reached the more radical part of his work: the rejection of any distinction between material and spiritual beings. He mocked Catholic apologists who described matter as inert and passive "the weakest and most imperfect of all beings." Brute matter, he countered, does not exist. Matter is active, its compounds continuously reorganized in a universal flux that involves all beings. Matter "can be characterized as brute and organized, passive and active, inert and living, insensible and sensible, depending on its different states." One should therefore talk of states of matter, not of matter itself. What need to be rejected are the "colored glasses of spirit and matter": they are useless for the investigation of nature.[68] If brute matter does not exist, neither does pure spirit. As Cestari explained in a later essay, the complete separation of body and soul that dominates modern philosophical systems is the result of semantic confusion and of a simplistic reception of Cartesianism. The mind can only exercise its faculties because is joined to the body: intellect and senses are no distinct powers. There is no thought without the body, and a person is not the conjunction of two opposite and mutually exclusive principles but rather a whole, "a third kind." Psychology, the study of the spiritual-intellectual-moral component of the human being, is therefore a delusional enterprise. What is real is the "science of man," which should be constructed bottom-up, starting from natural history and the study of animal sensibility, from which human sensibility and intellectual activity differ only in degree. Ultimately, the active principle should not be located in a faculty of the mind—the intellect or the will—but rather in the human being, "the one and only agent."[69]

Thus Cestari rebuked the Platonic-Cartesian spiritualism that characterized much continental philosophy, and that had continued to thrive even recently, often embedded in sensationalist perspectives. He did not shy away from the more dangerous implications of his monist position and reflected at length on Locke's famous reference to thinking matter. He concluded that there was nothing paradoxical in this notion. Quite the opposite: any operation of the mind, even the most sublime, depends necessarily on the material dimension; the mind, without the body, could not even have "consciousness of its own identity." And to those sensationalists who argued that the mind maintains its faculties in separation from the body thanks to God's direct intervention, Cestari replied that this argument was one of theology, not science.[70] Similarly, he held, to argue that the soul is spiritual and *therefore* is immortal is utter nonsense. Considerations about the soul's immortality and destiny belong to religion, not psychology and science. One might well argue for the materiality of the soul but not be an atheist if one recognizes, via an internal sentiment, the existence of a just and providential God. This position is the one that Cestari himself considered the most plausible. Don't be afraid of materialism, he wrote ironically; it's not that "horrendous specter" that many fear.[71]

Cestari, like most Neapolitan Jacobins, believed that the preeminence of dogmatic metaphysics over the sciences had simply been a matter of historical contingency. Emancipated from these transcendent and ultimately theological principles, the sciences would be free to flourish and foster universal social progress. Cestari even envisioned a "federation" of all universities and academies of the "learned nations," which would organize the "scientific system," sustain education, and sponsor scientific-literary activities. To those who might argue that this vision was a utopia, Cestari responded that there was so much "electricity" in this plan that its supporters could surmount all obstacles. Cestari's universalistic ideal had clear Enlightenment roots; it also echoed the vision of another priest who joined the 1799 Republic, Onofrio Tataranni, who had written about the unification of all the peoples of earth as they pursue peace, public happiness, and universal harmony.[72] Finally, "the great tree of the sciences," wrote Francesco Lomonaco, a Jacobin who defended Naples in 1799, "will form leafy branches, which will provide a restful shadow for a humiliated humanity."[73]

Lives and Deaths of Jacobin Mathematicians

Lauberg and Giordano were only two of many mathematicians who played important roles in the conspiratorial and revolutionary activities of the 1790s.

In 1799, most of them gathered in Naples one last time to assume key political and military positions under the tricolor flag—blue, yellow, and red—of the Neapolitan Republic. The fall of the republic would put an end to their careers and, for some, their lives.

Who were these Jacobin mathematicians? The evidence is scant. Many had studied with Abbé Nicola Pacifico (1734–1799), who had been trained in mathematics and natural sciences at the Archiepiscopal Seminar of Naples. Pacifico's personal library was always open to his students. He became a member of the Royal Academy of Sciences in 1779, in the class of Natural History, and was a Freemason, affiliated with the Great National Lodge and, later, the anti-despotic lodge of the Illuminati. His research at the Royal Academy around 1780 included a study of the Calabrian earthquake as well as new "methods for the study of convergent series, for the integration of differential formulas, and extending of their use." He also planned to extend "his philosophical considerations to agriculture and trade." Not surprisingly, mathematicians and reformers like Filangieri, Cirillo, and Caravelli were among his "dearest friends."[74]

One of these friends was Nicola Fiorentino (1765–1799), who had argued for the reliability of differential calculus, which he considered "exact, and free from any imputation."[75] Fiorentino was known primarily for his essay on the public economy of the Kingdom of Naples, written in 1794 when he was a professor at the Royal College of Bari. Following the reformist tradition, Fiorentino had offered an assessment of the present conditions of the country, explaining the difficulties experienced by both farm workers and entrepreneurs in terms of a profound ignorance in technical and scientific matters. To increase national wealth, well-to-do families, he argued, should direct their children toward scientific studies, and investments in agriculture should be encouraged. Other claims concerned the importance of a free-trade legislation and the abolition of import duties in order to increase internal competition. Too many young men were entering the legal profession in Naples, he argued, while the scientific study of agricultural techniques and machinery languished, due to "the great force of prejudices." New universities, tribunals, and schools should be opened across the provinces rather than concentrated in the capital, and chairs and academies of agriculture should be established. Fiorentino also argued for the introduction of a new legal code that would facilitate private investments in agriculture, trade, and manufacture. The link between science and economy was crucial to Fiorentino: trade and navigation depend on mathematics, physics, and chemistry, so the nation needed institutions to foster collaborations between practitioners of each.[76]

Vincenzo de Filippis (1749–1799) also knew Pacifico well. After arriving in Naples from Calabria, he studied law and found brief employment as administrator of a large estate. Soon his inclination for the sciences brought him to the University of Bologna, where he graduated in philosophy. In 1779, de Filippis joined the Royal Academy of Sciences. We also know that in that year he bought the works of d'Alembert in a Neapolitan bookshop. In the 1780s he participated in the debate over economic reform in Calabria. He hailed the suppression of the monasteries in the area hit by the 1783 earthquake and matched his physical analysis of the phenomenon with a study of its social and economic consequences.[77] De Filippis visited Naples again in 1786 to meet Donato Tommasi, and through him he was probably affiliated with the anti-despotic lodge of the Illuminati that same year. In a manuscript well received by his Neapolitan friends, he describes mechanics as "the most important and useful science of nature," the source from which every other physico-mathematical science originates. He believed that in mechanics one can reach mathematical certainty, given that its conclusions are derived with mathematical reasoning from a few evident principles. Not surprisingly, he read Lagrange's mechanics with great enthusiasm shortly after its publication in 1788. Between 1787 and 1792, De Filippis taught mathematics at the college of Catanzaro in Calabria. His lectures included practical sections in which he discussed how science can be applied to the solution of economic and social problems.[78] In the late 1780s, religious authorities accused de Filippis's close friend and fellow mathematician Gregorio Aracri (1749–1813) of atheism. Aracri, a Capuchin monk, had argued that moral laws are innate and grounded in human nature, and he defended a hedonistic ethic.[79] He lost his teaching position, and one of his books ended up in the *Index*. Formally, the reprimand was a procedure internal to the Catholic Church, but it sent a clear message to the lively group of Calabrian reformers and Freemasons. In 1792, de Filippis too left his teaching position, and we lose track of him until 1795, when he was under investigation in relation to the Jacobin conspiracy. He was indicted for Jacobinism in 1797, but the outcome of that proceeding is unknown.[80]

The biographies of these men are fragmentary but reveal a pattern. Young men from the provincial bourgeoisie are sent to Naples to study law or medicine or to enter the church or the military. During their studies they develop a passion for the modern sciences, in particular the mathematical sciences and their application to practical problems, from mechanics to agriculture. They all, at some point, teach algebra and infinitesimal calculus. They believe that the certainty of mathematics can be exported to other fields of human action, once those fields have been properly reorganized. They are fascinated by the rational construction of Lagrange's mechanics—where the entire science of mechanics is analytically derived from a few fundamental principles—and they adopt his algebraized version of calculus. They continue their studies abroad or, if in Naples, with unorthodox teachers such as Gerolamo Saladini and Vito Caravelli, either privately or at the Military Academy. Some end up teaching at the Military Academy, while others obtain teaching jobs in provincial colleges. Few have links with the University of Naples. The older ones take part in the reform movement of the 1770s and 1780s and join the egalitarian Masonic lodges. In the 1790s, they turn into a target for conservative elements within the Neapolitan Church, who are now closer to the government. Some of them are accused of religious indifference or atheism. In 1794, many of these mathematicians are involved in the Jacobin conspiracy: they will be incarcerated or flee abroad as a consequence.

*

But all was not lost yet. By late 1798, it became clear that the French would not miss an opportunity to occupy the southern kingdom. As the commander-in-chief of the French army in Italy, General Barthélemy Joubert, began planning the invasion, he asked Lauberg to join General Championnet in Rome—the starting point of the new campaign. Lauberg headed a commission of Neapolitan exiles charged with preparing the terrain for the French invasion and occupation. The Neapolitan army posed little more than a symbolic resistance to the invaders. Instead, there were met by a remarkable and unexpected popular resistance.[81]

What was called the "anarchy" of January 1799 tells us much about the increasing social fragmentation of late-eighteenth-century Neapolitan society. In December 1798, with the French crossing the northern border almost undisturbed, the court fled to Palermo. This sudden collapse of the monarchy had long been in the making. The combination of a deepening economic crisis and mounting military expenses proved unbearable for the crown, as its potential supporters were being alienated by unprecedented despotic measures, not least financial ones. Between 1788 and 1796, total bank deposits had fallen by 70 percent, due almost entirely to crown's appropriations. In the words of William Hamilton, the British Ambassador, "few, should it come to trial, would think such a government worth fighting for."[82]

A royal delegate with full powers signed an armistice with the French on January 12, but the City of Naples—an ancien régime corporation that represented the great aristocracy of the capital—refused to recognize his authority, claiming its own right to rule the nation in the absence of the king. Breathing life into medieval institutions, representatives of the main aristocratic families met in their own parliament, which in the last century had been reduced to a harmless political curiosity. Trying to control popular unrest, representatives of the City distributed food and set up gallows in the main piazze. They also began negotiating secretly with the French and the local Jacobins. The main religious orders were all out in force, walking the streets with their bare feet and exhibiting holy relics to the crowds. Serious rioting began on January 15 and targeted mainly the palaces of the Francophile aristocracy—some well-known figures were dragged out and butchered in the street. Things precipitated between January 20 and 21, as the arrival of the French triggered a massive popular insurgence. The city was devastated as rioters pursued alleged French sympathizers while fighting fierce door-to-door battles with the invading troops. General Championnet eventually entered the city, thanks also to the artillery of the fortresses, which had been seized by local Jacobins. On January 23 he promptly established the Provisional Government of the Republic of Naples, and Carlo Lauberg was nominated its president.

Lauberg's government was filled with fellow Jacobins back from exile and others just freed from local prisons. Vincenzo de Filippis became minister of the interior; Annibale Giordano was in the Military Commission; Nicola Pacifico was a captain in the Republican Guard. Many of those involved in the 1794 conspiracy entered state commissions or the republican army, which was led by some of Lauberg's former colleagues at the Military Academy. Giuseppe Cestari was in the commission for the internal administration. Like Lauberg, he was a radical on the feudal question and argued for the abolition of feudal rights without compensation. He also ordered many monasteries suppressed. Gennaro Cestari was in the ecclesiastical commission, whose goal was to prepare a "catechism of morals" comprehensible to the people but free from "superstition and error." The catechism made it clear that "only under a democracy does man enjoy these rights [equality, freedom, property, safety], which had been given to him by the Creator, and taken away by tyranny"; that "the people are the true sovereign"; and that "Christ recommended democracy." It is unclear how the catechism was received—Cestari's order to confiscate the churches' abundant silverware to finance the republican army was certainly unpopular.[83] Among the many Jacobin priests who supported the revolution from the capital's pulpits was Gregorio Aracri, whose sermons praised the newly acquired liberty and argued for the abolition of feudal abuses over the peasantry.[84]

The life of the Neapolitan Republic was only about five months. Few laws were approved, but one, written by Giuseppe Cestari and Mario Pagano, was extremely significant: the abolition of feudalism.[85] Pagano acted also as chair of a committee that produced the project for a constitution that was ready for discussion right as the republic fell. In it, references were made, implicitly, to Locke, Vico, Genovesi, the Neapolitan reformist tradition, the American and French constitutions, and the radical experience of Oneglia. Equality was its foundation: the first article stated that "all men are equal." Equality, it held, is not itself a right but the source of all rights, a relationship described as a law of nature and a principle of reason, which to Pagano were one and the same thing. This natural equality—wherein all men have the "same physical and moral faculties"—is the basis for civil and political equality. Sovereignty is an inalienable right of the people because it antecedes any form of social organization. The people, who enjoy an unprecedented number of rights in this constitution, need to understand what their "real interests" are. Hence the crucial role of a public education aimed at "shaping the mind of each citizen" by exposing them to the correct ideas, circumstances, and examples, according to the tenets of sensationalist pedagogy. Eleonora Fonseca de Pimentel, editor in chief of the main republican newspaper, called for a plan of national education to transform "the plebs into a people."[86]

Lauberg's political life as president was difficult, as he had to mediate between the envoys of the French Directory and local Jacobins who strove for the autonomy of the newborn republic. General Championnet was recalled to France in late February and arrested for disobeying orders, while a political commissar sent from Paris dismissed Lauberg in April and created a new government. Among other things, the Directory disagreed with Lauberg's radical position on the feudal question. Lauberg left the country soon afterward, never to return. As in earlier Italian republican experiences, the feudal question revealed the lines of social conflict within the democratic front, separating radicals, who pushed for land redistribution; moderates, closer to the line of the French Directory; and former aristocrats, whose anti-monarchic sentiments were more trusted than was their commitment to the abolition of feudalism.[87]

Meanwhile, the military situation was deteriorating. The Bourbon government had dispatched Cardinal Fabrizio Ruffo to Calabria to enlist volunteers to fight for the king under the crossed insignia of the Holy Faith (*Santafede*). It's instructive to consider the politics of the sacred images and symbols Ruffo chose for that army. Saint January, the most famous among Naples's patron saints, had not been enlisted: he was suspected of collaboration with the Jacobins. The Jacobins had done a good job of presenting him to the Neapolitans as a revolutionary saint, and the miracle of the blood had saluted the French's arrival. Ruffo had chosen Saint Anthony instead. Also, Ruffo's banners were invariably Christocentric, above all the cross and the sacred heart—a feature that the *Santafede* shared with the two other main

European counterrevolutionary movements: the Royal and Catholic Army of the Vendée and the Tyrolean Rebellion.

Relentlessly, the crusading army occupied one republican town after another. Bourgeois landowners and professionals who had planted liberty trees were massacred, their properties sacked. The religious fervor of the *Santafede* inflamed the Romantic imagination and remains to this day an emblem of the irredeemable backwardness of southern Italy. At a closer look, however, it's clear that this religious fervor is largely a historiographical myth. As shown by John Davis, Ruffo was far from a bloodthirsty fanatic. Instead he was a good commander who cleverly exploited existing conflicts and rivalries—between social groups, towns, and regions—to project the image of a single invincible counterrevolutionary army while in fact he was struggling to keep together a series of fairly disconnected conflicts and insurgencies. He exploited the peasants' hatred of the "Jacobin gentlemen," but he also cajoled the philo-republican aristocracy and provincial bourgeoisie and made it as easy as possible for them to recant. He recruited tactically, making the best of secular rivalries between provincial towns, so that in reality the composition of the army kept changing. Members of different communities joined the army temporarily, when its action aligned with their particular interests—often their goal was settling old grudges. Not surprisingly, Ruffo rarely trusted those who fought for the *Santafede*, and he seems to have doubted that faith was a significant motivation for most of the crusaders. His main concern—by far—was to control popular violence and *restore order* after each military victory.[88]

It's revealing that contemporary commentators close to the action described these events as a "great anarchy," recognizing similarities between the *Santafede* and older anti-seigniorial peasant rebellions. That order within his own ranks was Ruffo's primary concern heightens the significance of the choice of Christocentric images for his army: Christ the King imposed order, coherence, and direction on large movements that were continuously at risk of splintering and getting out of hand. By contrast, the fragmented and uncontrolled popular insurgences of the highlands almost invariably raised the insignia of the Madonna. If Christ's sacred heart was the symbol of the legitimist crusade for the defense of throne and altar, the Madonna was the flag of the mountain guerrilla. Her images, like those of minor local saints, were resources for communities who felt insurgence was their best option. This feature was not new. Peter Burke has commented on how Christocentric devotion was used by the Neapolitan clergy to restore order after the 1647 anti-Spanish revolt, which had taken the Madonna as its symbol: "order had to be restored in the supernatural as well as the natural world." David Blackbourn noticed how the Madonna chose carefully the sites of her apparitions in early modern Germany: she avoided the plains and chose the rocky uplands instead, "areas of sheep and goats," and in so doing "most clearly established her[self] as a distinct supernatural personage." Michael Broers aptly commented that this behavior was quite fitting for the "patron of *guerrillas*." The same pattern is visible when comparing the Mexican experience of the Hidalgo revolt (1810–1811), which raised the banner of the Virgin of Guadalupe, to the massive and coordinated uprising of the *Cristiada*.[89]

When the *Santafede* met the Russian and Turkish forces that had landed on the coast of Apulia, the siege of Naples could begin. Meanwhile, the French army had retired northward, due to the victories of the Austro-Russians in the Po Valley, leaving the Republican Guard alone to protect the capital. On June 13, 1799, all those defending the Republic were at the southern gate of the city, where the attack had begun at dawn. Ruffo had waited for the day of Saint Anthony of Padua, patron saint of the *Santafede*. Around ten thousand Republican troops were facing seventy thousand attackers supported by Russian artillery and by the guns of a British naval squad. After a full day of fighting, the attacking troops stormed one of the gates and entered the city. A week later, the surviving Jacobins entrenched in the city fortresses surrendered on the condition of reaching France safely. The allied generals agreed but later recanted as, they argued, the word given to rebels against a legitimate king need not be kept. The republican admiral was summarily tried on Admiral Nelson's flagship, hanged, and thrown into the sea.[90] Most prisoners were put on show trials, and some 120 were executed. By September, eight thousand trials against former republicans were in process across the kingdom. Giuseppe Cestari was killed in action. Mario Pagano and Domenico Cirillo were hanged, along with Eleonora de Fonseca Pimentel and three men who declared themselves mathematicians in front of the special tribunal: Vincenzo de Filippis, Nicola Fiorentino, and Nicola Pacifico. Lauberg's student Ettore Carafa was beheaded, a privilege reserved to the aristocracy. Annibale Giordano managed to escape from prison right before the execution and fled the country. Gennaro Cestari and many others continued their political and scientific battle in northern Italy or in France. In Naples, analysis had been defeated.

Empire of Analysis

If the guns of the Holy Alliance had wiped analysis out of Naples, French guns brought it back. In 1806 a French army crossed the northern border and occupied the continental part of the kingdom. The "French decade" (1806–1815) had begun. The new regime pursued an ambitious plan of modernization: the semifeudal kingdom was to be transformed into a centralized state under a new administrative and judicial system. The economy would also be completely reorganized, thanks to massive land redistribution and the construction of new infrastructures. A newly constituted corps of civil engineers was charged with implementing key elements of this project. The new regime's most important promise was a promise of order—order through modernity.

Once again, analysis was a key component of the culture that legitimated and oriented these reforms: it was seen as their necessary condition and its veritable engine. Aspects of this reform campaign endured well beyond the 1815 restoration of the Bourbon monarchy. Although reactionary pushback was strong, the monarchy continued to support institutions and projects that suited its centralizing and absolutist ambitions. The age of Restoration saw indeed a continuous tension between modernizing and reactionary forces, which almost invariably played out as *technical debates* between representatives of the corps of civil engineers and those who opposed their action.

If the French decade was the moment of the empire of analysis, it was also the moment of analysis for the empire. Analytical resources were mobilized to serve the urgent needs of the French imperial machine. The 1799 Jacobin revolution looked distant now, and few were even willing to mention it. Analysis too had changed. The tradition of eighteenth-century analysis as a set of resources for understanding and transforming reality survived largely as part of an allegedly apolitical techno-scientific instrumentality. Science was still understood in terms of analytic method, sensationalism, and ideological theory, but its reach had been severely reduced: Catholic dogma and universal metaphysical principles could not be questioned anymore. This configuration afforded new compromises with dogmatic academic philosophy, whose representatives increasingly adopted the ideological and sensationalist idiom.[1] This was indeed the idiom of numerous technocratic projects designed to support the action of the absolute sovereign, be it a French ruler or the restored Bourbon king.

During the French decade, economically active social groups from across the kingdom were brought into political life to address the question of modernization. The post-feudal agrarian elites' plea to reduce the central government's "despotism" would culminate in the constitutional revolution of 1820–1821. Their moderate political liberalism was being expressed scientifically in the language of analysis and found philosophical grounding in the theories of ideology and conscientialism (the theory that all knowledge is grounded on consciousness, the internal perception of the self and its modifications). Theirs was an active, autonomous reason that discovers its own essential and inalienable liberty, and legitimately resists the abuses of absolute power. By the 1840s, though, these positions were losing ground to Vichian historicism and German idealism. By the eve of 1848, the political and economic imagination of the agrarian elites who had fought for order and for the constitution seemed exhausted.

But not all defenders of analysis had embraced its imperial version. Through the existential trajectory of Ottavio Colecchi we follow lineages of practitioners who saw their work as directly connected to the Jacobin experience and the revolution. They envisioned a radical social transformation that could only be realized in the context of a unified Italian nation. This strain of patriotic liberalism included democratic and republican components and had deep roots in the military and in civil-engineering schools. Many of their members fought the patriotic wars of the Risorgimento and became active in the political life of the new nation. To them, the Risorgimento of the Italian nation was a process of liberation from both political and scientific oppression.

The French Strike Back

In February 1806 a French army entered the kingdom, heading straight for its capital. This time, the French did not find any opposition. Much had changed since the triumph of *Santafede*, and the policies of the first restoration had failed to meet the expectations of the loyalist groups that had brought the

Bourbons back into power. The royal family, accompanied by loyalist courtiers, fled to Palermo, and from there they organized a network of clandestine resistance with the support of the British fleet. An attempt to replicate Ruffo's crusade, however, failed. In March 1806, Joseph Bonaparte, brother of Napoleon, was proclaimed king of Naples. He formed a government composed of French officers and members of the Francophile aristocracy. A consultative chamber, the *Consiglio di Stato*, was created with a Neapolitan majority. When Joseph left Naples to accede to the throne of Spain in 1808, Joachim Murat (1767–1815), a famed cavalry commander and the emperor's brother-in-law, was crowned. Like the empire's other satellite states, the kingdom was supposed to support the imperial military effort, enforce the Continental blockade, supply France with raw materials, and open its market to French products—in fact a position of colonial subordination. Unlike those parts of the Italian peninsula that had been annexed to the empire or the Kingdom of Italy, however, the southern kingdom retained a certain degree of autonomy from Paris, which increased under Murat and turned into defection in January 1814.

North and central Italy formed the core of French hegemony, the areas where French power reached throughout its domains with the least mediation, while in the south the space for various forms of mediation and resistance was more significant. The French attitude toward the locals was not much different from what Michael Broers has detected in other parts of the peninsula and has characterized in terms of inter-European orientalism and cultural imperialism. The elites were perceived as sophisticated but morally and materially corrupted, an unreadable spectacle of effeminate men, cunning priests, and scheming women, whose salon life and artistic tastes were utterly incompatible with those of their Parisian counterparts. The French saw these elites as a world in long-term decline, enervated by a baroque and Jesuitical religiosity and at odds with the masculine world of the Napoleonic polity. The inhabitants of the peripheral highlands were, instead, uncivilized, threatening, and savagely violent, a view rapidly established after French military engagement with local insurgencies. "It would be very difficult, I can assure you," wrote a French general, "to take these troops back into that country since they have the greatest distaste for the cruel and barbarous war in which they have been engaged with the peasants." Former Jacobins did not fare well with French authorities either. The emperor himself had always had a deep mistrust of Neapolitan Jacobins—whom he had seen in action in the democratization of north Italy—and of the rebellious people of Naples in general. Thus he advised his newly crowned brother: "Arm your forts! Disarm, DISARM, the Neapolitans!"[2]

The costs of empire were massive for the southern kingdom: from 1806 to 1812 direct fiscal revenues grew by more than 50 percent, and, by 1812, 80 percent of those revenues went to military expenditures. Meanwhile, the prolonged state of war and the blockade had devastated trade and most economic activities. The kingdom, however, managed to achieve a remarkable degree of autonomy from Paris. Attempts to mediate the impact of imperialism intensified under Murat's reign, when French elements were gradually excluded from both government and administration. By 1811, only Neapolitan citizens could become civil servants. As the fortunes of the empire declined, Murat's bargaining power increased.[3]

The new regime's modernizing project was ambitious. The "gothic palace" of the ancien régime state needed to be reconstructed along simpler and more rational lines.[4] If fully realized, the new administrative structure would have transformed it into a highly centralized administrative machine operated by efficient, well-trained civil servants. Within months of the establishment of the new regime, feudalism was abolished. All forms of baronial and private jurisdiction were abolished together with feudal tributes and services, establishing the absolute sovereignty of the state. Every town and property was to be ruled according to the same law. Feudal lands were divided among landowners and communes, while the sale of common lands aimed to create a new class of landowning farmers. The national debt was consolidated through the sale of ecclesiastic and royal lands. The Napoleonic civil code reformed the juridical system, while a standard land tax and a new cadastre reformed the tax system. Military conscription was introduced. Overall, the project was a reformist effort that had few comparisons across Napoleonic Europe. The regime managed to push through the most urgent reforms by 1809; then, their difficult and contested implementation began.[5]

The new regime identified its key interlocutors and supporters in the professional and landed bourgeoisie: in the words of a French adviser, the *ordre moyen*, those who defended the *lumières* in Naples.[6] As we have noted, the lines that separated social groups were not so sharp in the southern kingdom, where former feudal landowners were jealous of their jurisdiction and monopolies but keen on the free market when it came to selling and buying land, and where gentlemen entrepreneurs found attractive low-risk investment opportunities in the legal and economic intricacies of the feudal world. And yet it is true that, especially in the provinces, the regime was able to mobilize social groups that, up to that point, had been active economically but not politically. The reforms offered plenty of new employment opportunities: the government needed to staff local institutions, administrative and juridical; and a season of public works was inaugurated to facilitate internal

communications and the movement of troops. The new bureaucratic machine, with its promise of fast careers based on merit rather than fortune, also
contributed to the creation of social consensus in spite of massive military
and financial problems.[7]

The promise of modernization was based on the establishment of a centralized administrative system. At its heart was the newly created Ministry
of the Interior, which combined numerous older institutions and, through a
rigid hierarchical system, directly controlled its peripheral branches as well
as prisons, hospitals, hospices, public works, and public education, while directing the development of agriculture, trade, and manufacture. The civil administration, linked to the executive power, was thus separated from the administration of justice. The country was divided into provinces, districts, and
communes; the system's elementary component was the communal council,
which represented the interests of local landowners and professionals. The
provinces were ruled by the *intendenti*, who reported directly to the Ministry of the Interior. Through this vertical structure, provincial elites could
participate in the exercise of political power, albeit in the context of a centralized and autocratic system. The government also extended the passive local
electorate to artisans and professionals, a move that sealed its alliance with
these groups.

The enemies of reform were many and powerful. Endemic brigandage,
insurgencies, and the reluctance of large sections of the provincial elites to
delegate power disrupted the new administration. Rather than passive remnants of the feudal order, these phenomena were forms of resistance of post-
feudal elites who had amassed vast estates through the enclosure process and
were now negotiating their position in the new order.[8] As resistance grew
more intense, the Ministry of the Interior and the *intendenti* were frequently
accused of plotting for an "agrarian revolution" aiming at an equal redistribution of land.[9] Absentee landowners and the Neapolitan tribunals formed a
power block that had for centuries resisted the land claims of both local communities and the central government. These long-term conflicts had formed
the background of the late-eighteenth-century crisis of the monarchy; in the
French decade they entered a new and acute phase. The antagonism between
executive and juridical powers was now continuous and pervasive in the capital as in the provinces, wherever a local authority conflicted with a justice of
the peace. These conflicts, which often played out as technical controversies,
were shaking the political and economic lives of most local communities.
But reforms did not threaten only the interests of absentee landowners, the
capital's tribunals, and the Church: the investment strategies of part of the
southern bourgeoisie were also endangered, especially those investments that

had flourished in the shadow of a feudal system that favored well-connected investors.[10]

Critics of reform lamented the excessive bureaucratization of the new administration, which they portrayed as too expensive and detached from the concrete reality of the country. To them, reformist tenets were French and alien, guided by abstract egalitarian norms that conflicted with both local traditions and common sense. Provincial borders had been redrawn in ways that ignored local history and customs. The separation between juridical power and executive power, which deprived the lords of their feudal jurisdictions and the old tribunals of their administrative role, was seen as undermining the sacrality of power by violating its inherent indivisibility. The newly centralized, vertical administrative structure was accused of suffocating local autonomy and dismantling the autarchic economies of rural communities, to which critics clung as the ideal form of social life. The myth of the organic nature of the feudal-communal system, within which orders and guilds coexisted harmonically, gained traction in early-nineteenth-century political and cultural discourse, typically in opposition to modern society as a machine, a mechanical aggregate of egoist individuals. Moralistic literature flourished, opposing the corruption of riotous urban dwellers to peasant wisdom and the values of uncontaminated country life. In 1808, academic philosopher Paolo Nicola Giampaolo wrote an entire book on the virtues of farming, exhorting his fellow countrymen to "look to the countryside for wealth, tranquility, and innocence: there one does neither hear seditious voices nor see intrigue against the state."[11]

Analysis for the Empire

Given its subordinate position within the imperial system, it's not obvious why the kingdom should be the site of such an ambitious and all-encompassing modernization project. The historiography has long presented that modernization as a top-down process, the imposition of foreign elements on a backward and unreceptive social reality. The republican fervor of a few French generals and administrators, however, is unlikely to offer the only explanation for the ambitions and transformations of the French period. Instead, some scholars have rightly emphasized the role of social groups who seized the opportunity to shape and direct the modernization project along lines and agendas that originated in local reformist and Jacobin experiences, and that responded to local aspirations. The ancien régime had been long imploding in Naples, but a new post-feudal order had yet to emerge: the new rulers' promise of order through modernity did not go amiss.[12]

The local roots of the new reformist wave become apparent when considering the large number of administrators and advisers who came from the ranks of Bourbon anti-feudal reformers, the progressive Masonic lodges and, even more visibly, the Jacobin network. That this last group should have played a prominent role in the new regime is remarkable. One factor that played in their favor was the arrival in Naples of Antoine Saliceti (1757–1809), a Corsican close to Napoleon and Joseph Bonaparte who became the powerful minister of police and war. A *robespierriste* who had survived Thermidor, Saliceti had provided vital intelligence to the Directory and to Bonaparte during the north-Italian operations. A *Jacobin à outrance*, he had contacts with most Italian republicans—in particular, he had worked with Buonarroti and the Neapolitans in the electrifying days of the Republic of Oneglia.[13] Facing the urgent need to staff the administration, Saliceti called back a large number of Jacobin exiles. They knew the terrain, needed a job, and had no patrons. He thus constructed a formidable personal network while filling every branch of the administration, central and local, with loyal and overly committed men who "saw themselves as missionaries of the new order." Whether the modernity that they envisioned for their country was in line with Paris's plans was, of course, a different matter.[14]

Among the returning Jacobins were many technicians and men of science who had studied with Lauberg and Giordano. One was Ferdinando Visconti, the military engineer whom we met as one of the 1794 conspirators: he became the director of the Topographical Office in Naples. The influx of exiles taking up relevant technical positions, and bringing analysis back with them, was perceived by the synthetic mathematicians who had gathered around Nicola Fergola as a foreign invasion meriting resistance.[15] The synthetics' fears were not unfounded: General Director of Public Education Matteo Galdi, himself a former Jacobin and a student of Lauberg, advocated openly for an end to the synthetics' domination of mathematical education in public institutions, and for the need to bring analysis back to Naples. Interestingly though, he did not push for a complete elimination of synthesis from the schools as his maestro once had. Rather, he argued for a *compromise* in which analysis and synthesis would coexist, with different functions, in mathematical teaching. We shouldn't pursue extreme solutions, he wrote, but aim for "an intermediate way." The former Jacobin and *anarchiste*, once arrested in Milan for his opposition to the French Directory, had apparently come to terms with the Napoleonic regime. After years of service in north Italy and in the Batavian Republic, Galdi's interests had shifted from political theory and radical egalitarianism to the question of technical and economic modernization. This shift showed in his politics as well as in his mathematical pedagogy.[16]

During the Jacobin triennium Galdi had written a programmatic booklet on "revolutionary public education" wherein he stated that education is a necessary component of any revolution, as changing government is far from enough: the old religion, customs, and ideas must also be replaced with new ones "more suitable to the new order." A revolutionary public education must, above all, "teach the democratic principles to the mass of people."[17] Ten years later, Galdi wrote a substantive plan that would guide the reorganization of all levels of education in the Kingdom of Naples well beyond the French decade. This plan was not pedagogy for revolution though; it was pedagogy for the new imperial order. The book addressed the "princes" and referred to Bonaparte simply as "the GREAT." Education is important because the stability of the "modern monarchy" depends on it, Galdi held; ignorance, by contrast, is a cause of disorder and anarchy, which were plentiful in the feudal world. The reforms of the new regime had inaugurated a "new Enlightenment" (*nuovi lumi*); what was needed next was the pacification of Neapolitan society and the active commitment of agrarian elites, the main agents of reform.[18]

That Galdi now thought of education in terms of political stability is clear from his treatment of the pedagogy of philosophy and theology. At a first glance, his language and concepts look familiar: sensationalism, the connection of the physical and the moral, philosophy as ideology, the equality of all intellects. And yet, it doesn't take much to see that the philosophical curriculum now had a metaphysical core, a "transcendental metaphysics" that provided ontological canons to all sciences and humanities and must frame moral and political action too. Emblematic of this return to metaphysics was the repositioning of logic and metaphysics at the center of colleges' curricula. Theology was also back as a legitimate component of the world of science, reflecting Galdi's new understanding of the role of established religion in the management of a modern state. The fear of another *Santafede* insurgency was palpable, and schools might not be enough to prevent it. As an 1810 memorandum put it: "No matter how many schools, grammar schools, and colleges Your Majesty may build, these will never bring instruction to the most numerous classes of the Nation." For shaping the masses' behavior, religious sentiments were of the essence.[19]

Education was not only about political stability: in a modern monarchy the curricula of schools, universities, and special schools must be designed to serve the state's needs. Hence the centrality of political economy and statistics, which are deeply intertwined and should inform policies relative to commerce, agriculture, finance, and population management. The mobilization of curricula did not ignore the humanities, which were seen as key to

instilling civic virtues and, as in the case of Eastern languages, directly relevant for the economy of the kingdom.[20] In mathematics Galdi promoted the
pedagogical combination of analysis and synthesis, a *via di mezzo* between
the extreme positions that had characterized mathematical life in Naples up
to the gallows of 1799. Galdi rebuked foundational fears and anticipated that
analysis's recent developments would find solid foundations in geometry;
however, he also argued that those who abandoned synthesis to pursue purely
algebraic infinitesimal methods were gravely mistaken. Galdi's language is
clearly borrowed from the synthetics: there had been an "abuse" of formulas and calculus in the mixed mathematics; the analytics were "ungrateful"
and disrespectful of deep thought; algebra's "bold flights" were driven by the
imagination rather than reason. An effective mathematical pedagogy must
focus precisely on controlling and channeling the student's imagination. By
contrast, geometry is the only God-given source of certainty that humans
have—apart from metaphysical axioms—and Fergola had shown the way to
expand legitimately the limits of the mathematical sciences. If ever Lauberg
read his former pupil's essay in his Parisian exile, he must have choked at this
point.[21]

Galdi's political and pedagogical trajectory was far from unique. Many
former Neapolitan Jacobins were coming to terms with the Napoleonic normalization.[22] The watershed of 1799 should not be taken as a complete break
with revolutionary and democratic experiences, but the protagonists of that
season were now reflecting and acting under very different conditions.[23]
While reactions to the new situation varied, there is one visible pattern that is
relevant for our story: the turn to modernization as a *technical problem*, one
to be addressed with the tools of mathematics, statistics, and political economy. The new political framework excluded revolutionary and democratic-
egalitarian solutions, but it afforded the reconceptualization of modernization as a technocratic transformation of society. There was continuity with
Jacobin revolutionary mathematics: analysis remained the privileged tool for
implementing social change. But there was a crucial difference: analysis was
now a neutral tool, the distinctive expertise of a technical elite in the service
of executive power. The *direct connection* between mathematics, egalitarianism, and republicanism had been severed, and with it vanished the very possibility of a revolutionary mathematics.

*

Most technicians close to the new regime agreed with Galdi that it was time
to move on from the heated mathematical controversies of the recent past.
Luca de Samuele Cagnazzi (1764–1852), a professor of political economy and

a key adviser to the government, arguably the most prominent statistician in Italy at the time, wrote an essay on analysis and synthesis in mathematics that was an invitation to overcome "the spirit of controversy" and look at these methods from the point of view of their technical meaning and utility. He too thought that students should learn both methods, and that while the synthetic method was "luminous" the analytic was a powerful "mechanism." The two should be integrated, and geometrical intuition must be, ultimately, the foundation of infinitesimal calculus. Teaching calculus only through "abstract ideas" and without its geometrical foundations is a "deformity," as the students would act like "automata," without understanding what lies behind each formula.[24] Turning modernization into a technical problem meant detaching mathematics' formal tools from their original source of legitimation—analysis as a universal logic and language. Therefore, mathematics now needed a new legitimation, a foundation. The political reframing of modernization had produced a profound and pervasive *mathematical* problem.

Fergola and the synthetics had been pondering the question of foundations early on, and they had a clear answer: purity. Hence it should not surprise that, among the administrative elites that supported the Napoleonic order, we encounter functionaries who celebrated Fergola's school as an autochthonous mathematical tradition worth defending, and argued for its integration in a new system of public education. One such functionary was Vincenzo Cuoco (1770–1823), one of the key figures in the genealogies of the moderate patriotic liberalism that would become hegemonic in nineteenth-century Italy. In 1801 Cuoco wrote a celebrated history of the Neapolitan Revolution that launched the *topos* of the "passive revolution," later re-elaborated by Benedetto Croce and Antonio Gramsci: the Jacobins ignored local traditions and popular sentiments, imposing a foreign constitution on a reality that could not receive it. Cuoco was resolutely anti-Jacobin, and his keen awareness of local differences wreaked havoc with enlightened universalism, but he was also distant from reactionary nostalgia. He celebrated the organic development of political institutions with historicist sensibility, and his pages are rich in Vichian-inflected insights about popular spontaneity as the veritable engine of history; the people are "the great and only agent of revolution and counterrevolution."[25] But the ideas of the people are not necessarily anchored to tradition; they keep changing, and the role of political elites is precisely to control and guide these profound and mostly irrational transformations.

Cuoco was read as a theorist of the subaltern relationship of the vast postfeudal masses to the propertied elites in a framework of economic liberalism and a modern monarchy. He became a key political actor during the French decade, when he was charged with implementing anti-feudal, financial, and educational reforms. He would support Murat until the very end, in 1815, when the ruler of Naples played his last desperate card as the champion of Italian national unification.[26] Cuoco's main critique of the Jacobins of 1799 was that they had not been able to control and channel popular energies. He remained always wary of the masses. "All our political errors," he wrote, reminiscing about his participation in the short-lived Republic, "derived from sharing with many ideas that should have been for the few."[27] Within the Napoleonic order Cuoco turned the political problem of the failed revolution into the pedagogical problem of a new system of public education. Celebrating the ancient Pythagorean heritage of southern Italy, Cuoco described it as a source of scientific wisdom but also of an elitist worldview: that true knowledge is for the few; what is truly essential in a body politic is that "everyone knows their place, that they work and that they love order. To obtain these things, both science and subordination are necessary." Like other members of the French decade's technocratic elites, Cuoco believed that the main goal of popular education was not political emancipation but the consolidation of a new post-feudal and nationalist order centered on the dominance of agrarian elites. To this widely shared perspective he would add an original powerful narrative of the nation's cultural primacy and of its mission.[28]

Cuoco, like Galdi and Cagnazzi, argued for an integration of analysis and synthesis in that strategic sector of public education that was the pedagogy of mathematics, overcoming "the great dispute" over methods. Overall, Cuoco supported an instrumental conception of the sciences, which he considered primarily as tools of government—hence his emphasis on statistics. But when it comes to pure mathematics, to the "metaphysics" of mathematics, he described in some detail the mathematical pedagogy of "our illustrious Fergola," whose texts he obviously knew, and celebrated the incomparable *perspicuity* of synthetic procedures. If analysis was a useful heuristic method for the professional mathematician and for the technocratic management of the state, synthesis should be preferred for its metaphysical significance, which made it especially suitable "to form the mind of the young." Restoring order along moderate liberal lines after Jacobin and counterrevolutionary anarchy meant, first of all, restoring order in mathematical education.[29]

*

This sociotechnical effort to reshape the kingdom was framed by a consistent and pervasive philosophical perspective. It was a sensationalist and ideological epistemology that subordinated knowledge to physiology and sensibility, and encouraged an empirical approach to the study of nature and the human mind. Various ideological systems describing the workings of the mental faculties were devised in those years, always moving from experience, which meant from sensations (external) or reflection (internal). Invariably, these systems included advice on how to shape the ideas and sentiments of the Neapolitan people at a time when the government urgently needed to impress its final "civilizing" push. In France, after Thermidor, *idéologie* as the grammar and logic of all possible sciences had turned into an all-comprehensive discourse that connected the natural sciences, the science of man, and the science of society, in defense of the liberties conquered with the Revolution and in support of a new order—for Cabanis, what was most needed in 1798 was *bon sens* and *l'amour de l'ordre et le gout des utiles travaux*.[30] The marginalization of the *idéologues* under the consulate, however, marked the crisis of their programmatic conjunction of scientific and political discourse, and of their vision of a harmony of the body politic based on the balance of powers and the active role of intellectuals. The collapse of this political project made Cabanis's epistemological connection of the physical and the moral obsolete. It was replaced by a new division of competences and a new institutional geography of knowledge. As noted by Jean-Luc Chappey, the crisis of the ideologues shows how Bonaparte overcame a significant political opposition thanks to the reorganization of the "order of intellectual production," namely the purification of science—its separation from politics—the specialization of knowledges, and the constitution of new elites based on competences at the service of the state.[31]

While the French *idéologues* were influential in Naples—Cabanis's main work was translated in 1807—the Neapolitan ideologists saw themselves as the heirs of an uninterrupted sensationalist and empiricist tradition that ran through Neapolitan philosophy, starting at least from Genovesi.[32] An important link between Genovesi and the philosophy of the French decade is Melchiorre Delfico (1744–1835), whose long career spans from the high point of Bourbon reformism well into the Restoration. Like many other provincial reformers, Delfico participated in the anti-feudal campaign of the 1780s, arguing for free-trade policies and for the modernization of agricultural and breeding techniques. A moderate, he preferred not to get involved in the experience of the 1799 Republic, though afterward he spent a few years abroad for good measure. Instead, he took up major administrative and political positions during the French decade. It was probably Saliceti who invited him to be part of the *Consiglio di Stato* in 1806; nine years of intense ministerial work followed—he was even interim minister of the interior for a few months. He was on innumerable commissions, covering a range of areas from economic and juridical matters to education. And yet he did not give up his scholarly work, which focused on the basic mechanisms of intellectual and social life. The titles of his publications were concise manifestoes: *Thoughts on History and on Its Uselessness and Uncertainty*, or *On the Necessity for Physiological Knowledge to Precede the Study of Intellectual Philosophy*.[33]

Unlike his contemporary Pagano, Delfico showed little interest for palingenetic catastrophes and political revolutions. Instead, he continued to demonstrate an unshakable faith in the gradual progress of science and the human spirit. He wrote often on the inutility of the discipline of history, but in reality he aimed to replace military-diplomatic history with a "progressive history of knowledge, experiments, observations, discoveries."[34] He wielded his emphatic and provocative anti-historicism as a weapon to dismantle feudal institutions and the feudal legal infrastructure: tradition in juridical matters, he argued, is an endless source of confusion and abuse.[35] Instead, laws should be grounded on the universal features of human nature. Human beings are social because of their "organic constitution": every moral phenomenon depends on it. Intellectual and moral ideas are always, in principle, reducible to physiology.[36] If sociability is based on organic dispositions, then the gradual emancipation of humanity—which includes women's emancipation and universal peace—can be pursued by eliminating all historical sediments that have concealed the universal principles of the science of society.[37]

Historians of philosophy have not been kind to Delfico. His exhibited lack of historical sensibility fared poorly in idealistic quarters, while twentieth-century commentators did not relate to his Panglossian social views. Yet Delfico's philosophical and political longevity are striking and need to be accounted for. A personal friend of King Ferdinand and the philosophical light of the French decade, he was a man for all seasons, someone who could be hailed as the last heir of the Neapolitan Enlightenment as well as the father of a new scientific age. Delfico's militant epistemology became the framework of the French-Neapolitan modernization project, and continued to be the hegemonic language and theory of knowledge of Neapolitan liberal elites up to, at least, the constitutional government of 1820–1821. It's not difficult to see why the new regime would find interesting a theory of behavior that promised to explain empirically how sensations and sentiments—rather than ideas—modified people's behavior and habits, offering potential techno-scientific solutions to the problem of social order in that most unruly corner of Europe.[38] Above all, the case of Delfico illustrates the interpretive flexibility of ideology and of its analytical toolbox. Its key elements had been constitutive of radical political projects: the Neapolitan Jacobins had imagined their revolution through the language and concepts of sensationalism and analysis. After 1799, the political realignment of most intellectuals with Bonaparte and the liberal agrarian elites began with a denunciation of the utopian, abstract nature of revolutionary praxis. In epistemology there seemed to be continuity: sensationalism and analysis were not abandoned. On the contrary, ideology consolidated its position as a coherent framework that brought together the culture of sensibility, the combinatorial analysis of ideas, and the deployment of purely algebraic-analytic mathematical methods. But the continuity was only apparent.

The position of Galdi, Cagnazzi, and Cuoco on mathematical methods and pedagogy is instructive: the ideological perspective that dominated official Neapolitan philosophy in the first quarter of the century had been neutralized, deprived of its revolutionary potential. The analytic study of ideas was disconnected from broader assumptions, and its scope was strictly limited to the empirical and operative dimension. A rigid dualism excluded a priori any monistic, materialistic, and atheistic outcomes. A chasm separated the empirical study of the mind from the axioms relative to the essence and destiny of the soul; the laws of physics and physiology from the metaphysical principles that inform the created universe. Even the most radical physiological reductionism could thus easily coexist with metaphysical evidence of the existence of God and recognition of the divine cosmic order. Which explains how ideology became an almost universally accepted common terrain, the meeting point of academic philosophers, who could update their methodological equipment without renouncing their metaphysical commitments, and the technocratic elites of the French decade, who increasingly fashioned themselves as the neutral, disinterested technicians who alone could implement the regime's modernization project. Ken Alder coined the term "Techno-Jacobins" to refer to those French Jacobins who conceived their revolution as, at once, a political and a techno-scientific process. He noticed how, after Thermidor, the survivors refashioned their personae and social function in terms of scientific neutrality and technocratic efficiency.[39] In Naples, it was the gallows of 1799 that marked—spectacularly—that watershed.

Training Engineers

The governments of the French decade saw techno-scientific training as essential to the success of their modernization project. In 1808 the minister of the interior reorganized the Royal Academy of Sciences as part of the Royal Society of Naples, designed on the model of the Institut de France. The ministry also supported a Royal Society for the Encouragement of the Natural and Economic Sciences, whose goal was "promoting public and private economy, agriculture, arts, wealth, and prosperity in this part of Italy, by means of

mathematics, chemistry, natural history, medicine, and veterinary science."
The society published a journal and awarded prizes to those citizens whose
discoveries had enhanced agriculture, husbandry, and manufacturing. The
most distinguished Neapolitan men of science were invited to join the society, including Fergola. He, however, refused membership, alleging health
reasons.[40]

The discussion over public education was intense during the French decade, and it produced a wealth of literature—much richer than the contemporary literary and philosophical production. An overall reform began to
be implemented with the 1811 Organic Decree for Public Education, which
divided it up into primary, secondary, and higher levels, and put it under
the direct control of the Ministry of the Interior. Primary schools were to
be created in every town and village, and at least two secondary schools in
each province, one for boys and one for girls. Three thousand free primary
schools had opened by 1814. Together with Italian and Latin grammar, students were taught arithmetic, good manners, catechism, and practical agriculture.[41] In 1814 there were also eleven colleges, whose curricula focused on
Latin, Greek, mathematics, and philosophy. The reform plan for the university saw a reduction of theological chairs in favor of scientific ones. Control
over the university was taken away from the major chaplain and given to a
king-appointed prefect.[42]

The reform of the university ended up less radical than it looked on paper, while the new Royal Society turned out to be primarily a celebratory
space. The government promoted scientific disciplines but seemed unwilling
to challenge the powerful professoriate; rather than trying to transform traditional institutions, it decided to bypass them. Above all, the regime needed
an efficient corps of engineers to build essential infrastructures. In November
1808, the Ministry of the Interior established the Royal Corps of Engineers
of Bridges and Roads, possibly the single most important institution created during the French decade. A French general of the Military Engineering
Corps, Jacques de Campredon (1761–1837), prepared a plan for the new institution based on the decree that had reformed the analogous French corps
in 1804. The plan, approved in 1809, segmented the territory into divisions
headed by inspectors; each division consisted of departments headed by chief
engineers. It was also decided that engineers should reside in their respective
departments and that they were under the authority of the *intendente*—thus
linked directly to the executive power.[43]

When the Corps was created, Campredon chose the engineers personally,
mostly from the ranks of civil architects, but it was clear that the engineers
were to be trained in new schools that would suit the needs of a modern

administrative monarchy. And indeed the 1811 decree also completely redesigned the curriculum of civil engineers. Campredon and General Giuseppe
Parisi—a former patron of Carlo Lauberg—transformed the Military Academy, now called the Polytechnic School.[44] Its four-year course included mathematical teaching in analytic geometry, descriptive geometry, and mechanics,
and its goal was to train army officers and prepare students to enter more-
advanced schools where they could specialize in fields like civil engineering, military engineering, artillery, or maritime constructions. The school
published the lectures of its professors under the collective title *Course of
Mathematics* (1813–1815). The lectures of analytic geometry and calculus were
written by Ottavio Colecchi (1773–1847), a former Dominican monk who cultivated analysis with unusual fervor and who would leave his mark on Neapolitan scientific and political life.[45]

Graduates from the Polytechnic School could then seek admission in the
School of Application for the Corps of the Engineers of Bridges and Roads,
whose four-year course cranked out "aspirant engineers" who competed to
enter the lowest level of the rigidly hierarchical Corps. The School of Application was designed as a primary center of techno-scientific education and
research. Financed generously by the government, the school was a serious
threat to the prestige of the university, where civil architects had formerly
earned their degrees. The school offered its professors higher salaries, and
enrollment increased significantly when Neapolitan tribunals began granting
expert qualifications exclusively to Corps engineers. Candidates at the school
were selected on the basis of an examination that included arithmetic, geometry, trigonometry, calculus, analytical geometry, design, French, and Latin.
This carefully designed examination offers a clear insight into the specific social groups that the Corps was mobilizing—mostly landed, professional, and
commercial elites from the provinces. The endless controversies and negotiations over the structure of this examination reflected broader struggles over
the role and authority of the Corps, and would continue well into the Restoration. Once accepted, students received an eminently practical education in
which firsthand experience was valued more highly than reading books or
listening to lectures.[46] Professors also drew attention to the natural connections between disciplines and to the overall unity of the curriculum, which
emphasized the great applicability of mathematics and the role of analysis as
its unifying language.

The four-year course consisted of two cycles: the first was devoted to the
study of mathematics, mechanics, hydraulics, and architecture; the second focused on applications to the various dimensions of civil engineering, including
construction, machines, chemistry, and agronomy. At the end of the course

the students were ranked on the basis of a final exam, and high-scoring students could enter the Corps of Engineers. During this one-week examination, students were expected to solve problems in mechanics, descriptive geometry, constructions, agronomy, and chemistry, and to complete an architecture project.[47] The School of Application's pedagogy reflected an instrumental view of science, which valued above all its capacity to transform reality, and embodied the epistemological doctrines of Delfico and other pedagogists who emphasized the active role of the student. Machines were key pedagogical tools: it was the continuous verification of the mathematical formalism's function in technological applications that gave it meaning and legitimacy in the eyes of the students. All courses integrated theory and applications and were structured according to analytical principles.[48]

Carlo Afan de Rivera (1779–1852), director of the Corps of the Engineers from 1824 to 1852 and the most eminent champion of modernization in Naples, made it clear that the study of the sciences and political economy were closely related. "Our scientists," he wrote, echoing a famous Galilean passage, "are afraid of being degraded by getting too close to the workshops, and to offer guidance to practice with their theory; so they neglect to apply the sciences to the arts." Instead, under his direction, engineers were engaging with land restitutions, the establishment of new manufactures, and the construction of new roads and harbors. For the first time in Naples, Rivera claimed proudly, the scientific method was being applied to the administration of the state.[49]

The *Analytic Library*

A new periodical celebrated the return of analytic savoir-faire in science, economics, and politics: the *Analytic Library* (1810–1823). Its articles were mostly in line with the cultural vision of the government and sustained its attempt to consolidate and shape the "public spirit," the social backbone of the new regime. The ministry of police and the provincial *intendenze* actively promoted the diffusion of the periodical among government officials, jurists, professionals, academics, and students.[50]

The *Analytic Library* is also interesting because it gave space to the voices of those few who had not given up the republican and democratic ideals, revealing political tensions within the capacious world of analysis. Those views were not expressed in openly political terms, but rather through scientific and mathematical positioning, most visibly in the rejection of the pedagogical compromise between analysis and synthesis. As a consequence, the periodical did not have an easy life. As early as 1811 it was temporarily closed

down due to the Royal Academy's pressure on the government because one of
its articles contained a vehement attack against Nicola Fergola and his school.
In 1823, at the height of political reaction, it was shuttered for good. Its mission, however, was continued by the *Progress* (1832–1846); it was a moment of
renewed liberal ferment and this journal defended free trade, political liberalism, Rivera's engineers, analysis, and the advancement of science.[51]

The *Analytic Library* opened its first number with a manifesto that called
for the unification of all branches of knowledge in a sensationalist, sentimental, and ideological perspective. The editor argued that human reason is well
equipped to grasp the basic structures of reality and to effectively guide human action. All sciences are related, grounded in sensations and processes
of abstraction, to the point that "matter cannot be detached from the most
sublime contemplations." Artificial boundaries between pure and mixed
mathematics are therefore pointless. The wording is reminiscent of Lauberg
and Giordano: "Doesn't every human craft [*arte*] aim to give us knowledge
about the extension, nature, and quantity of things?" Physics, mathematics,
politics, and morals share the same goal: "the construction of a Universal
Mathematics." The periodical reviewed books from all disciplines, reporting
with particular enthusiasm on those that educated the lay public in the recent
advancements in French mathematics, and those that applied mathematics,
especially probability calculus, to social and economic matters.[52]

The editor of the *Analytic Library* was convinced that moral and social
laws could be rationally derived from the material constitution of human beings, their needs, and their sentiments. In turn, the science of economics "depends entirely on morals." "It has been argued that administrators only need
good sense and matter-of-fact knowledge," the editor quipped, "but what is
this good sense, after all?" In reality, "economic truths" are "necessary consequences of the nature of things." A clear lineage was traced that linked the
present moment to Genovesi, Delfico, and other champions of Neapolitan reformism, who together transformed economics from an aggregate of empirical observations to a science characterized by "connections between ideas,
universal principles, and certain deductions."[53]

As one would expect, much space was devoted to the physical sciences,
where "the solution of problems depends more on the perfection of analysis
than on the exactness of observation." Laplace's *Mechanique celeste* and Lagrange's mechanics were taken as paradigmatic. Quoting Laplace, the editor
argued for the necessity of integrating empirical observation with imagination: the mere collection of facts would produce a "sterile nomenclature," not
a science. It is "higher mathematics" applied to "the results of experience that
can perfect the work of physics by expressing through calculus the universal

laws of nature." Analysis is not an artificial, external apparatus superimposed over nature but rather the only instrument that can capture with precision its universal laws.[54]

In mathematics, "preference should be certainly given to the analytic method" and to the "method of projections." The application of algebra to the theory of curves was "one of the most fruitful connections ever made in science." As for calculus, it did not need any geometrical foundation: Lagrange had reduced its techniques to the rules of algebra, so that its rigor was "not inferior to that of the ancient demonstrations." The science of calculus is certain and "applies to everything," its language being constructed of a "universal grammar"—hence the certainty of its statements. Those who claimed that the new analytic methods were uncertain were not just wrong: they were "hindering the progress of science."[55]

The *Analytic Library* saluted the advent of a new scientific age, one in which reason, senses, and imagination would cooperate to produce reliable and useful knowledge. Mathematicians, in particular, should not be mere contemplators of "intellectual truths" but should rely on their imagination as well: theirs is an active, productive reason. The true core of science, then, is not a set of absolutely certain principles: the sciences study the relations between things, and the most general philosophical truths are those related to the analysis of human cognition—"the procedures of our intelligence." The traditional view of logic as the art of deducing consequences from otherwise known principles was replaced by the field of ideological-transcendental studies.[56]

The main philosophical novelty of the French decade was the engagement with the work of Kant and German idealism, which was understood at first as a useful resource to update the ideological conception of mind and argue for a reason that is *autonomous and legislative*. This shift to a newly active mind is visible, for example, in the account of geometry as "an entirely human creation." From Genovesi onwards, Neapolitan sensationalists and ideologists had grappled with the problem of assigning to the mind some kind of active, reflexive faculty without reintroducing Platonic and Cartesian innate ideas. In the 1810s, Kantian criticism seemed to provide a viable solution: one could thus integrate the analytic and sensationalist perspective with a clearly defined organizing function of the human mind. For Pasquale Borrelli (1782–1849), thought is a self-regulating "force"; for Francesco Paolo Bozzelli (1786–1864), the faculties of the mind are "activities"; and for Pasquale Galluppi (1770–1846), all philosophy is grounded on "consciousness," the internal perception of the self and its modifications. Galluppi's conscientialism was a clear expression of this rejection of dogmatic academic philosophy in the name of an active and transformative mind endowed with analytic and synthetic faculties. Philosophizing begins with consciousness, which becomes the fundamental mechanism of all the intellectual operations described by ideology.[57] These transcendental critiques of traditional ontology—and of the supreme criterion of logical evidence—would be part of the philosophical infrastructure of the liberal and constitutional movement of 1820–1821.[58] Already in 1812, the editor of *Analytic Library* had announced that the Republic of Letters would soon be replaced by "the universal monarchy of the exact and natural sciences."[59] That monarchy, it was officially proclaimed in 1820, should be a constitutional monarchy.[60]

The concrete nexus between ideologically inflected conscientialism and political constitutionalism was the notion of "liberty of conscience," which Galluppi characterized as "an inalienable human right." No just government can deprive its citizens of their liberty of conscience. The philosophical discovery and articulation of the active powers of the mind, their epistemological significance, was also, at once, the recognition of the active power of volition, which "manifests the existence of our liberty." It was a liberty of the will that has been restricted by an "absolute power" and needed to be rescued.[61] The battle for autonomous reason and the battle for political liberty was one and the same. During the nine-month-long constitutional government, the kingdom experienced a freedom of speech that had not been seen since 1799: provincial agrarian elites voiced their resistance to the "arbitrary" central government, while military and professional elites who had thrived under Murat expressed distrust in the king as an agent of that modernization in which they were still invested. They all converged toward the constitution as the solution to the problem of post-feudal order. Their hopes, however, would be crushed in March 1821, when an Austrian army would defeat the constitutional army, thereby making it possible for King Ferdinand to revoke all concessions.

Risorgimento Analysis

By the early 1830s a new king was ruling, and the anti-liberal push seemed to be weakening. In reality the government was simply revamping the modernizing plan of the French decade, strengthening its administrative structures and intervening more decisively in economic matters. Once again it was technical elites who were charged with delivering modernization within the framework of an autocratic state. Technical officers from the army, along with Rivera's engineers, were tasked with implementing change. By offering prestigious administrative positions to high-ranking officers, the government also aimed to break up the dangerous alliance between military and agrarian elites, which had precipitated the crisis in 1820–1821. Meanwhile, pro-constitution liberals cautiously intensified their campaign through journals like *Il Progresso*, under the banner of classical economy, free trade, industrialization, technical expertise, Galluppi's conscientialism, and Bozzelli's theory of the "social forces."[62] Theirs was far from a revolutionary discourse: as in the "orderly revolution" of 1820, what they were asking for was social order and the recognition and protection of their political and economic rights in a legitimist perspective, mainly through the strengthening of local autonomies.[63]

In the 1840s, patriotic (Italian nationalist) and democratic political aspirations intensified, further complicating and destabilizing the political situation. The government took another sharp reactionary turn. As the supporters of unmitigated political reaction and Bourbon legitimism gained space, their views found philosophical expression in the newborn Neo-Scholastic philosophy, which took its first steps in Naples but would become, by the end of the century, the hegemonic Catholic philosophy worldwide. Old champions of moderate liberalism like Galluppi were now part of the academic establishment, and their work turned in an apologetic and spiritualist direction. The common enemy was "rationalism," a term that referred to the historical, economical, and philosophical productions of a composite group of students of Vico and Kant, who had turned to Hegel attracted by his dialectical theory of history and his absolute idealism.[64] Their reading of the German philosopher left no space for ambiguities: "the Idea is the rationality of revolution." Reason rules the world, and it realizes itself as liberty. It also leads to democracy. Philosophy must come down into the world and fight "for the entire humanity, for its needs and sorrows, for its desires and hopes." In 1843 the Neapolitan police reported about the activity of local revolutionary-patriotic groups and possible infiltrations of Mazzinians and members of the "Communist Sect." As Bozzelli would have said, the "social forces" were once again changing.[65]

We can conclude our survey of the trajectory of analysis in Naples with some brief considerations on the period leading up to 1848. Our guide is Ottavio Colecchi, a champion of democratic liberalism, Kantian philosophy, and analytic mathematics. He was indeed the last notable protagonist of the mathematical controversy between synthetics and analytics. The great rift that had divided mathematical life in Naples for over fifty years was finally losing its significance, but it had not disappeared yet. Colecchi came from a modest provincial family and had entered the Dominican order in 1794. After earning a doctorate in theology he had devoted himself completely to mathematics and philosophy. In 1809, the French-Neapolitan government suppressed most monastic orders, and Colecchi became a secular priest, though he continued to wear his black-and-white Dominican habit. By that time he was already "a very famous predicator"—and a troublemaker, as shown by an 1807 police search for weapons in his monastic cell.[66] In 1810, Colecchi was in the capital, where he published a mathematical article against the synthetic school. He was knowledgeable about recent developments in algebraic analysis and accused Fergola and his school of making few original contributions to this field and thus impeding the progress of Neapolitan mathematics. Unlike those who were calling for moderation and compromise in the controversy over mathematical methods, Colecchi condemned the synthetic method resolutely:

> I fear that Euclid, with his great power, can cause to mathematics the same damage Aristotle caused to philosophy. I agree that we must respect the father of geometry, and every other ancient geometer; but the exaggerated deference to synthesis, the servile attachment to the ancient constructions, can indeed hinder the progress of these sciences. In fact, it seems that the damage is becoming evident. While in France Laplace writes the *Mecanique Céléste* and the *Exposition du Systéme du Monde*; Monge writes the *Geometrie Descriptive* and the *Analise Geometrique*; Puissant writes the *Geodesie* and the *Recueil de diverses propositions*—where, through the method of coordinates, he solves the hardest problems with unparalleled simplicity and elegance [. . .] here in Naples one discusses the "problems of contacts," and a new property of the triangle; the method of inscribing a triangle in a circle whose sides pass through three given points; and one writes with didactic rigor a memoir on rational functions and their reduction to partial functions, enriching these and other similar jokes with plenty of scholia and notes.[67]

Colecchi delivered a scathing critique of Fergola's problem-solving methods and of synthetic mathematical culture in its entirety. The article appeared in the newborn *Analytic Library* and earned him a teaching position at the Polytechnic School but also many powerful enemies. It resulted, as we have seen, in the temporary closure of the periodical. By the end of the French decade, Colecchi was the charismatic leader of a group of young engineer-mathematicians who would be active as private teachers and in the military and engineering schools of the kingdom through the Restoration and up to the national unification of Italy (1861). Among them were Salvatore de Angelis (1789–1850) and Francesco Paolo Tucci (1790–1875), both teachers at the School of Application and the founders of a prestigious private school that trained a new generation of analytic mathematicians.[68]

Tucci's mathematical and professional trajectory is remarkable: he had studied with Fergola between 1808 and 1811 and was part of the synthetics' inner circle. By 1812, however, he had manifested his inclination for Lagrange's analytic geometry. A nineteenth-century historian speculates that Tucci's "temperament" induced him to leave synthesis for analysis. A publication of 1812 reflects nicely the shift. Up to that point he had solved problems with purely geometrical methods, in a way that delighted the estimators of the geometry of the ancients. That year, however, in a synthetic piece originally read at a local academy, he provided a solution to problems relative to certain conic curves and to the surfaces originated by their rotation. Tucci showed that he was able to master the techniques of geometrical analysis, as should be expected from a pupil of Fergola, while simultaneously performing a subtle apology for algebraic analytic methods by providing in the footnotes the relative analytic solutions, which he deemed simpler, very general, and applicable to any kind of curve. The reader would thus conclude that geometrical analysis is a beautiful exercise whose results are, however, rather limited.[69] Later, as a professor at the Military School, Tucci continued to argue for the superiority of modern analysis in geometrical problem solving, commenting also on a French debate on geometrical methods that had gained visibility in 1817–1818.[70]

After the 1815 Bourbon restoration, Colecchi left the country. That he had to leave while many other protagonists of the French decade maintained their positions shows that he was perceived as distinctively dangerous and impossible to integrate in the new regime. Thanks to the Dominican network, he traveled across Europe and ended up at the Dominican priory in Saint Petersburg, where he was responsible for the pastoral care of Italian Catholics in the region and taught mathematics and philosophy at the Main Pedagogical Institute (later the University of Saint Petersburg). Colecchi preached in the Church of Saint Catherine of Alexandria, on the Nevsky Prospekt, and gave private lessons to the children of the Russian aristocracy, including the sons of Alexander I. He is recorded as a member of the Imperial Academy of Sciences.[71] Colecchi returned to Italy via Königsberg, and in 1819 he was back in the Kingdom of Naples as a professor of analytic mathematics at the provincial college of Aquila. The reactionary turn that followed the constitutional government of 1820–1821—which he had supported—did not spare him. The bishop of Aquila considered his teaching dangerous for public order: "Colecchi," the bishop wrote to the special tribunal in Naples, "has shown a decided inclination for the pestiferous and abominable philosophy of Gante [Kant], which is subversive of all morals." To the bishop this charge meant that even if Colecchi wasn't a liberal conspirator (*carbonaro*), he certainly thought like one. Colecchi lost his position and for the next ten years he could only teach semi-clandestinely and far from the capital.[72]

In 1831, Colecchi was authorized to teach privately mathematics and philosophy in Naples, but a couple of years later Bishop Francesco Colangelo, then minister of education, managed to have the authorization revoked. Colangelo was a prominent representative of the reactionary Catholic group that had reshaped Neapolitan culture after 1821. To him Colecchi's mind was "disordered and dangerous." He reported to the police that Colecchi was an atheist, as he claimed that the existence of God could not be demonstrated. Such claims were "dishonorable," Colangelo commented, especially for a mathematician, who should know better about the role of the creator in the universe.[73]

Critics associated Colecchi's alleged irreligiosity and his democratic aspirations with his enthusiasm for Kant's three critiques, which he had read during his 1818 stay in Germany. As a consequence, Colecchi spent the 1830s under strict police control. About thirty thousand Neapolitan citizens lived in similar conditions. Meanwhile, taxes on foreign books increased, university students were required to participate in "spiritual congregations," and all privately owned books had to be reported to a special commission for approval. In 1843, Colecchi tried to publish his main philosophical works in a three-volume collection, but the general inspector for public education intervened to prevent the publication of the third volume, wherein Colecchi had planned to discuss the philosophy of Hegel and Spinoza. A couple of anonymous satirical sonnets from this period portray Colecchi as an "arrogant mathematician," an "apostate," and an atheist. But Colecchi, like the Jacobin priests before him, always defended his faith, and when the time came for him to dictate his epitaph, he began with the line "Ottavio Colecchi, priest." He separated, however, the sphere of religious sentiments from that of knowledge, wherein individual reason is sovereign. Like Lauberg and the Jacobins, Colecchi had defended philosophical sensationalism and mathematical analysis on the assumption that human reason is self-sufficient in making sense of empirical reality and discovering the true laws of nature and society. Kantian criticism, of which he was the first competent specialist in Italy, offered him powerful resources to argue for the autonomy of reason, replacing old-fashioned ideological analysis and philosophies of consciousness.[74]

Through Colecchi's experience we see the tension between liberal and democratic fronts. In philosophy, Colecchi disputed fiercely with Pasquale Galluppi, whose reading of Kant ultimately subordinated human reason to religion, arguing that reason can only recognize—not create—moral values. Galluppi's philosophy had gradually replaced previous academic traditions during the 1830s. In 1831 the Minister of the Interior had called Galluppi to the highly symbolic Chair of Logic and Metaphysics. Like other moderate liberals, he had turned to Christian apologetics and the defense of the Bourbon status quo. In 1820–1821, his experimental philosophy of consciousness and liberty expressed the views of the agrarian elites who had supported the constitutional movement. Leading up to the 1848 revolution, the spiritualistic core of his philosophy, which set clear boundaries for human cognitive faculties and the scope of the sciences, functioned as a bastion against atheism, materialism, and the ontological pretensions of a new generation of students of mathematics and philosophy, many of whom had gathered around Colecchi.[75]

But the strongest attacks against Colecchi came from reactionary Catholics. Right after the publication of his works was interrupted, the editors of the periodical *Science and Faith* accused Colecchi of numerous "metaphysical and moral" errors.[76] They did not reconstruct Colecchi's selective reception of Kant but remarked—correctly—that he accepted the idea that any sound science must be grounded on synthetic a priori judgments. This idea, they continued, was precisely what "our illustrious" Galluppi had always fought against, "making it impossible for Kantianism to consolidate its positions in Italy." According to Colecchi, the notions of space, time, substance, and cause are not derived from experience and are sufficient (he rejected Kantian categories) for the synthesis of sensible intuitions. Moreover, the three "ideas of reason"—the soul, the world, and God—do not provide us with concepts of knowable objects. Individual human reason, on the other hand, is "autonomous and legislative." It is self-sufficient in its construction of the entire edifice of knowledge and is the source of all natural and moral laws. To the editors of *Science and Faith*, these positions were all "grave and fatal errors," which turned Colecchi into an enemy of Italian and Catholic philosophy.[77] In the Neo-Scholastic perspective of *Science and Faith*, individual reason is passive and ultimately powerless vis-à-vis nature, political authority, and God. Political dissent and pro-constitutional protests were no more epistemologically legitimate than Kant's moral "monstrosities"—or the imaginary constructions of the arrogant analytic mathematician.[78]

Excluded from the official education system, Colecchi lived his last years in dignified poverty. But he was intellectually active, and he published numerous articles on philosophy and mathematics in periodicals such as the *Progress*. Topics ranged from the philosophy of Victor Cousin to problems in descriptive geometry. Mathematics remained a concern for Colecchi, and he did not miss opportunities to duel with Fergola's students, whose approach was still hegemonic at the university and in the most prestigious scientific academies. In philosophy, Colecchi argued that Kantian transcendentalism had finally dissolved that opposition of analytic and synthetic methods that was still a problem for Galluppi and the ideologists. Transcendentalism finally offered a way out of the dilemma between empiricism and metaphysical rationalism. Interestingly, Colecchi was convinced that the transcendental approach also made him tackle certain mathematical questions more effectively. He argued, for example, that mathematical induction is a perfectly legitimate method, and that it makes for rigorous demonstrations.[79] Colecchi further argued that mathematical induction is not based on experience, but is rather "an operation of the self" and as such can be known "exactly." He admitted that while the algorithm behind the pattern can be clear to our intelligence, we lack an intuition of the entire pattern when it's infinite. But this lack should not pose a problem, he continued, as not all evidence is based on intuition. Whether this formalist approach can be squared with Kant's view of mathematics is debatable, but that's not our concern here. Rather, note how Colecchi was using a transcendental perspective to question, once again, Fergola's purist and intuition-based approach, which had banned from the realm of legitimate mathematics all those techniques that depended on nonmathematical considerations and which were not entirely perspicuous.[80]

Colecchi passed on his uncompromising analytic faith to his numerous students, who were part of a new generation of engineer-mathematicians based at the School of Application and the Military School. Among them were many patriots who supported the process of Italian national unification, and some revolutionaries—"skilled at haranguing the crowd," according to a police file. Faculty and students from those schools joined the insurrection of 1848 in large numbers, and many of them served in the ensuing constitutional government.[81] Meanwhile, Colecchi's philosophical competence and intransigent character had also earned him the admiration of the Neapolitan neo-Hegelians, a group of philosophy students whose revolutionary and patriotic plans were soon to surface. That Colecchi was close to democratic, socialist, and Mazzinian conspirators jailed in Naples was no secret, as he frequently visited them. When he died in 1847, the newspapers did not mention his death, so as to avoid turmoil—but to no avail. His funeral was the occasion for one of the most significant anti-Bourbon demonstrations yet. The eulogy was given by Bertrando Spaventa (1817–1883), a student whose name opens the histories of Italian Marxism. To democratic and republican revolutionaries like him there was an obvious connection between their struggle and that of the revolutionaries of 1799.[82]

4

The Shape of the Kingdom

To the engineer-mathematicians of the School of Application, analysis was a transformative force, the force of modernity. Analytic practices aimed to transform not only scientific and social life but also the very *shape* of the kingdom—the way of organizing its infrastructures, administrative boundaries, and agrarian landscape. And indeed landscape, both human and physical, was at the core of disputes over modernization. The Corps of Engineers, whose vast responsibilities included managing natural resources, constructing and maintaining large infrastructures, fostering manufacture, innovating agricultural techniques, reclaiming land, and founding new towns, played a major role in all such disputes, as did their distinctive knowledge. Analysis, the engineers argued, would provide them with the correct way of thinking and would guide their action unfailingly.[1]

The engineers' analytic tools shaped how they represented and modified the landscape. The ancien régime shape of the kingdom was the outcome of specific relations between urban areas, plains, and highlands—the Apennine periphery—and was constitutive of a political culture based on corporatism, autonomy, and the reciprocities of clientage and patronage in areas where the authority of the state could seldom translate into permanent territorial control. Historical geography, noted Michael Broers, is essential "to the shape—quite literally—of the history of modern Italy."[2] To the French, their allies, and those who continued to support the Napoleonic model of the state during the Restoration, this culture represented an irrational, uncivilized way of organizing political life; nothing could be salvaged from its distinctive way of constructing a public sphere or of its baroque collective rituals. A civilizing mission, revolutionary regeneration, and modernization all invoked a literal *reshaping* of the kingdom.

The engineers' enemies were many and powerful, and it is impossible to reconstruct the engineers' vision and action without acknowledging the fierce resistance they faced. In part, this resistance was similar to the resistance faced by enlightened entrepreneurs and reformers like Grimaldi. There were clear continuities between earlier modernization projects, and therefore continuities in the resistance against them should be expected. I use the term "resistance" here to refer to a broad spectrum of phenomena, a set of what should be more properly called "resistances," which include forms of protest against the destruction of the moral unity of the peasant community, the defense of secular local autonomies, and the protection and consolidation of the interests of part of the trading and agrarian elites, all the way up to the articulation of the discourse of political reaction.[3] These phenomena were often but not necessarily related, and when their interests aligned—as in the case of the *Santafede* crusade—this alignment was based on contingent strategic considerations rather than on some shared essence. I use "resistance" to refer to this constellation of phenomena, and "reaction" to refer, more specifically, to the self-conscious articulation of discourses that opposed social change and enlightened modernization in their Jacobin-revolutionary and Napoleonic-technocratic forms.

While highland peasants resisted new taxes and conscription, many members of the elites across the Italian peninsula described the Napoleonic campaigns as the new barbarian invasions, and found the methods of the new regime alien and repugnant. Carlo Botta, a disaffected *bonapartiste* from Piedmont turned Italian patriot, was expressing a common view when he described the Italian Jacobins of the triennium as "infatuated by certain geometrical governments that they had not yet experienced."[4] Debates on the legitimate use of mathematical tools to represent and transform the physical and human landscape of the kingdom were particularly intense in key areas like statistics and cartography. These modes of representation changed profoundly during the French decade, and it's instructive to reconstruct how their technical features embodied the aspirations of social groups that aimed to impose a new order on the post-feudal world. During the Restoration, critiques like Botta's were raised against those who continued to defend the "modern monarchy," its idea of the state, and its attempt to transform the relationship between center and periphery. The 1830s exchange between the long-term director of the Neapolitan Corps of Engineers and a leading conservative politician was emblematic of this momentous fight over the Neapolitan landscape. Through these two contrasting gazes I reconstruct the engineers' technocratic project and the vision that opposed it. The first embodied a narrative of progress, analysis being, once again, one of its key components.[5]

The second was an expression of a reactionary utopia that, in the name of tradition and common sense, rejected not a particular reform but the very possibility of social change.

The tenets of this reactionary utopia found a sophisticated and successful visual representation in the productions of a local group of landscape painters, the school of Posillipo. The Restoration was a golden age for landscape painting in Naples, characterized by the rapid rise of a new figurative naturalism, a term that referred to both the subjects of the paintings—uncontaminated nature—and to their pictorial style, which was anti-academic, free, and "full of sentiment." The 1820s and 1830s saw the codification of a series of visual tropes about the Neapolitan landscape that, I argue, embodied a reactionary utopia. Nature in these paintings is perfect and timeless; the ancient artifacts that emerge from the lush vegetation are themselves naturalized; they belong to a landscape that needs no change, that *cannot* change. When human figures do appear, they are "innocent peasants" whose way of life is as ancient and unchangeable as the pine trees or Vesuvius. The contrast with the analytic and future-oriented representations of the engineers could not be starker. And yet, the *posillipista* gaze is not less modern than that of the engineers: there would be no untouched nature without the engineers' iron bridges.

Engineers and Their Enemies

After 1815, the restored monarchy continued to support the process of administrative centralization. Significant reactionary forces, however, had come into power. The 1820s and 1830s saw an intense conflict between the champions of progress and a fierce political reaction at both the central and local levels. The newly centralized structure of the state suited the absolutist ambitions of the crown, and Vienna clearly indicated the need to retain existing administrative expertise. Prime Minister Luigi de' Medici opened up the Restoration age by attempting a compromise, which was called "amalgam politics"—a choice fiercely contested by the reactionary faction.[6]

Some institutions, however, were so emblematic of the French period that they could not go unchallenged. The Corps of Bridges and Roads and its School of Application were both abolished in 1817, replaced by a General Directory of Bridges and Roads, largely restoring the pre-1806 situation.[7] The new institution was characterized by the suppression of the inspector, a drastic reduction in personnel, and a redefinition of the engineers' juridical and economic status, which became more precarious with the implementation of short-term contracts. The function of the Corps was reduced to controlling selected public works, while funding and planning depended entirely on the Provincial Deputations for Public Works, which expressed the interests of local elites. This shift in the decision-making process clearly diminished the Corps' autonomy and reach.[8] The engineers' distinctive expertise and epistemological authority were undermined too. During the French period, the figure of the civil engineer had been vested with unprecedented prestige: engineers had become officials equal to those from rhetorical-juridical backgrounds and were considered essential to the national interest. Things changed after 1817. The School of Application reopened, but with a shortened course, and with many of its teachers on temporary contracts. The Corps was a lighter, more flexible institution that engaged problems as they arose, depending on the financial situation, while a general plan of public works for the entire kingdom had been shelved.[9]

Carlo Afan de Rivera, an army officer who became the Corps director in 1824, forcefully argued for the necessity of restoring the Corps' original features. Supported by administrative and military elites who longed for a return to the technocratic *dirigisme* of the French period, Rivera's saw many of his ideas implemented with the reform of 1826, which restored the stability and dignity of Corps positions, and returned to engineers control over the technical dimension of public works across the kingdom. New uniforms, designed by Rivera himself, signified their renewed status. The Corps, however, remained an anomaly and the source of endless tensions in the Bourbon administration.[10]

Under Rivera's long directorship (1824–1852), the Corps and its school developed and implemented a vision that was largely an extension of the French period. Elaborating on traditional associations between engineering and artisanship, Rivera considered his engineers at once scientists and artisans (*scienziati artisti*) who could reshape the kingdom through the application of modern scientific theories to architectural, infrastructural, and socioeconomic questions. They were in an ideal position to do so, Rivera argues, because they were *disinterested*. They were impartial and independent and looked at problems "in the abstract"; that is, not from a "private" point of view. The lamentable state of public works in the kingdom, according to Rivera, did not result solely from incompetence but from the excessive pressure of "particular interests." To resist such interests, engineers needed to turn themselves into a well-structured and disciplined group. For the traditional engineer and architect, who acted in isolation, it was virtually impossible to resist the pressure of powerful actors, be they absentee landlords or financial speculators. Only the Corps could function as a reliable cognitive and moral actor in the exclusive interest of the state.[11]

According to Rivera, the moral and epistemological virtues of the engineers enabled them to tackle problems systematically. In his writings, the metaphor of a social group as a machine resurfaces again: the Corps is an efficient machine and the kingdom itself a "great machine," whose parts interact in precise and calculable ways. Public works are the cogs and wheels of this machine, connecting its parts and facilitating the movement of goods, people, and information throughout the kingdom. To think of public works as isolated tasks is to miss their true meaning. The main limit of those engineers and architects who had worked for the Neapolitan crown before the French period was precisely that they did not see public works *as a system*.[12]

Unlike their ancien régime predecessors, Rivera continued, his engineers were well versed in applied mathematics, hydraulic architecture, general constructions, and the physical sciences; they could "investigate and calculate with precision the local circumstances and their relations to the branches of political economy." Therefore, they could "plan and supervise skillfully the most useful public works." It is not coincidental that, in Naples as in France, what Antoine Picon called the "invention of the modern engineer" took place contemporaneously with the gathering of an unprecedented amount of statistical information and with systematic cartographical efforts.[13]

Rivera guided his engineers toward three main integrated areas of intervention: transportation, land reclamation, and reforestation. He argued that massive deforestation, in addition to depriving the kingdom of construction wood, was linked to the gradual but steady migration of peasants from the plains to the hills as they tried to escape malaria. In turn, deforestation had only worsened the hydrogeological disaster and favored the expansion of swamps at the bottom of valleys and along the coasts. Such a vicious circle had to be addressed systematically. To Rivera, land reclamation in the plains was the first necessary step for modernizing agriculture—making it intensive and highly productive. This new agriculture, freed from feudalism's remnants, would attract new investments and innovate its techniques. Reclamation must be accompanied by the construction of new towns and roads, the relocation of entire populations, and the establishment of state-supported economic initiatives. Rivera referred to this set of related operations as "integral reclamation" (*bonifica integrale*), which would reshape the physical landscape as well as the ways of life of its inhabitants.[14]

Rivera was well aware of the opposition his plan faced. The single most insurmountable problem, he wrote, was not technical but "the interest of those who profit from the evils that afflict society, and defend their rights in the name of private property." The interest of the state was often at odds with those of individuals and institutions but, he thought, the state must prevail.[15]

When Rivera was writing, new concentrations of land and power had replaced the configuration of the feudal-communal world. These agrarian elites had benefited from the enclosure of common lands during the French period, and had developed highly exploitative labor systems that guaranteed them social and political hegemony. Throughout the Restoration, they revolted whenever they felt threatened by centralizing projects. As noted by John Davis, the southern landscape would remain essentially unstable and "disordered," as the socioeconomic tensions produced by modernization processes could not find viable political solutions.[16]

The new landscape envisioned by Rivera had distinctive physical and social traits. He argued that extensive cereal farming and transhumance were inefficient and worked against the public interest: they must be replaced by intensive rotation farming and breeding. To do so meant breaking down former feudal and communal lands into small and medium farms controlled by entrepreneur-investors.[17] The transportation system—vital to the military, administrative, and economic needs of the kingdom—had to be redesigned entirely. The existing star-shaped road system, centered on the capital while underdeveloped in the peripheries, was to Rivera a clear sign of the inferior mathematical training of eighteenth-century Neapolitan engineers. In fact, the system reflected the ancien regime relationships between urban areas, plains, and mountainous regions, and was functional to the economic interrelation between the center and its highly autonomous peripheries. Here as elsewhere, Rivera turned aspects of the fundamental sociopolitical problem of reshaping the kingdom—in essence, reordering its center-periphery relationships—into manageable and rationally solvable techno-mathematical problems: problems best expressed and tackled through the abstract language and methods of analysis.[18]

In Rivera's plan, roads should connect the main ports and innervate the most productive areas in a polycentric system that reflected and was functional to a new administrative and economic order. He believed that traditional paths were often badly designed and made the movement of goods and people too slow and expensive; they needed to be abandoned. For example, traditional roads often ran through too many towns and were too steep in mountainous regions. Supported by the "evidence of calculation," Rivera argued that it made economic sense to lengthen roads to follow valleys and rivers and thus keep lower gradients—no more than 5 percent.[19] Like all projects by Rivera and his engineers, this one too included a detailed analytic description, a calculation of the necessary work and materials, and a cost evaluation.[20]

The formidable opposition to Rivera's plans has been described as the outcome of the limited "awareness with which the ruling class of south Italy faced the problems of modernization."[21] This resistance should be interpreted in terms of the concrete interests of a restricted financial and commercial elite that controlled the kingdom's economy. The post-feudal conditions of the provincial economy offered attractive low-risk investments to a few well-connected merchants and financiers. The interests of these investors aligned partly with those of absentee landlords and those of provincial elites who resisted losing their power to the central government. This combination constituted a daunting obstacle for the supporters of modernization, who were mostly professionals, entrepreneurial landowners, military officers, and former members of the Napoleonic administration. As for the crown, it was chronically indebted to Neapolitan bankers, and this weakened its reformist resolve, particularly in times of crisis.[22]

Clashes between supporters of a strong central government and defenders of local autonomies included endless controversies on the technical-mathematical expertise of engineers and the scope of their legitimate action. The technical was the political, and no relevant actor seems to have missed it. Not surprisingly, the School of Application was under constant pressure, and faced numerous attempts to close it down. In 1835, for example, Rivera organized a one-week public exhibition of models and projects to showcase the skills of its students. The *Progress* reported about it with an apology for the engineer as civil servant. Engineering, the article claimed, needed substantial governmental support, as it required not only "talent and study" but also resources "that a private individual could hardly afford." Apparently, the "competent and distinguished public" was impressed. The article concluded on an optimistic note: these students would realize "marvelous artifacts," bringing Naples up to par with "the most educated and richest regions of Europe." But the exhibition signaled that pressure upon the school was mounting: the Ministry of Finance had asked Rivera to organize it as a clear and public demonstration that the school was not a waste of money.[23]

Since taking up the directorship of the Corps, Rivera had emerged as champion of the engineers and of a vision of the country that aligned with the project of a centralized administrative monarchy and a dirigiste economy. He faced powerful opponents, including first-rank politicians such as Giuseppe Ceva Grimaldi (1777–1862), whose writings and political action reveal some key strategies of those who resisted modernization. Ceva Grimaldi, an erudite aristocrat with literary ambitions, had kept a low public profile during the 1799 Revolution, but in 1800 he left the administration of his feudal estate to sit on the committee for the restoration of public education. He abandoned political life during the French period but was nominated *intendente* at the return of the Bourbons, charged with controlling some of the provinces considered more at risk of liberal conspiracies. In his 1817 book *Reflections on the Police*, Ceva Grimaldi discussed Beccaria, Helvétius, Montesquieu, and Voltaire, only to conclude that the police should force the "public spirit" back to "the old way of thinking." After the constitutional insurrection of 1820–1821, Ceva Grimaldi's career advanced rapidly. In 1830 he became minister of the interior and of public education. His political action was mainly directed at decreasing the economic pressure on the popular strata through paternalistic policies such as lowering the price of wheat. Rather than arguing for a general reform of the Neapolitan economy, he lamented the abolition of the medieval guilds, which he saw as the workers' best protection against the greed of modern entrepreneurs. At the height of his controversy with Rivera, he was the president of the *Consulta Generale del Regno*, an important administrative and juridical institution, and in 1840 he became prime minister. He abandoned politics during the revolution of 1848, when King Ferdinand II conceded the constitution.[24]

Ceva Grimaldi constantly targeted the Corps, which had been designed precisely to subtract the control of the territory from the intermediate bodies that he defended so vehemently as a legacy of ancient wisdom. In his pamphlets, he describes the Corps' autonomy from the rest of the administration as a monstrosity and a threat to the government's authority. He also kept reminding his readers that the Corps' members were all too keen on liberal ideas, and many of them had been involved in liberal conspiracies. Ceva Grimaldi openly declared his *distrust* for engineers and their machinations: theirs was, at once, a political and technical subversion.[25]

In 1832 Rivera published a selection of studies on political economy and some ambitious technical projects with the title *Considerations on the Means to Revalue the Gifts Given by Nature to the Kingdom of the Two Sicilies*.[26] Grimaldi's response came a few years later with *Considerations on the Public Works in the Continental Kingdom from the Normans until Our Days*.[27] These two books can be taken as manifestoes of two conflicting visions of the southern kingdom. From their titles it is clear that they belong to separate conceptual worlds. Rivera emphasized the wealth of natural resources available, paying little attention to their former modes of exploitation: he was convinced that nothing useful could be learned from the economic and technological practices of the past. He was confident that the powers of modern science and analysis alone could transform the kingdom into a modern state. Analysis had unified the fragmented realm of human knowledge; now it would unify the nation as well. Ceva Grimaldi's considerations, by contrast, focused on the long tradition of public works under the dynasties that had ruled southern Italy. He went all the way back to the Norman conquest of the tenth and eleventh centuries, aiming to prove that a solid local tradition in public works had always existed. To Rivera, the present was the starting point from which a scientific vision of the world could transform political and economic life. To Ceva Grimaldi, the present was the outcome of a historical process through which existing social and political institutions received their meaning and legitimization. He found laughable the idea that "our civilization began at the time of the institution of the Corps of Bridges and Roads."[28]

For all Ceva Grimaldi's historical sensitivity, he seems to describe a world that has never really changed—in his pages even revolutionary upheaval vanishes into an undifferentiated past. His narrative offers, ultimately, a powerless contemplation of an immutable order. Time flows monotonously in this world, where the future is not a project but merely an emanation of the past. Human agency is marginal: men are temporary administrators of a reality that cannot be planned or even modified. Historical change does not—and cannot—transform the essential traits of this reality. Politicians and administrators face one choice: embracing tradition and defending the status quo or destroying it according to impious eighteenth-century revolutionary principles. Whatever decision they make, political life will always return to its natural condition, and any revolutionary aberration will eventually be forgotten and reabsorbed in a futureless present. The kingdom as described by Ceva Grimaldi is a reactionary utopia, a world in which traditional institutions function effortlessly, all channels are navigable, and all roads are well kept and safe. His is a sentimental journey through the thousand little *patrie*—cities, towns, local communities—that made up the ancien régime kingdom and whose very existence as autonomous entities was now threatened by ominous external forces.[29]

In this perspective, the Corps of Engineers was not only unnecessary but detrimental. For one thing, it was an overly expensive way of running public works. Furthermore, to grant the privilege to direct all public works to a single institution was dangerous; works should be planned by local authorities and executed by engineers of their choosing. As a general rule, Ceva Grimaldi argued, public works should be kept to a minimum. Railroads, for example, might be "advantageous for trade" but can also be "advantageous for the enemy" in case of military invasion—an argument that aligned with those of many local governments that feared railroad construction would accelerate the depopulation of their poorest regions. According to Ceva Grimaldi, the construction of railroads was fraught with "moral and political consequences," and he thought the plan to contract the construction of new roads to private companies that would be authorized to charge tolls would create a new form of "industrial feudalism," the outcome of letting laissez-faire principles "triumph over all the ancient customs and privileges." "Someone might find that I am too attached to the old ways," Ceva Grimaldi quipped, "and yet trying to retrieve what was good in them is less crazy than ditching them altogether."[30]

Rivera's technical plans were hardly comparable to Grimaldi's literary reflections. The concepts they deployed, it has been remarked, are so radically different that they seem to produce a "reciprocal incomprehensibility."[31] Yet both recognized a single object at the core of their dispute: analysis, the savoir-faire of the engineer. According to Rivera, the Corps was fully justified in making decisions about the shape of the Neapolitan landscape without consulting local authorities. Engineers' expertise, based on the universal language of mathematics and the "analytic spirit," separated them from the rest of the administration and from all others involved in traditional decision-making regarding public works. Only the problem-solving expertise and the rigidly hierarchical organization of the Corps could guarantee the central government an effective control of the territory and the implementation of its modernizing policies. Once decision-making was shifted to the technical level, it was no longer accessible to those who had not mastered the chief engineers' analytical expertise and mathematical language. One of the recurrent complaints local authorities voiced to the national parliament during the 1820 consultation was indeed that they had lost control over their own territories to the despotic Corps director.[32]

By contrast, Ceva Grimaldi aimed to obliterate the distinctive cognitive and institutional traits of the Corps, reducing it to a mere branch of the administration. He questioned the morality of the engineers and the disinterestedness upon which Rivera had grounded their cognitive authority.[33] With plenty of evidence that engineers could be corrupt, Ceva Grimaldi argued, how could they be trusted? Similar attacks were forceful and repeated. In 1831, for example, the attorney general of the Court of Audit questioned the engineers' morality in an attempt to cut public spending and decentralize control of public works.[34] The pressure was such that Rivera wrote up a memoir in defense of his engineers, arguing for both their technical skills and moral virtues, while still acknowledging real cases of corruption, which he blamed on excessively low salaries.[35]

Ceva Grimaldi also questioned engineers' expertise. Wasn't it ridiculous, he commented, to believe that Neapolitans had to wait for the arrival of the French to learn how to build roads and bridges? The necessary skills could easily be found outside the Corps: Neapolitan artisans and masons were heirs to a millenary tradition of public works. The absurd monopoly of the Corps should therefore be abolished, following the example of Great Britain and the United States. As for the School of Application, Grimaldi suggested it be made public, widening access and increasing the teacher-to-student ratio. Admission to the school should not be restricted to "a few initiated"—students who were selected on the basis of rigorous exams in analytic mathematics.[36] Grimaldi's attempts to open up the School aimed clearly to weaken its distinctive analytic training. To challenge the reliability of analytic methods in mathematics was to challenge the scientific authority of the engineers of the Corps and the political and economic choices legitimated through their expertise.

*

The controversy over the reform of the Neapolitan system of weights and measures provides an insight into the interplay of the political, economic, and technical dimensions. The route to reform had been long and difficult. Attempts to unify Neapolitan measures and weights in the 1780s had failed, as had those of the French period.[37] In 1830 Rivera adopted a nonmetric system based on the decimal progression to be used in the Corps and its School: it was limited to surface measures, and its terminology was traditional, based on the Neapolitan palm. This system, Rivera claimed, was easier to use than the existing one, "and we cannot understand how people could be more inclined to calculate in what is, in fact, a more difficult way."[38]

Following this pilot project, a commission proposed to extend the use of this system to the entire kingdom. The administrative organ responsible for such decisions rejected the proposal in 1837. Ceva Grimaldi, the president of this organ at the time, in collaboration with synthetic school leader Vincenzo Flauti, defended his decision, accused the Corps and its school of squandering public money, and openly questioned its technical expertise. Flauti had the book published in his private printing house, along with his introduction and notes. He had read the manuscript with "incredible avidity" and supported it enthusiastically. He was confident that Grimaldi had clarified once and for all "the real meaning" of the question, showing that "this matter falls within the competence of the wise economists, and not of mathematicians, as it is commonly believed."[39]

According to Flauti, mathematicians only began to deal—illegitimately—with the question of weights and measures in the eighteenth century, "when a furious freedom, aiming to renew the human species, tried to destroy all ancient customs and habits." In this "very unhappy period," a new system of measures was introduced together with such pernicious innovations as a new calendar or the decimal division of the circle. Of all these alleged improvements, Flauti continued, only the metric system survived, though limited to certain scientific purposes. Expanding the use of such a system to the whole society would mean changing inveterate laws and customs, producing end-less "disorder and confusion."[40]

The different systems in use should not be unified but rather integrated, argued Flauti. The decimal system was indeed excellent for "purposes of cal-culation" but was not equally suitable for the practical needs of trade and daily life because of its abstract nature. The problem of identifying the best system of measure was "underdetermined": the engineers offered one of many logi-cally possible solutions, but hardly the most appropriate, because it did not take into account the *concrete conditions* in which this system would be used. To Rivera, the abstractness of engineering methods was the source of their technical effectiveness, universality, and moral-epistemological superiority: they were disinterested and therefore objective. To Flauti, that same abstract-ness was the essential limit of the knowledge of the engineer.[41]

Flauti and Ceva Grimaldi were defending a situation in which multiple sys-tems were in use in the kingdom, and even the same unit could differ signifi-cantly from region to region. One of the reasons for these discrepancies were the disparate conditions found across the territory. For instance, traditional surface measures were related to the time or the number of people necessary to work the land, which would vary between the plains and the highlands. Techni-cal differences embodied different forms of labor and life. More generally, this dazzling variety of systems reflected the interconnected but highly autonomous condition of the kingdom's peripheries. Such discrepancies, Flauti observed, never caused "any harm to landowners, or the government." If mathematicians had a role in the debate on weights and measures, Flauti argued, it should be limited to questions like the exact determination of the length of the Neapolitan palm—without questioning its suitability for practical uses.[42]

The synthetics and their political allies had stalled standardization projects for decades.[43] By 1838, however, opinion within the Royal Academy of Sciences had shifted in favor of the engineers. Flauti was losing his grip on the academy and blamed the "mob" of "young and inexpert" engineer-mathematicians that had entered its ranks. Led by Ferdinando Visconti, now director of the Topo-graphical Office, they were contesting the predominance of synthetic mathe-matics within the academy; Visconti himself had designed the measure system supported by Rivera. Flauti repeatedly accused Visconti and his fellow engi-neers of introducing "mathematical abstractions" to matters of societal admin-istration, ignoring "the public interest." He noted maliciously that Visconti, of all people, should know well the difference between an abstract and a concrete solution to a problem in light of "his own fatal experiences." As we have seen, Visconti had been a Jacobin and a close collaborator of Lauberg; in 1794 he had been sentenced to ten years in the fortress of Pantelleria; in 1822 he had lost his position due to his support for the constitutional government.[44]

For Flauti and Ceva Grimaldi, the legitimacy of the traditional systems of measure was first of all historical: it stemmed from their "very ancient origin," a mythical golden age in which "men had faith in their own customs, and they did not think of asking scholars to establish their system of measures and weights."[45] The wisdom of the common people had designed a variety of units of measure to fit different places and usages. According to Ceva Grimaldi, the people have "the monopoly of good sense," and the government should take their instinctive resistance as a command. The long controversy came somehow to conclusion in 1840, when Rivera published the official tables for the newly rationalized and unified measure system.[46] In fact, the usage of the new measures remained limited; most farmers, traders, and landowners simply ignored it.[47] If reactionaries deployed the rhetoric of "immemorial tradition" and "popular common sense," Rivera's appeal to "logic" and "simplicity" was not less rhetorical. Consistent implementation of the new system would require significant investments and nonelementary levels of mathematical training. Above all, it would imply transforming the existing center-periphery relationships. It was a system, as Rivera at times conceded, well suited for the army, the scientists, the technicians, and the entrepreneurs; a system designed for a dirigiste techno-scientific management of the kingdom.[48]

Numbers and Maps

Analysis needed reliable data to be effective. Cartography and statistics had been essential tools for the Neapolitan Enlightenment's reformist efforts. Socioeconomic change, reformers believed, should move from and be guided by a detailed knowledge of the concrete situation of the kingdom. In the 1780s and early 1790s the government had supported, to some extent, the collection of information about the provinces, a genre well exemplified by Giuseppe Galanti's impressive *Descrizione* (1786–1790). "In Naples," Galanti wrote, introducing his work, "the state of the island of Tahiti is probably better known than that of the provinces."[49] Building on this tradition, the governments of the French period and the Restoration set up offices of statistics whose activity was characterized by a tension between centralizing ambitions and the resistance of local elites, who were often reluctant to disclose key information. The limited results of their data collection efforts were yet another outcome of a fundamental clash between different modes of conceiving the center-periphery relation and the nature and tasks of public administration.[50]

Silvana Patriarca has shown how, in the Italian peninsula, statistics was a mode of representation that came to embody different aspirations, from the enlightened rationalization of society to the patriotic cause of national unification. At the juncture that concerns us here, the emergence of the Napoleonic modern state and its adaptation to the Restoration age, statistics was seen, above all, as a body of knowledge functional to re-establishing social order, facilitating the exercise of power, and fostering production. Governments faced a fundamental problem of social order, and statistics was part of a possible technical solution—"statistical surveillance," as Patriarca calls it. Statistics also embodied the moderate liberal elites' aspiration to establish their hegemony in the post-feudal world by fighting socioeconomic "backwardness," clearing the remains of the feudal-communal system, and completing the process of enclosure of common lands and infrastructures. Through its detailed descriptions, calculations, and classifications, statistics promised to address the related issues of popular unrest and the effects of uncontrolled economic change. The alignment of these moderate liberal interests with those of the Napoleonic and Restoration governments opened up institutional spaces for the theorization and practice of statistics across the peninsula.[51]

What kind of statistics do we encounter in Naples? Theodore Porter describes the statistical science that emerged in parts of Europe in the 1820s and 1830s as a "numerical social science of facts," which expressed the anxieties of rapidly growing urban societies as industrialization was ramping up. At its core was the attempt to extract the "laws of society" from the wealth of newly available numbers. What we see in Naples is a particular version of this general trend, characterized by strong continuities with previous forms of descriptive statistics, which included numbers but also large amounts of qualitative materials, mostly descriptions of territorial and administrative entities. In this approach, shared by Rivera and his engineers, statistics was a governmental science complementary to political economy. To evaluate statistically the conditions of a region meant, ultimately, to be able to assess its degree of "civilization." And indeed statistical knowledge was considered distinctive of a "civilized" community, one that is settled over a territory, is open to free trade, and has a government to guarantee the security of property. Statistics thus offered a rationally ordered and comprehensive catalogue of the components of the civilized state, a description of those "forces"—territory, population, and government—that enabled administrators to "act with confidence in all sectors of public administration."[52]

The most prominent statistician in early nineteenth-century Naples was Luca de Samuele Cagnazzi, professor of political economy and adviser of government in matters of economics.[53] He published the first Italian book of statistics (1808) and defended the widely held conviction, embraced by Rivera, that the southern kingdom was graced by nature in all kind of ways, and that its mythical ancient opulence, swept away by centuries of ignorance, could be revived through free trade and rational administration. Cagnazzi, as we have seen, was part of a Napoleonic elite of state functionaries that rejected the political excesses of Revolution and supported an agenda of modernization along moderate liberal and dirigiste lines.[54] Other leading Italian statisticians shared a similar background and approach, the most influential being Melchiorre Gioia (1767–1829), whose quantitative and tabular statistics turned all possible objects of administration, public and private, into useful data, from the "degree of slope of hills" to the "quantity of manure scattered on a hectare," to the "foreigners who have no acquaintances and support."[55]

We have seen how Cagnazzi, dealing with the controversial question of mathematical methods, had opted for a compromise solution that turned analysis into a set of neutral techniques that should be judged only by their usefulness. However, he gave synthetic geometry epistemological and pedagogical priority because of the certainty and *perspicuity* of its operations: geometrical demonstrations can be grasped *at a glance*. The kind of statistics favored by Cagnazzi and Gioia had similar features: public functionaries, wrote one of their followers, should use statistical tables built in such a way that "nothing may escape their gaze," and to build them thus meant eliminating details as needed, "as in the making of pictures." Statistics, and especially well-crafted synoptic tables, should make it possible to grasp the situation of a town or region at a glance, revealing *immediately* the effects of "physical and moral causes" on production, commerce, and habits. Patriarca describes this use of tables as "a new technology simultaneously of knowledge and power."[56]

In order to be perspicuous, vast amounts of data concerning topography, population, natural resources, agriculture, crafts and trades, institutions, and customs needed to be arranged in an orderly and comprehensive description whose immediate goal was to analyze a region's sources of wealth. Statistics was guided by what Gioia calls a "descriptive logic." This logic embodied analytic ideals that we have explored in the first chapter, such as the ordering from the least to the most complex, the cause-and-effect sequence, grouping by analogy, and the extensive use of tables. Cagnazzi talked of this sequence as having ontological significance: the ordering of statistics should indeed reflect the "nature of things." Others commented on the "healthiest logic" that informed such statistical descriptions, where causes precede effects, and where "the very arrangement is a demonstration."

*

Patriarca notes how this insistence on the need for a well-ordered statistics marks an important difference with respect to the French models, which varied greatly in terms of the order of presentation and tended to cover vast and heterogeneous domains. By contrast, all Italian statistical descriptions share a similar structure, one that carried a precise normative message. First came topography, as land and climate were seen as exercising a decisive influence on economic, intellectual, and moral phenomena. A description of the population followed, considered both as a resource of the state ("a general force") and an agent of production. Then descriptions of the different forms of labor: agriculture, crafts, industry, and commerce. Government came next, described in its capacity for direction, relief, repression, and the financial apparatus through which it collected resources. Finally, the outcome of all the determinants was described: the character and habits of the population, or "public spirit."[57]

This ordering reflected a specific conception of the relation between state and society, one in which society was not subordinated to the state, as was the case in contemporary German-Austrian statistics. Gioia noted that describing the features of the government right after topography and population, as the Germans did, broke "the course of economic ideas," and the natural causal links, because one needs to consider the forces of production first. The liberal tenets of the separation of state and society and of the autonomy of the economy were balanced, in the Italian perspective, by a pragmatic support for industrial protectionism, which was considered necessary especially in the southern kingdom, due to its structural weaknesses. State power was ultimately an instrument for development and the growth of productivity—the engine of civilization.[58]

Once again, this time through the prism of statistics, we recognize the aspirations of those agrarian elites who had consolidated their position during the French period and were eager to establish their hegemony in the post-feudal world. The state, in their view, was an agent of rationalization that should support production without interfering with their entrepreneurial action, without upsetting the "natural order of things." Statistics thus became a site for their collaboration with the monarchy, as each party tried to impose order on the turbulent body politic of the kingdom. Patriarca notes how statistics increasingly entered the literature for the popular classes, especially in the 1840s, with the clear pedagogical goal of "teaching the laboring classes about their place in the world and to remind 'dreamers' of all kinds about the need to study 'the facts' in order to avoid proposing 'inept'—as well as dangerous—schemes of social and political change."[59]

By the 1830s and 1840s, statistics had emerged as a well-structured body of knowledge that was reliable, effective, perspicuous, and, above all, *neutral*. A reviewer of an essay by Cagnazzi made it clear that statistics is "impartial"; indeed, it's the "cold witness" of the country's progressive developments. Its relation to politics was that of a "humble" adviser to governments.[60] Technicians like Cagnazzi had been key actors in this process of purification of statistics. In his pioneering work of 1808, Cagnazzi grounded his own technical apparatus ("exact and truthful") into a familiar sensationalist framework, but he also warned his readers about the risk of "abusing" science. Reason is a sublime divine gift, he argued, but it's also "very weak" and cannot guide us to the discovery of the fundamental causes and laws of the "universe's complex machine." The ultimate workings of this machine will always remain "an impenetrable arcane." Worse still, our reason can doom us when it falls prey to the imagination, a faculty that lightens passions and vices. When this happens, politics and religion end up determining scientific opinions—which therefore are distorted, sophistic, and abstract. Thus these Italian statisticians, who regarded themselves as the heirs of Enlightenment, were trying to strike a balance between what they saw as the quantifying excesses of the French tradition and the full-fledged rejection of quantification voiced by reactionary authors.[61]

In his quest for a purified and objective statistical knowledge, Cagnazzi and his followers consistently deployed the "analytic method," but warned about earlier abuses by "the builders of utopias," those who thought that society should change as the scene changes on a theater stage—the Jacobins. Instead, statistical descriptions should be disinterested and impartial; they should be precise not in a merely numerical sense, but in the way a portrait can be a precise representation of the sitter. Statistics must remain truthful to its own internal logic, and to its bound and concrete nature. There was a distinctive caution in Italian statisticians not to reduce "civilization" to purely materialistic indicators and therefore numerical terms. They tended to be wary of *cifromania*, the obsession with numbers. Numerical indicators are ambiguous, these statisticians argued, requiring philosophical interpretation and an understanding of local conditions, and cannot be quickly generalized. One needed to paint the "portrait" of a society via such indicators rather than simply making calculations, as civilization meant more that the output of the productive forces of a nation. It had to do with "the liberty and security of all its inhabitants," the "freedom of contracts," the "force of honor, of religion, of national pride," and many other factors that could not be easily quantified. In conclusion, statistical descriptions were certainly useful tools, but it was the government's *wisdom* that, ultimately, must decide the best course of action.[62]

*

Topography was an essential component of any statistical description. Parallel to the emergence of statistics as a professional body of knowledge and a tool of government, Naples saw the emergence of a new standardized and militarized topography. Genovesi, Grimaldi, Galanti, and many other reformers had argued that a precise physical description of the kingdom was a sine qua non of any significant reform project, and the third quarter of the eighteenth century saw the production of two impressive maps that tried to address this question, covering both the kingdom and the city of Naples. If these maps had originated from the initiative of individual *illuministi*, the project of a great new map of the kingdom launched in the 1780s was the brainchild of Ferdinando Galiani, who invited a prominent cartographer, Giovanni Rizzi Zannoni, to lead the initiative.[63]

In a significant coincidence, Rizzi Zannoni arrived in Naples from Paris in 1781, the same year in which Caracciolo was trying to convince Lagrange to join the local Academy of Sciences. In both cases, local reformers aimed to bypass the cumbersome presence of local academic figures whom they perceived as obstacles to reform and scientific progress. At the time Nicola Fergola wasn't yet a powerful actor in Neapolitan academic life, but to both Caracciolo and Galiani, his reluctance to embrace analysis and its technological incarnations was clearly a problem. In particular, Galiani knew that Fergola had presented to the Academy of Sciences a cartographical project to produce partial chorographic maps of the provinces while carrying out numerous astronomical and geographical measurements.[64] This project was far too limited in Galiani's view, both technically and conceptually, as he aimed for a complete mapping of the territory of the kingdom and of its surrounding— strategically relevant—areas; above all, he aimed for a large-scale rationalization and standardization of representational techniques. Skillfully, Galiani masked his project as a mere revamping of an older map and thus managed to sidestep the academy's project and set up what was in fact the first Italian topographical office, under Rizzi Zannoni's lead.[65]

The Topographical Office became a site where forms of expertise merged and transformed: mechanics and clockmakers turned into makers of topographical and astronomical instruments, while engravers were hired among the specialists who had produced vast numbers of prints illustrating in detail the archeological findings of Herculaneum. The maps produced by the Office would soon be in high demand, especially during the war years. Even

Cardinal Ruffo, in 1799, used Rizzi Zannoni maps to guide his *Santafede* army through the provinces. In the French period, the Office's employees began to wear uniforms and were partially militarized. By 1812 the atlas of the kingdom was complete, and at that point its maps were playing a crucial function in the abolition of feudality and in the selling of ecclesiastic and common lands.[66]

Rizzi Zannoni died in 1814 and was replaced by Ferdinando Visconti. The Topographical Office moved from the area of the Royal Palace to that of the Military Academy, a relocation that signified its reinvention as a militarized and purely technical institution. Like many other technicians of the French period, Visconti maintained his position after the Restoration, although the Topographical Office was under close scrutiny and saw its capacity reduced. In 1817, Visconti launched an ambitious project for a new map of the kingdom, one that would respond to new political and military needs while addressing new potential users—an embodiment of what has been described as a new "cartographic eye." This ferment coincided with significant technological change (as lithography began to be used extensively), new fundamental geodetic and topographic measurements, and the preparation of a politically sensitive new cadastre. Visconti was removed from his position after the constitutional government of 1820–1821, which he had openly supported. He would return to the direction of the Office only in 1835, when the debate on weights and measures was gearing up. He was indeed the author of the compromise solution that, with Rivera's support and against Flauti's and Ceva Grimaldi's protests, would be eventually approved in 1840. Shortly after his death, many of his former employees participated in the insurrection of 1848, where—as a tribunal later attested—they distinguished themselves especially in the construction of barricades.[67]

The territory of the southern kingdom was the object of many different gazes, among them the gazes of the royal inspector, the Napoleonic military officer, and the technician of the modern state—statistician, civil engineer, and cartographer. Rivera and his engineers turned these modes of representation into resources for their analytic renderings of the Neapolitan territory through technical drawings integrated with analytic tables and calculations. The function of these representations was to guide and legitimate their transformative action. We now conclude our considerations on the shape of the kingdom by considering a set of visual renderings of the way of seeing that we have associated with such enemies of the engineers as Ceva Grimaldi and Flauti. This way of seeing found expression in a new wave of pictorial depictions of the Neapolitan landscape. These evocative representations codified a series of powerful yet seemingly harmless tropes. They embodied, I argue, a reactionary utopia.[68]

Return to Nature

Historians of art describe the 1820s as the years of Neapolitan artists' discovery of landscape painting. During this period, unprecedented attention was directed to the natural beauty of the southern regions and to the traditions and the values of country life. These were summed up in the classical topos of *Campania Felix* (prosperous countryside), an idealized, organic rural society that follows the laws and rhythms of an immutable nature.[69] Remarkably, there was no established tradition of landscape painting in Naples, as the genre had hitherto been considered inferior and unworthy of patronage. To be sure, during the seventeenth century Naples was one of the main attractions for the grand tourists, and its famous surroundings were codified in a repertoire of typical images. These Neapolitan vistas (*vedute*) were composed according to the well-defined conventions of the genres of urban vista, topographic landscape, and idealized landscape. Through the eighteenth century Naples remained, with Venice and Rome, one of the main Italian centers for the production of vistas, but the genre was mainly practiced by foreign painters who worked in the southern capital for short periods on behalf of British, German, or French patrons. Neapolitan painters became seriously interested in portraying the natural beauty of their country only in the 1820s, when the landscape genre began to enjoy an unprecedented success that far surpassed that of the vistas or the cheap gouaches.[70]

The birth of a Neapolitan school of landscapists sparked a revolution in the local artistic scene. It was a revolution in pictorial techniques as well, centered on a new anti-academic use of oil paints and the introduction of watercolors. Change was led by the Posillipo school, a somewhat heterogeneous group of painters who started to portray the surroundings of Naples en plein air.[71] Posillipo was a fishing village often portrayed by these painters, and the use of this name was initially meant to be ironic.[72] However, the name stuck. The painters embraced it, and under it they achieved unrivaled popularity, as well as academic positions and the patronage of wealthy collectors, including the Neapolitan royal family. What caused this sudden change in the local artistic taste? During the seventeenth and eighteenth centuries local painters had remained substantially unaffected by the presence of prominent foreign landscapists. Only during the Restoration did foreign stimuli and suggestions become suddenly interesting, originating a new artistic movement centered on the figures of Anton Sminck Pitloo (1791–1837), Dutch but Neapolitan by choice; and Giacinto Gigante (1806–1876).[73]

The sources for the technical, stylistic, and thematic changes introduced by the *posillipisti* can be identified with relative ease. Of notable influence

were foreign visitors who had worked in Naples in the past century, such as
Gaspar van Wittel, Thomas Jones, Carlo Bonavia, Antonio Joli, Philipp Hack-
ert, John Cozens, and the French expedition of Abbé Saint-Non, who ordered
a large number of engravings of Neapolitan and Sicilian views to illustrate his
literary description. More immediately influential were painters who stayed
in Naples during the school's emergence in the 1810s and 1820s: Joseph Re-
bell, Johann Christian Clausen Dahl, Joseph Turner, Richard Bonington, and
Jean-Baptiste-Camille Corot. In short, the Neapolitans became interested
in those artists who rejected the traditional structure and language of the
vistas, who adopted an unprecedented freedom of composition—choosing
unusual subjects and small canvas—and whose touch had become quick and
imprecise as colors turned more intense and brilliant. But why did this "total
renewal of every Neapolitan convention in landscape painting" happen only
in the 1820s?[74] If a century of continuous presence of the greatest European
vedutisti in Naples had not contributed to create a local school of landscape
painting, why did the relatively short stay of a few artists in the 1820s have
such a profound effect?

To answer this question we must consider, first of all, the patronage system. In eighteenth-century Naples the church, the feudal aristocracy, and
the royal family fostered an unusually high demand for decorative academic
paintings and frescoes. The Fraternity of Saint Luke, which was founded in
1664 and for a long time remained the only Neapolitan institution devoted to
the encouragement of painting, was linked to the Jesuits, for whom visual language was key to disseminating their Counter-Reformation religiosity. The
local aristocracy had generally been conservative in its artistic taste, as had
the recently installed royal family, which was strongly committed to a cultural
politics of self-legitimation. The classicist language of Rococo style, rich in
mythological allegories, was well suited to exalt and glorify the virtues of the
dynasty. Thanks to the works in the Neapolitan churches and in the magnificent royal palace of Caserta, at the end of the eighteenth century there was
still a rich production of frescoes, a technique that had declined elsewhere.
While foreign *vedutisti* were spreading images of pine trees and Vesuvius all
over Europe, the Neapolitan masters sustained the Baroque tradition of Luca
Giordano (1632–1705) and Francesco Solimena (1657–1747) for the whole century and, starting in the 1730s, developed local forms of Rococo.[75]

At the turn of the nineteenth century, however, wars, the abolition of the
feudal system, and the suppression of a great number of religious orders all
but dismantled this system of patronage. In this period, as an observer aptly
noted, "the Saints went out of fashion."[76] And they were not to return, not even
with the Bourbon restoration. Commissions for religious and mythological

paintings or frescoes dried up, and the best painters turned their attention else-where.[77] Francis Napier, a British diplomat stationed in Naples, noticed how the cultivation of landscape painting began receiving unprecedented attention around 1815. To him, this moment marked the beginning of "modern paint-ing" in the southern capital.[78] This new era was characterized by the work and teaching of Anton Sminck Pitloo, a Dutch-born painter who had studied in Paris and Rome before settling in Naples. Pitloo's style intercepted the changing taste of the local elites, and in 1824 he became a professor of landscape painting at Royal Academy of Fine Arts. Instrumental to this appointment was a small oil painting that marks a turning point with respect to theme, composition, and technique. The painting, probably presented at the first biennial exhibition held at the Bourbon Museum in 1826, was an original view of a park near the royal residence of Francavilla. This work exemplified Pitloo's new interpreta-tion of landscape, particularly in the free, sketchy, and unclassical rendering of the vegetation in the foreground. Also unusual was the chromatic impasto used to represent the dense tangle of the garden vegetation. By contrast, the old houses in the background were portrayed with realistic precision, according to the late-eighteenth-century conventions of the vista.[79]

The new style quickly became popular and around Pitloo formed the Posil-lipo school. "The great merit of Pitloo," Napier explained, "consisted in his love of nature"; the modern way of painting "is not very careful and scholastic, but

FIGURE 1. Anton Sminck Pitloo, *Il boschetto di Francavilla al Chiatamone*, 1824. Oil on canvas, 44 × 75 cm. Naples, Museo di Capodimonte, collezione Banco di Napoli. © 2022. DeAgostini Picture Library/Scala, Firenze.

full of sensibility."[80] Pitloo was modern in the eyes of connoisseurs because nature alone was his source of inspiration. The Posillipo school defined itself as a naturalistic reaction against the conventionally artificial and academic landscapes of painters like Salvatore Fergola. Compared with the works of the *posillipisti*, Fergola's landscapes and scenes of the Bourbon court suddenly look artificial, rigid, and, ultimately, ugly. By the 1830s, they appeared irremediably outdated. Napier wrote that Fergola's court scenes are populated by "manikins" who were "not picturesquely distributed"; his landscapes had neither "the dignity of the old" nor "the verity of the modern."[81] "Academic" was thus opposed to "modern" as "convention" to "reality." Modern painters broke the artificial constraints of tradition to capture the verity of landscape.

Pitloo's break with academic painting was apparent in both themes and techniques. Nature was the absolute protagonist of his paintings. If human figures were represented, they were always at a distance, rapidly sketched, their role being that of "an agreeable adjunct to the inanimate scene."[82] A great effort was made to perfectly reproduce the colors of the landscape, particularly those of the Mediterranean atmosphere. This new attempt to render the atmospheric unity using chromatic impastos was linked to another stylistic innovation, the elimination of drawing, whereby a quick, sketchy execution captured the impression of a vanishing moment. To this end, colors were distributed in counterpoised masses rather than filling predefined areas. This style's best results were achieved in small, preliminary sketches taken en plein air.[83]

Equally anti-academic was Pitloo's predilection for unusual composition and his choice to portray landscapes never painted before. His sites are often difficult to identify, a rejection of another important aspect of the traditional *veduta*—its informative function. What was important for Pitloo was not the topographical correctness of the representation but the fact that it portrayed a landscape filtered by the artist's sensibility. Optical precision, which was still evident in the houses in *Francavilla Woods*, was absent in paintings like *Sunset at the Castle of Baia* (1833).[84]

Admiring connoisseurs described the *posillipisti* as being close to nature as no painter before them had been. Describing the work of Gabriele Smargiassi, Napier noted that he explored "the remote, untraveled by-ways of the country, to discover and record the charms of a virgin landscape." There was nothing conventional in his paintings, which "faithfully reflect[ed] the hues of the forest and the sky."[85] In this quest for reality, Smargiassi adopted the new *macchia* technique and constructed his landscapes with daring impastos. The habit of traveling in the countryside in search of unconventional views was common to most *posillipisti*. Vincenzo Franceschini specialized in the

FIGURE 2. Anton Sminck Pitloo, *Tramonto sul Castello di Baja*, 1833. Oil on cardboard, 22 × 32 cm. Sorrento, Museo Correale di Terranova, inv. 2844. (@MondadoriPortfolio).

study of forests, storms, and natural calamities and was appreciated for his rendering of foliage; Théodore Duclère was known for the neatness and fidelity of his landscape oils and for his early use of watercolor.

The production of the Posillipo school was centered on the problem of realistically representing southern nature. This central concern did not prevent most *posillipisti* from dealing with other related themes, such as domestic animals or the simple life of peasantry. In fact, they did not perceive any discontinuity when moving from landscape to popular scenes. Peasants and their animals seemed to emerge naturally from the countryside, so that inanimate nature and its inhabitants formed a sort of organic unity, best represented through the *posillipisti's* quick touch and vibrant impastos. Thus Raffaele Carelli had no rivals "in depicting the sentiments and actions, the mirth and labours, the devotions and the dances of his humbler countrymen." He favored the figures of peasants, washerwomen, fishermen, and tarantella dancers, and he turned conventional costume painting into "a living, spirited, and interesting reality." Among the representatives of this rustic style in the Posillipo school were the Palizzi brothers, Giuseppe and Filippo. The former specialized in representing cattle, whereas the latter (who, according to Napier, "drew from nature without instruction") favored domestic animals, peasants, and local landscapes. Giuseppe was known for compositions "that reproduce the pastoral repose and primitive rustic manners and costume of a stationary peasant life." What he presented was not simply animals or peasants but "a condition of existence." Achille Vianelli too was attracted to popular manners and religious piety, rendered in watercolors and sepias that traded optical precision for an interest in the disposition of the masses and the use of light.[86]

The *posillipisti* portrayed mostly anonymous portions of the countryside; some preferred inanimate nature or marinas, others the expressions of an allegedly simple and unchanging peasant life—popular religiosity, dances, and markets. They rejected the artificialities of the academic tradition and relied only on the direct observation of the rural world and its inhabitants in order to represent them faithfully. Modern painters wanted to offer truth to their public; the truth they represented was *Campania Felix*.

Campania Felix is best portrayed in the watercolors of Giacinto Gigante, the most successful among the *posillipisti*. The son of a fresco painter of the old school, Gigante, originally a draftsman, began studying oil painting with Pitloo in 1825, turning to nature and sentiment. According to Napier, "his disposition was rather impatient and solitary than academic; he revolted from the formality and tedium of the schools, and having mastered in less than two years the technical difficulties, he thenceforward threw himself entirely on Nature as his mistress and his guide."[87] Gigante's studio was "far from the bustle and animosities of the Academy and the town"—and indeed he sought out nature, especially where it was secluded and calm, ditching traditional subjects and searching the hills for unusual views.[88] He drew sketches en plein air and later completed them with ink or watercolors; sometimes he experimented with mixed techniques, such as watercolor and tempera or white lead. He accompanied his drawings with plenty of notes about the weather, the light, the tones of the colors, and his own feelings. The sketch was only a component of a complex operation of appropriation and representation of natural reality.[89]

Gigante was well aware of the cultural and pictorial tradition that had flourished around the myth of *Campania Felix* and consciously re-elaborated its tropes. His transformation of the *veduta* is well illustrated by his early watercolors. The drawing, taken from reality, was accurate, topographical. Then, in the atelier, it was transfigured based on the notes that had accompanied its execution: the lines of the original drawing were neglected and colors distributed in *macchie*.[90] The result was an organic composition in which each constituent was treated according to its specific representative function. So, for example, in *Marina di Mulo*, *Marina grande di Capri*, and *Veduta di Ischia*, Gigante used topographical precision to represent the ancient urban structures of villages and their architectonical peculiarities in ink, whereas

FIGURE 3. Giacinto Gigante, *Via dei Sepolcri a Pompei*, first half of the nineteenth century. Watercolor, 21 × 31.7 cm. Naples, Museo Nazionale di San Martino, inv. 23105 (@MondadoriPortfolio).

the surrounding landscape was sketched and macchiato in watercolor.[91] Similarly, in his many views of archaeological sites such as Pompei or Pozzuoli, his reproduction of the imperial remains is extremely accurate, apart from a general increase in the size of the artifacts—a common expedient to emphasize the unreachable greatness of the classical age.

Gigante's watercolors are also interesting for what they do *not* represent. Consider, for example, a couple of watercolors both titled *Maddaloni*.[92] This is a small town not far from Naples, whose name was immediately associated with a famous aqueduct, one of the main achievements of eighteenth-century Neapolitan engineering. Gigante ignored this emblem of the enlightened Bourbon period and focused instead on the medieval structure of the town. His views of Maddaloni are celebrations of the Lombard tower and the architectural peculiarities of the old houses. Similarly, when drawing the river Garigliano, Gigante renders the contorted vegetation on the shores but omits the much-celebrated suspended iron bridge that Rivera's engineers had completed in 1832.[93]

And yet, it would be misleading to simply contrast the engineering gaze and the mechanical realism of modern topography to the *posillipisti*'s lyricism and subjectivism. These two modes of representation-appropriation were both *modern* phenomena, and were connected in interesting, unexpected ways, both conceptually and technically. A piece of technology favored by Gigante and other *posillipisti* shows this connection nicely: the camera lucida. Patented by William Wollaston in 1806, this instrument had quickly

FIGURE 4. Giacinto Gigante, *Maddaloni*, 1844. Ink, watercolor, and white lead, 54.5 × 21.5 cm. Naples, Museo di Capodimonte, inv. 5214. © 2022. DeAgostini Picture Library/Scala, Firenze.

reached Naples. Ferdinando Visconti, director of the Topographical Office, bought at least eighteen of them in 1818–1819, and local production began shortly after. Gigante, who had already seen it used by artists in 1816, probably became familiar with this instrument in 1820, when he was collaborating with the Office. The Office employed a number of draftsmen who intervened at various points of the mapmaking process; while topographers relied on plane tables for their surveying work, the artists used a variety of techniques and instruments including the camera lucida. The plane table expressed the features of the territory through geometry; the camera lucida, not to be confused with the much older camera obscura, created only a fleeting illusion of objective vision, leaving to the artist a crucial role in selecting and emphasizing what was relevant: lines, details, light effects, and the relation between outline and masses of color. In the words of a British artist of the period, the camera lucida was "consistent with the most perfect freedom of execution in the hands of those who possess taste or capacity to represent nature with spirit."[94]

In Gigante's hands, this versatile technology became a key element in a complex process of representation-appropriation of the landscape. Landscape was something experienced by the artist, filtered through their sensibility, and therefore natural and authentic. The mechanical-mathematical and the experiential were two modes of representation of landscape that emerged simultaneously in Naples and, to a significant extent, within the same technical cultures and institutional spaces. The ways of seeing that they embodied were in tension but also synergistic: there could be no *posillipista* vibrant sensibility without the standardized and mathematically accurate representations that draftsmen like Gigante had contributed to construct.

Gigante was less interested than other *posillipisti* in costume scenes. Still, his landscapes are populated by figures that are taken from the *posillipista* repertoire. Proud peasants contemplating Roman ruins or the spectacle of the natural scenery; villagers resting in the shadow of medieval buildings; colorful groups of women in traditional dresses.[95] The inhabitants of Gigante's watercolors invariably transmit a sense of tranquility. Their attitude toward the landscape is respectful, and their presence seems natural and ancient, like that of the imperial ruins emerging from the sultry vegetation. These watercolors do not simply portray *Campania Felix*; they are also a celebration of the people who had been living on this fruitful land since remote ages.

The myth of *Campania Felix* pervaded Neapolitan restoration culture well beyond painting. As we have seen, much contemporary literature and philosophy contrasted peasant wisdom and virtue with the social disorder and moral corruption of urban life. Peasants were often portrayed as custodians of tradition and religion. The moral superiority of the inhabitants of the countryside over urban dwellers was demonstrated, according to their apologists, by the scarce success of the atheistic and revolutionary doctrines outside the main cities, and by the fact that they had constituted the backbone of the 1799 *Santafede* army. On this ground, the defenders of altar and throne constructed the persona of the innocent peasant who lived in a condition of true happiness, unaffected by the moral corruption brought about by eighteenth-century philosophy and by disruptive industrialization.[96]

Visual and literary representations of *Campania Felix* contrasted starkly with the real conditions of peasant life in the first decades of the nineteenth century, a period characterized by the consolidation of agrarian elites who had profited from the redistribution of ecclesiastic and feudal lands. These landlords brought no radical change to land administration: most of them were absentee landlords who rented their land to farmers who in turn employed day laborers. In the 1820s and 1830s, the situation of these workers was remarkably precarious. The enclosure process and the 1806 abolition of feudalism had deprived the great mass of peasants of their traditional rights and pushed them out of rented allotments. To complicate things further, prices fell following the dismantling of the French empire, generating a long-lasting agrarian crisis that hit hard both day laborers and small landowners.[97] John Davis notes that the political innovations of Bourbon reformism and the French decade "had left southern Italy caught uneasily between an older agrarian order that had been seriously destabilized and newer forms of bureaucratic power that were in large part still to be created." As a result, the rural world of the Restoration age was characterized by social and physical degradation. The landscape was scarred by massive deforestation and hydrogeological disasters. Such degradation only accelerated under anti-feudal laws. As historian of landscape Emilio Sereni remarked, "it cannot be argued that, overall, such a process of degradation . . . was accompanied by a visible process of reorganization in new and superior forms." The landscape, dominated by the fallow system, became, according to Sereni, "shapeless" (*informe*).[98]

Yet Gigante and his fellow *posillipisti*, the champions of naturalistic representation, saw *Campania Felix*. Whose way of seeing was that? Foreign demand certainly played an important role in shaping the *posillipista* sensibility, but its influence should not be overestimated. The painters were strongly committed to their own anti-academic program of renewal of Neapolitan figurative art and were generously patronized by Neapolitan elites. Smargiassi, for example, met the taste of the Bonapartist aristocracy in Paris and the Russian royal family but also "shared the patronage of the sovereign and the royal family in no common degree"; he worked constantly for "the Count of Aquila who has manifested the most convinced preference for his productions." Similarly, Gigante had patrons in France and Russia, but also became the king's favorite landscapist and the royal children's private art teacher. Raffaele Carelli, who was engaged by the Duke of Devonshire for a tour of Sicily in 1834, enjoyed the patronage of the Count of Montesantangelo. Gonzalvo Carelli, who had numerous English patrons, worked for the Duke of Terranova and for the Neapolitan royal family. By and large, the local aristocracy and the royal family remained the main source of patronage for Neapolitan painters through the first half of the nineteenth century. The success of the Posillipo school therefore signals a profound transformation of the taste of this social group.[99]

Gigante's naturalistic landscapes are filled with historical memories, literary references, and cultural values. Raffaello Causa, a prominent historian of this artistic movement, described well how Gigante's paintings express an entire way of life: the artist "looks, interprets, and reacts with an eye and a sensibility shared by his clients."[100] The results are images of Neapolitan landscape as *Campania Felix*, codified in a series of tropes whose presence in the *posillipista* production is at once imperceptible and constitutive. Imperceptible in that the myth did not enter the representation in the way traditional conventions of idealized landscape did. Its presence is subtler and therefore more effective. The pretension to faithfully represent the countryside through nothing but the filter of artist's sensibility was itself a mythical construction, which allowed the naturalization of certain pictorial statements about natural and social reality. Gigante's representations, as those of other *posillipisti*, tell us of a rural society that is ancient, still, and unchanging as southern nature itself, a society whose traditions, customs, and religion are rooted in the classical period and in the early Middle Ages and follow respectfully the structures and rhythms of nature. Far from showing any sign of physical or social degradation, the *posillipista* landscape told of an eternally flourishing Neapolitan countryside and of a rural society that was not in need of change. As in Ceva Grimaldi's literary descriptions, in a *posillipista* painting change could only happen as a temporary aberration, as violence against nature.

Algorithm or Intuition?

What does it mean to provide a solution for a mathematical problem? Is it a matter of applying a universally valid algorithm, a procedure that could, in principle, be mechanized? Or is it a matter of relying on some form of intellectual intuition, which might vary greatly from individual to individual? And would a reliance on intuition mean that problem-solving procedures cannot be fully codified in a set of explicit instructions? In the Kingdom of Naples this set of questions was so salient that in 1839 it originated a public challenge. Vincenzo Flauti, an eminent professor at the local university, invited his colleagues to solve three geometrical problems. The problems were hardly new, but the point of the challenge was clearly methodological: the prize would go to the most elegant and direct solutions. Flauti was confident that the outcome of the challenge would demonstrate once and for all the superiority of purely geometrical methods, and the non-algorithmic nature of geometrical problem solving. He was convinced that mathematicians should continue to look at the classical tradition to nourish their intellectual talent, rather than adopt handy, ready-made algebraic techniques. The challenge did little to clarify the question. Flauti claimed that the analytic solutions were not solutions at all and disqualified them, proclaiming one of his students the winner. Memory of the controversy vanished rapidly, its traces incorporated into a reassuring narrative of mathematical advancement. It's clear, however, that these mathematicians did not share basic criteria to assess whether a given solution to a geometrical problem was elegant, intuitive, or economic—in fact, they did not even agree on what "solving a geometrical problem" actually meant.[1]

This challenge was the last noteworthy episode of the fifty-year debate between the Neapolitan synthetic school and its analytic opponents. The synthetic school defended the use of purely geometrical methods in mathematics, while the analytic school promoted the systematic use of algebra. The scope of the controversy extended to geometry, calculus, and other branches of mathematics. Debates between supporters of geometrical and algebraic methods in mathematics were far from uncommon in early modern Europe and were often construed as an opposition between the techniques of the ancients and those of the moderns. The respective merits of these methods were the subject of debates in which aesthetic, practical, and philosophical criteria were used to support one method against the other, or some combination of the two. The debate invested foundational issues as well: were the modern algebraic methods simply an abbreviated and more manageable form of the ancient ones, or were they independent from geometrical truth? If they were independent, what was the source of their reliability? Providing answers to these questions required taking a position on broader epistemological matters such as the nature of knowledge and human reason. This mathematical version of the *querelle des anciens et de modernes* constituted the conceptual framework of much early modern mathematics, from the rediscovery of the classical tradition in the sixteenth century and the emergence of algebra to the transformation of the discipline in the early decades of the nineteenth century. Only against this background can we understand the historical development of the long-term processes that came to define modern science, from the mathematization of natural philosophy to the algebraization of geometry and calculus.[2]

The classical tradition of Greek geometry and the modern tradition of analytic geometry that emerged with the works of François Viète and René Descartes were powerful resources that could be set against each other or integrated: neither choice was self-evident. In practice, many early modern geometers found it useful to mix the methods. For instance, synthetic methods were used to teach elementary geometry because of their exemplary rigor and clarity, whereas algebraic methods were used in research because of their heuristic power. The Neapolitan controversy, however, was characterized by an intransigence that did not allow for such flexibility. Its protagonists saw the two methods as divided by a veritable epistemological and moral chasm: only one deserved priority. From a structural and modern axiomatic point of view, the question of the priority of one method over another makes little sense. And yet, for half a century, Neapolitan mathematicians, engineers, and politicians perceived this controversy as one of vital importance not just for mathematics but also for the future of their country.

At the core of the controversy were the research and teaching of Nicola Fergola, the most prominent Neapolitan mathematician of his generation. Appointed a professor at the University of Naples in 1789, Fergola ruled over Neapolitan mathematics for about thirty years: the golden age of the synthetic school. Under his guidance, a generation of students discovered the geometrical works of the ancients and was trained to emulate their clarity, elegance, and rigor. Fergola's teaching ability was renowned, and his lectures were packed with students from across the kingdom. Among them were future men of science, politicians, administrators, and ecclesiastics. After Fergola's death in 1824, the synthetic school lost its unitary character, but part of Fergola's program continued to be pursued under the lead of Vincenzo Flauti. By that time, Flauti had accumulated an impressive number of administrative responsibilities within the university, the Royal Academy of Sciences, and the Ministry of Public Education. Flauti's control over scientific productions is well illustrated by his virtual monopoly on mathematics textbooks: his own edition of Euclid's *Elements* went through no fewer than twenty-two editions and granted its author fame and revenue.

The synthetic school's hegemony over mathematical life in Naples was challenged by a small but combative group advocating a purely analytic approach to mathematics exemplified in the works of Joseph-Louis Lagrange. Among those who championed analytic mathematics in Naples were the protagonists of earlier chapters: Carlo Lauberg, Annibale Giordano, Ottavio Colecchi, and the engineer-mathematicians of the School of Application. By the time of the political unification of Italy in 1861, this once acrimonious controversy had sunk into oblivion. The erasure had been so effective that historian of mathematics Gino Loria declared his complete ignorance of it when he first came across some of Fergola's books in the late nineteenth century.[3] To him, such a debate over methods could not but appear misguided, and even more obscure were the accusations of moral corruption and anti-scientific attitude that Neapolitan synthetics launched against those using analytic methods.[4] The analytics' scathing judgments on pure geometry and their attempts to ban Euclid's *Elements* from the schools of the kingdom were equally puzzling.

Analysis and Synthesis

We know already about the eighteenth-century meaning of analysis, and about the many and diverse embodiments of the analytic reason. It's now time to learn more about what these Neapolitans meant by "analysis" and "synthesis" as specific mathematical methods. These terms have a long history and have been used in a number of different ways in mathematics, philosophy, and other fields of inquiry. They stem from ancient Greek as composed terms: "analysis" can be translated as "loosening up," "dissolution," or "resolution," the standard Latin translation being *resolutio*. In Greek geometry "analysis" also meant "back from solution" and referred to working back from an assumed solution to more fundamental statements, until something already established was reached; the geometer would then reverse the process and move toward the solution of the problem or the demonstration of the theorem. These two basic meanings of analysis as "decomposition" and "regression" coexisted and were still relevant in the eighteenth century. "Synthesis," on the other hand, can be translated as "to put together" or "to compose" (standard Latin translation: *compositio*) and referred, among other things, to the process that followed the regression of geometrical analysis and concluded with the actual establishment of new geometrical knowledge. Together, analysis and synthesis have often been used to define and contrast styles of mathematical and philosophical reasoning and to clarify the nature of knowledge in general.[5]

Let's take a closer look at the ways these methods work. Consider the following problem: trace the tangent *t* to a given circle *C*, at the point *P*. One could proceed by performing a geometrical construction, tracing the radius *OP*, then tracing the line *t* passing through *P* perpendicular to the radius *OP* (fig. 5a). However, one could also solve it with an algebraic algorithm. First, one must translate the terms of the problem into algebraic form by choosing a system of coordinates: in this case the one at the circle's center. With respect to these coordinates the equation of the curve is

(1) $x^2 + y^2 = r^2$

and the point *P* will be individuated by the couple *a,b* (fig. 5b). Then one must record the equations of the curves whose points of intersection are relevant to the solution of the problem, in this case the equation of the circle and the general equation of a straight line passing through the point *P*:

(2) $y - b = m\,(x - a)$

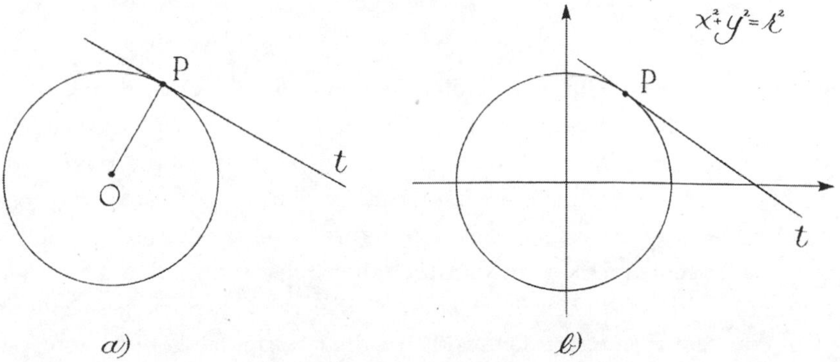

FIGURE 5. One problem, two solutions. Drawing by Monica Maldarella, 2022.

to calculate the gradient m of the tangent line t. The solutions of this system are the points of intersection between the straight line t and the circle (which, in the case of the tangent, do coincide). Solving the system, one obtains the following equation of the tangent line:

(3) $ax + by = r^2$

The Neapolitan synthetics believed that the first procedure was the fundamental method for solving geometrical problems. They referred to it as the synthetic method (*metodo sintetico*) or composition method (*metodo di composizione*). The analytics, on the contrary, attributed epistemological priority to the second procedure, the analytic method (*metodo analitico*) and to what they called "two- or three-coordinate geometry."

Although this example does not illustrate the sophisticated practices of the two groups or the complex problems that attracted their attention, it highlights an important difference between the two methods: their degree of generality. The first problem-solving procedure relies on a specific feature of the circle—namely the fact that the tangent to a circle at a point P is always perpendicular to the radius OP. If one were to consider a different kind of curve, a specific feature of the curve in question could be used to solve the same kind of problem. For instance, to find the tangent at a certain point P in the case of an ellipse, one should consider the following property: the tangent in P is a bisector of the angle between the straight lines connecting the two foci to P. On the contrary, the *same* analytic procedure is applicable to both a circle and an ellipse. In this sense, the analytic method can be described as easier and more general. Nevertheless, when using this method, knowledge about specific properties of the relevant figures becomes not only unnecessary but also much more difficult to establish. So the fact that the tangent in P is perpendicular to OP can be established only by recognizing the validity of the equation

(4) $b/a \, (-a/b) = -1$

where b/a is the gradient of the radius OP and $-a/b$ is the gradient of the line t. Thus, while the analytic method delivers the solution to the problem, it fails to provide information that is immediately evident when using the synthetic method. Also, the synthetic method requires that the geometer look constantly at the diagram to *see* the relations between the relevant objects (in this case, the tangent and the radius). Although geometers can achieve a certain degree of generality in their considerations, they cannot detach them entirely from concrete geometrical figures. With the analytic method, by contrast, one achieves immediately a much higher degree of generality by translating the conditions of the problem into algebraic language. Variables and parameters are introduced precisely in order to make one's considerations more general, so that all the possible cases falling under the initial conditions of the problem are considered.

Schematically, the basic steps of the synthetic method are the following: the geometer assumes the tangent as already traced; sees that for every possible position of P, the tangent would be orthogonal to the radius in P; knows already how to trace a radius in P; constructs the radius in P; then traces the tangent t in P. With the analytic method, the geometer considers which quantities would give, if known, the solution to the problem; takes these quantities as unknowns; expresses the conditions of the problem in algebraic equations; introduces parameters to consider all the possible cases; enters the equations in a system and solves it to find which values satisfy all the conditions; and traces the tangent translating the numerical values back into geometrical representation.

An important meaning of "synthetic" and "analytic," and the one that occurs most often in the Neapolitan controversy, refers to the use of geometrical or algebraic reasoning in problem solving. Synthesis was virtually synonymous with purely geometrical procedures, and analysis with algebraic and infinitesimal procedures. Another important meaning of analysis was derived by the twofold method of classical Greek geometry briefly described above in terms of regression. It was, properly speaking, a "geometrical analysis," a procedure that does not rely on algebraic tools. In Greek geometry, solving a problem requires constructing a figure that corresponds to a specified description. Basically, a problem is solved by reducing it to one that has been already solved. The twofold method includes the following process of analysis: first, the desired figure is posited as if it had been already constructed; then a series of properties is deduced from it until some element is reached that is known to be constructible from previous results. Then the process of synthesis begins, starting from the elements to which the original construction has been reduced and, following approximately the reverse order of analysis, leading through a series of deductions to the desired construction. One can now recognize a simple case of this twofold process at work in the example of the tangent to the circle. The geometer starts by assuming they have already traced the tangent, sees that it would be orthogonal to the radius, something they already know how to construct, then traces the radius and the tangent as orthogonal to it. In other words, analysis moves from the unknown as if it were known to its possible antecedents until arriving at premises one recognizes as known; synthesis moves from the known to what one must construct to solve the problem.[6]

Expressions such as "Greek geometry" and "the classical tradition" of problem solving need to be qualified. The Neapolitan synthetics considered themselves the legitimate heirs of an ancient geometric tradition—the "Greek school," as they called it. In their historical reconstruction of the evolution of mathematics, they were the last link in a long chain connecting the Greeks to Renaissance geometers to the moderns—especially Galileo, Viète, Descartes, and Newton. Newton was to them the last great champion of the Greek tradition that had entered a period of decline and corruption in the eighteenth century. The idea that such a heterogeneous lot of mathematicians shared a unitary conception of mathematics and a common problem-solving method is hardly plausible. Historical scholarship has challenged the belief that even Greek geometers shared a single, well-codified method.[7] In fact, the procedures adopted by the Neapolitans were derived from a composite set of texts that spanned at least six centuries and included works by Euclid, Archimedes, Apollonius, and Pappus.[8] This corpus contained mostly synthetic textbooks, such as Euclid's *Elements* or Apollonius's *Conics*—texts in which results are presented systematically via geometrical synthesis and that became exemplary tokens of demonstrative rigor. However, these texts did not exhibit the original reasoning of the geometers and the rationale for their choices, so solutions and demonstrations contain seemingly arbitrary steps. Greek geometers also wrote texts that illustrated their geometric analysis, such as Euclid's *Data*, but few survived. This dearth of extant texts contributed to the consolidation of an image of the ancient tradition as one defined by compilations of theorems ordered according to the rigorous deductive steps of the synthetic method. Neoplatonic authors of later commentaries, such as Pappus and Proclus, reinforced this image. They were interested in mathematics as a purely abstract and rational science, hence their predilection for synthesis, theorems, and formal proofs over the heuristic power of geometrical analysis as a problem-solving tool. According to them, mathematics deals with the world of eternal and immutable essences, and the statement of a theorem is the description of an eternally true state of ideal things. Problem solving, an activity in which the geometer performs operations with and on geometrical entities, modifying them and bringing new entities into existence, was thus inferior.

These considerations can help us understand the scarcity of texts of geometrical analysis and the emphasis on the theorematic rather than problematic form of geometrical knowledge in the ancient corpus.[9] It was this image of the ancient tradition as one primarily interested in demonstrative rigor and logical deduction that seventeenth-century geometers found unsatisfactory.[10]

The use of algebraic and infinitesimal methods to solve geometrical problems was in part a response to the limited heuristic power of the traditional techniques: a replacement of the lost geometrical analysis of the ancients with the new algebraic analysis of the moderns. But this approach was not the only way to overcome what was perceived as a most serious impasse in the development of mathematical knowledge. Some seventeenth-century geometers tried, despite the sparsity of surviving records, to *revive* the original practice of geometric analysis, reconstructing its procedures in a long series of "divinations" of the lost analytic texts. This practice declined sharply across Europe in the eighteenth century, but it was revived one last time by the Neapolitan synthetic school, in its puzzling and untimely fight against the mathematics of the moderns.[11]

Synthetic, Cartesian, and Lagrangian Methods

What did synthetic and analytic solutions to geometrical problems look like? Here is a fine synthetic solution from a textbook by Fergola: *Let ABC be a triangle of given kind and magnitude. Inscribe a square in it whose side lies on the base BC.*

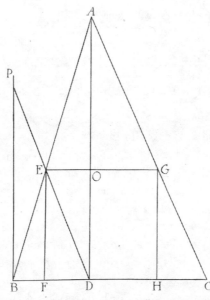

FIGURE 6. A synthetic solution, adapted from Fergola, *Della invenzione geometrica*, 198–99. Drawing by Monica Maldarella, 2022.

Geometrical analysis, step 1: supposition of the fact.

> Suppose that the square EGHF is already inscribed in ABC, so that the
> side FH lies on BC and the angles FEG and HGE lie in the sides AB and
> AC. Or, alternately, suppose that O is that point of the triangle's height
> AD such that tracing through it EG is parallel to BC, and for E and G the
> perpendiculars are EF and GH; then EF=GH=EG.

Geometrical analysis, step 2: consequences of the supposition.

> If EG=EF, then EG:AO=DO:AO. [1]
> This implies, for the nature of the given triangle, that EG:AO=BC:AD. [2]
> Therefore, BC:AD=DO:AO.

Geometrical analysis, step 3: reduction of the problem.

> Given that DO:AO=BC:AD, and that BC and AD are of given magnitude,
> the problem can be reduced to dividing AD according to a given ratio or
> to finding two segments whose sum is given and whose ratio is also given.

Geometrical composition, step 1: construction.

> Trace from B the perpendicular BP, equal in length to BC. Then trace PD
> to find the point E on AB and EG, parallel to BC. For E and G, trace the
> two perpendiculars EF and GH. EFGH is the square the problem asked for.

Geometrical composition, step 2: proof.

> PB:EF=BD:DF, or EO
> and BD:EO=BC:EG
> then PB:EF=BC:EG
> so EF=EG, and EGHF is a square, inscribed in ABC.[12]

This solution illustrates nicely the three-step procedure of geometrical analysis. First, suppose what is asked in the problem ("supposition of the fact").
Then, develop the consequences of this supposition: look both at the consequences of the conditions of the particular problem (hence, in the example,
the proportion [1]) and at the consequences of the nature of the figures involved (hence the proportion [2]). Finally, find one consequence that is constructible, that reduces the problem to one previously solved. Solutions like
this one are, according to Fergola, simple, elegant, and economic.

As was customary in synthetic textbooks, Fergola enriches his pages with
plenty of historical considerations. Endorsing an interpretation that went
back to Diogenes Laertius, he attributed geometrical analysis to Plato himself. In fact, he judged it to be Plato's greatest philosophical achievement,
one worthy of "an angelic mind." Fergola aimed to reconstruct this alleged

Platonic heuristic, summing it up in a few informal "heuristic canons" that codify the three-step procedure shown above. The geometrical analysis of a problem is an "ontological principle of reduction," and "a rich source of truths and methods."[13] Fergola also referred to it as "the art of inventing" (from the Latin *invenire*, to discover), which he asserted to be one of the most beautiful intellectual virtues, as opposed to empirical or practical. This ability is innate but must be nurtured through assiduous training and "infinite effort."[14]

Carrying out the analysis of a problem, however, is only half of the problem-solving process. After the reduction of the problem, which leads to some previously known geometrical truths, it is necessary to perform its "geometrical composition."[15] The materials for the composition of the problem are already contained in the geometrical analysis by which the problem has been solved. But the elegance of the composition depends entirely on the way these materials are chosen and rearranged by the geometer—therefore, on "skill and ingenuity." The construction of the problem generally proceeds in the reverse, starting with known geometrical truths and progressing to the construction required by the statement of the problem. However, reversing the analytic process is not always possible or desirable. The geometer may decide to employ indirect forms of reasoning or to introduce some independent geometrical truths—in the form of lemmas—in order to make the composition more direct and elegant. There may be no strict correspondence between analysis and composition, and knowing the composition of a problem may be insufficient for reconstructing its original analysis—hence the difficulty in reconstructing the lost analytical works of the ancients. In the last example Fergola preferred not to move backward from the reduced problem, as he had found a simpler and quicker way to compose it. This example showed students that a skilled geometer is only loosely bound by methodological principles: the art of inventing is not algorithmic. Finally, a proof was required to allow construction. In practice, the constructed solution is turned into a theorem for which a proof is then provided: hence the awkward expression "the proof of a problem."

Fergola devoted a significant part of this textbook to modern algebraic problem-solving methods, which were introduced as legitimate and useful techniques. For Fergola there were two ways to use algebra in problem-solving procedures. One—legitimate—was analytic ("Cartesian") geometry, where the use of algebraic algorithms is subsidiary and coexists with a belief in the essentially geometrical nature of problem-solving procedures. In principle, any such analytic passage can be replaced by the correspondent geometrical reasoning. Moreover, to properly solve a problem, the geometer must perform a final construction of the solution to translate the analytic solution back into geometrical form. The second way of using algebra he called the Lagrangian or "very modern" (*modernissimo*) method, in which problem solving is not introduced by any preliminary construction but by a set of general formulas that can be thought of as expressing metric relations between kinds of geometrical figures. The geometer manipulates these formulas to obtain the desired results and does not need to inspect figures. Thus, not all analytic reasoning is necessarily translatable into corresponding geometrical steps. Eventually, finding the possible solutions for the equation means solving the problem itself. There is no need to perform a final geometrical construction, and in any case it is taken for granted that equations are always constructible. In this method, analysis is not a subsidiary tool but an independent and fully legitimated way of reasoning that allows geometers to discover new truths in a mechanical way.[16]

The structure of a solution in the Cartesian method and in the purely geometrical method is the same: preliminary construction, analysis, construction of the solving equation, and proof. According to Fergola, synthesis and Cartesian analysis are the "two wings" that make geometers advance in their discoveries. Students needed to learn how to move back and forth between algebra and geometry—which can be far from obvious.[17] Thus synthetics valued and developed techniques that most of their foreign colleagues would consider antiquated and pointless. Such was the case of the techniques for constructing equations, a field of study whose popularity had greatly declined during the eighteenth century, but one necessary to geometrically construct problems solved through algebraic analysis.[18] Outsiders must have found Fergola's emphasis on the necessity of "proving a problem which has been analytically solved" quite peculiar. Basically, students were asked to go backward from the roots of the solving equation to the initial conditions of the problem. They would then replace algebraic expressions with equivalent geometrical magnitudes, and equations with proportions, translating in synthesis each analytic passage of the original solution. "A geometrical problem that is solved analytically," Fergola stated, "cannot be considered properly solved if it is not constructible."[19]

By contrast, to see what a purely analytic solution looked like, we can examine a textbook by a leading Neapolitan analytic mathematician, Fortunato Padula.[20] Padula's choice to include in his collection the kind of plane- and solid-geometry problems favored by the synthetics was openly polemical: he wanted to show how algebra could be effectively deployed in all kinds of geometrical problem solving, even those considered best treated with synthesis. To do so, however, Padula had to compromise the generality of his analysis. The extreme generality of the Lagrangian method is missing: each solution begins with an inspection of the figure, and Padula even provides the final composition of the problem. This hybrid textbook aimed at confuting the synthetics' claim that the use of indeterminate equations made constructing a solving equation difficult, if not impossible. Padula provided constructions for each and every problem. At the same time, he insisted that a problem can indeed be considered solved *without* a final construction, let alone a final proof. Padula's collection is characterized by a methodological uniformity that was supposed to express the spirit of the analytic approach. Readers were invited to appreciate its generality and effectiveness, which transcended the particular nature of the problems. The difference from synthetic textbooks was stark, as synthetics differentiated, classified, and solved problems using a large number of different ad hoc strategies. By contrast, Padula adopted an algorithmic approach: again and again, he used the same set of instructions to put problems in equation form and solve them. But the solution to even the simplest problem in Padula's collection occupies a few printed pages—an aspect of the analytic method that the synthetics took as a sure sign of its inferiority.

In these solutions, Padula first assumed as unknowns the coordinates of that geometrical object that, if known, would determine all the other quantities, thus solving the problem. Then he expressed these quantities as functions of the unknowns and expressed the conditions of the problem in algebraic language, forming an equation. "It is certain," Padula remarked, "that with a little practice in the application of this method, and with the comprehension of its spirit, it will be possible to put any problem in the form of an equation, as soon as we read its conditions. We cannot imagine how this could happen using geometrical analysis." Not only did the new method allow geometers to immediately put every problem in equation form; it also guided them mechanically to the discovery of all solutions to the problem itself, by considering every possible case that fell under the initial conditions. In commenting on a problem that had a large number of solutions, Padula declared that "the true solution of a problem is its solving equation." One could then perform particular constructions from the solutions of this equation through the rules of algebra, "without referring any more to particular considerations about the original figure." Padula also remarked that from the equation of the general case, a much higher number of possible constructions can be derived than through the synthetic study of the figure. It often happens that, thanks to algebra, one recognizes a given problem as a particular case of another one, "whereas to those who are only familiar with geometrical procedures [they] would seem completely different." In other words, the mechanical procedures of algebra bring out relations that the synthetic geometer simply cannot *see*.

"From this point of view," Padula continues, "one cannot but recognize the superiority of the algebraic method over the method of the ancients."[21]

The algebraic method brought uniformity and generality to the practice of problem solving: it turned it into a mechanism, an unremarkable algorithmic practice that did not posit any particular intellectual challenge. In their respective self-representations, the synthetic mathematician was a gifted contemplator of geometrical truths, an otherworldly genius.[22] The analytic mathematician, by contrast, was an efficient and well-trained engineer who was able to solve all kinds of problems through the disciplined implementation of appropriate algorithms.

An Epistemological Stroll

The problem-solving procedures sketched above were clearly informed by different epistemological assumptions. Both the synthetic and analytic methods operated by reducing the original problem to a simpler task. In the synthetic method, analysis reduced the construction sought in the original problem to a known construction. The reasoning remained bound to geometry: it was about geometrical objects, their properties, and the ways in which they could be manipulated and modified. In the analytic method too, analysis reduced the construction sought to a simpler task, *but the nature of this task was not geometrical*. Instead, it involved solving a system of equations whose solutions were, in algebraic terms, solutions to the original problem. The connection between geometry and the problem-solving process was severed: algebraic language was not an aid to geometrical reasoning but a legitimate method in itself, which happened to be applied to the solution of a geometrical problem.[23] Lagrange, and Padula after him, had unbounded confidence in the generality of formulas and aimed to demonstrate the power of analysis as a universal way of reasoning. The use of algebraic algorithms in geometry was above all an expansion of the empire of analysis, the last in an impressive series of achievements obtained through its application. Padula and the Neapolitan analytics did not even conceive of themselves as geometers but as engineer-mathematicians. What unified their diverse productions was the faith in analysis as a universal language.

The mathematical culture of the synthetics gave priority to local conditions, while the analytics' mathematical culture prioritized general ones— two kinds of knowledge that cannot be achieved contemporaneously. If one decides to solve problems by means of the accurate inspection of particular figures, it will be difficult to move toward the highest levels of generality. If one decides to solve problems through the manipulation of general formulas, it will be difficult to acquire knowledge of the particular figures under consideration. In the synthetic approach, the preliminary construction sets the stage for the entire problem-solving process. Geometers reason on particular figures and perform particular constructions. They might be interested in generalizing their reasoning to include other classes of figures and other possible spatial configurations, but in the end, the solutions are all derived from the original case, maybe with some extensions. Thus, one could not only provide a solution for the original case, let's say a triangle, but could also generalize the result to a polygon. Neapolitan synthetics described the scope of their method as limited to specific geometrical objects: different problems required different constructions, with no underlying universal procedure. Their method relied primarily on the intellectual intuition of the geometer, who must be able to see the relations between relevant figures and choose the most suitable among the many possible constructions. It was a method that required ingenuity, skill, and experience. Synthetic pedagogy was designed accordingly, allowing for long periods of training under the guidance of senior mathematicians. Students were not supposed to internalize a set of rules but to find their own way to contribute to a millenary tradition. The teacher did not provide them with algorithms but with exemplary works upon which to meditate. Solving problems and proving theorems was, for the synthetics, a "sublime art."

The analytic method works in the opposite way. In its purest form, it starts from the very general—the construction of sets of equations that can be seen as expressing metric properties of entire classes of figures—and then derives solutions to problems from these equations. In Padula's textbook, this method was applied to problems typically studied by synthetics, allowing for a much higher level of generalization and a higher number of possible solutions. He described this method as universal: ideally, every problem could be put into an equation and solved through a sequence of identical steps. By solving this equation the geometer obtains all possible numerical solutions for the original problem. Problem solving consists in following an algorithm, an explicit set of instructions, and in principle anyone, if properly trained, should be able to solve problems effectively.

The practice of solving problems thus had different meanings in the two schools. For the synthetics, problem solving was essential: it was the starting point of each and every geometrical investigation, the necessary prelude to the discovery of new geometrical truths. Mathematicians are primarily problem solvers, and their activity is nourished by a never-ending reflection on the problems of classical antiquity. For the analytics, problem solving as conceived by the synthetics was a secondary activity. Lagrange simply ignored it.

Padula indulged in it for polemical reasons, declaring that he found it utterly uninteresting. To him, the solutions of specific geometrical problems were simply byproducts of analytic reasoning, the core of any trustworthy mathematical and engineering procedure.

We can also characterize the two methods in relation to abstraction. The synthetics made local considerations about specific objects. They drew, inspected, compared, cut, and manipulated concrete geometrical figures. In keeping with the Aristotelian-Thomistic tradition, they understood their activity as dealing with "geometrical abstractions," meaning they did not consider a particular triangle but what was typical of every possible triangle.[24] In this sense, geometry is an abstract science because it does not depend on any particular instances and is centered on the scrutiny of typical figures. Not surprisingly, synthetics described the activity of the geometer through a family of metaphors that revolve around the faculty of vision. The geometers must be able to *see* that a problem can be reduced to another one, that it has a certain number of possible solutions, and that a solution can be constructed in a certain way. The connections between geometrical ideas are "luminous," and their apprehension is accompanied by intellectual satisfaction.[25]

At the same time, the synthetics criticized what they called the "abstract reasoning" of the analytics, where "abstract" implies a lack of intuitive content. Their main argument against the reliability of analysis in problem solving depended precisely on its being abstract in this sense. How can we know, the synthetics asked, that the analytics are actually saying something meaningful with their formulas if those formulas cannot be directly related to any geometrical state of affairs? When the synthetics tried to define the art of geometrical discovery and demonstration, they referred to it as a form of intellectual intuition—in fact, the highest achievement of the human intellect. This definition is invariably contrasted to the *mechanical procedures* of the analytics that, in keeping with the visual metaphor, are described as "blind." Algebraic calculus was therefore a blind mechanism. Pedagogically, a mathematical education not based on synthesis could only produce a "deformity": the students would end up "acting as automata," ignoring the meaning of the formulas they manipulate. Mathematics without synthesis turns its practitioners into soulless, machine-like number crunchers.[26]

By rejecting the unrestricted use of algebraic algorithms, the synthetics were defending a concrete, bound mathematics founded on the intuitive knowledge of the properties of geometrical objects. Any attempt to introduce nongeometrical methods into geometry or, conversely, to export geometrical methods to other disciplines was problematic and potentially misleading. The world of the synthetics was epistemologically stratified: what was true at one level was not necessarily true at another. By contrast, the analytics inhabited an epistemologically homogeneous world where languages and methods crossed disciplinary boundaries. Reliable mathematical procedures were simply expressions of a more fundamental way of thinking and acting that they called analysis, and every problem could be reduced to a matter of calculation. They were not afraid to "bastardize" or "alter the purity" of geometry; to them, mathematics should be studied primarily in view of its application to mechanics.[27] And solving geometrical problems did not require *seeing* anything: one simply followed the universal rules of algebra. Like the synthetics, Padula compared the analytic method to the clinking workings of a machine—but he did so enthusiastically: to him the machine was the best emblem of rational thinking.[28]

Threads and Traces

It is instructive to consider the historiographical destiny of the Neapolitan controversy. The story can be told briefly, as the controversy has attracted little scholarly attention. In his 1924 reconstruction, mathematician-turned-historian Federico Amodeo wrote off the controversy by assuming that some mathematicians are synthetically minded, while others are analytically minded.[29] In this perspective, history was not a relevant dimension, as the key explanatory factor was to be found in a timeless notion of individual mathematical style.

More articulated was a previous work: Gino Loria's 1892 study of Fergola and his school. Loria was respectful of the technical ability of synthetic geometers, yet puzzled by their apparent misunderstanding of the development of modern mathematics. He concluded that Fergola's overall project of reinstating synthetic geometry as the foundation of all mathematics was fascinating, technically sophisticated, but desperately anachronistic. In spite of their quality and elegance, Fergola's geometrical textbooks were forgotten because what he was trying to do was equivalent to "resuscitating a corpse," the corpse of Cartesian geometry. In other words, Fergola's mathematics was backward. He looked at the past rather than the future and did not understand the mathematical revolution that was happening all around him. The eventual failure of his program was therefore "ineluctable." As one might expect from a fin de siècle scholar, Loria was fond of the epigenetic metaphor and applied it liberally to the history of mathematics. He could thus claim that it was necessary for Neapolitan science to pass through the "Cartesian stage," which had not been fully institutionalized in seventeenth-century Naples. Even so, Fergola's program contained the "deadly germ that would bring it to the grave," and by the early 1840s, it was "a putrefying corpse."[30] These are revealing lines, and not just because of the morbid metaphors and decadent gusto.[31] Here Loria authoritatively endorses what will become the standard interpretive line: the *backwardness* of Fergola and the synthetic school, their mathematical practice as anachronistic. The synthetic school was thus normalized as a temporary aberration on the path to modernity. After all, hadn't the synthetics fashioned themselves as the staunch defenders of an ancient tradition against the rise of the modern methods?

That Fergola's program had *political* significance was a point first made in 1908 in an insightful essay by mathematician Vito Volterra. Volterra consciously superimposed the emergence of "Italian mathematics" through the nineteenth century on the contemporary independence wars, the political unification of the peninsula, and the appearance of Italy as a nation on the stage of international politics. He referred to the striking number of professional mathematicians directly involved in the Risorgimento and in the early political life of the new nation to construct a narrative of moral, political, and *scientific* liberation. Revealingly, he began his story in the midst of political reaction at the opening of the nineteenth century, a period that was perhaps "the saddest and most obscure" in Italian history. He argued that during this period "internal conflicts were almost mirrored by scientific intransigence and intolerance," thus making it clear that the Risorgimento was as much about science as it was about politics. Volterra was sensitive to such connections; as a politically engaged mathematician he experienced science as essential to the democratic life of the new nation. In his narrative, the Neapolitan synthetics, "men who were otherwise ingenious," are described as engaged in a mortal fight against "Lagrange's great discoveries and all that was modern and new in science." They aimed at "setting [science] back of a number of centuries," and their scientific intolerance exemplified scientific life in the despotic states of pre-unitary Italy.[32] Fergola's mathematics was a politically inflected aberration that slowed down scientific progress, much as the reactionary Bourbon regime had slowed down the process of national unification and the creation of the liberal Italian state.

In a much later survey, Massimo Galluzzi offered a useful European contextualization of the synthetics' productions. He related their attempt to restore the dignity of geometry to the revival of synthetic geometry across Europe after 1800. The fascination with Greek-like aesthetic canons for geometrical proofs was not peculiar to those working under Vesuvius. On the contrary, this age was one in which pure geometry flourished after a long period of decline. As noted by Michel Chasles, some Neapolitan productions had achieved that Greek-like style for which many were striving across the continent. Then, contradicting himself somewhat, Galluzzi offered another version of the backwardness argument: the Neapolitan synthetics' predilection for geometrical methods should be understood in terms of available cultural resources. They operated in a largely "anti-scientific" context and had only limited contact with mainstream European mathematics. Reading the classics was "the only scientific education that [Fergola] received," so he trained an entire generation to obstinately defend Greek geometry against mathematics' most recent developments.[33]

The analytics, by contrast, are presented as a group of younger and more internationally connected mathematicians who benefited from being trained in French-like engineering schools rather than at the local university. This reading was partially contradicted by the findings of Giovanni Ferraro and Franco Palladino, who pointed at manuscript evidence of Fergola's interest for and competence in the techniques of late-eighteenth-century analysis. The intransigence of the synthetics, and therefore what the two authors perceive as the involution of Neapolitan mathematical life, should not be attributed to Fergola himself but to his pupils—Flauti above all—who had betrayed their teacher's original, much broader research program.[34] Well aware of the way synthetic texts continuously blend politics with science, Ferraro and Palladino understand them as a distortion of scientific practice in the service of extrascientific goals. The marginalization of the analytics under the synthetics' mathematical regime should therefore be read in straightforward political terms: scientific arguments were a cover-up for political persecution.[35]

This brief historiographical overview offers insights and promising leads. In particular, it invites us to reflect on *backwardness* as an explanatory category in the history of science. Commentators agree that the synthetics' mathematics was anachronistic and therefore in conflict with that of the more up-to-date analytics. This backwardness depended on the age of the synthetic mathematicians and, above all, on their alleged cultural isolation and limited resources. This interpretive line, however, runs into problems quickly. First, there was no significant age difference between the representatives of the two groups. Second, Fergola was well aware of eighteenth-century technical developments in algebra and calculus. His unpublished papers reveal comprehensive knowledge of the works of Euler and Lagrange, among others. As we shall see, he had even engaged in a project aimed at grounding contemporary calculus on a more solid foundation.

Another reason to be wary of the backwardness interpretation is that there was no lively synthetic tradition in Naples when Fergola entered the scene. The city had been home to prominent geometers around 1700, but by the time Fergola was a student, in the late 1760s, there was neither a distinctive mathematical tradition to be defended nor a particular interest in synthetic geometry. In fact, the emergence and institutionalization of the synthetic school took place amid the growing popularity of French analytic methods. These considerations suggest that Fergola's synthetic program is best understood not as a defense of the mathematical status quo but as a new program, indeed a *reaction* against a perceived degeneration of mathematical practice. Synthetics were obsessed with lineages and understood themselves to be the last custodians of a millenary geometrical tradition that went back all the way to the Greeks. This tradition, they argued, had arrived in southern Italy with Greek settlers and had never abandoned the former Magna Graecia.[36] To the synthetics, their untimeliness was something to exhibit proudly.

Rather than take either the synthetics or their critics' claims at face value, I interpret their narrative of backwardness as a key component of a *new* image of mathematics that was in important ways closer to our modern mathematics than the analytics' boundless trust in the power of analysis. The notion of backwardness should not be taken for granted. Adhering to a past tradition is never just a matter of passive inertia; it requires resources and choices. Understanding why someone chooses to follow a tradition should be as interesting for historians as why someone decides to break with it—even more so if that tradition is largely imaginary.

But if Fergola's Greek-like geometry wasn't the residue of an established local tradition, how should we understand its emergence and, above all, its credibility? To provide a more satisfactory interpretation, we must embolden our historical imagination and raise our gaze from the techniques themselves. The meaning of a formal procedure and the understanding of its technical destiny involve more than mathematical symbols. Symbols themselves mean nothing if they are not interpreted, and interpretation is necessarily situated and historical. Competing mathematical cultures gave the Neapolitan problem-solving controversy its form and meaning by opening up certain technical possibilities and giving credibility to certain technical solutions while making others unviable. In other words, the techniques briefly discussed in this chapter would not have taken that precise form apart from concerns and beliefs that were not—strictly speaking—technical. How to best construct and perform a formal proof is a question that cannot be dissociated from assumptions about the nature of mathematics and rational thinking. Each step of a proof, for example, must be supported by a shared understanding of what counts as a logical deduction. These assumptions routinely remain unexpressed precisely because they're shared and unproblematic. But that was not the case in the Neapolitan controversy.

The Geometry of Reaction

Fergola and his followers attributed to geometry a priority that was at once historical and logical. They thought its lineage could be traced back to the pre-Platonic geometers of ancient Greece and saw synthetic geometry as the only legitimate foundation for all of mathematics. Fergola had come to believe that modern mathematics lacked sound epistemological foundations and aimed to provide them. In his early manuscripts, he explored the possibility of infusing calculus with "Euclidean rigor," which meant grounding it in rigorous and purely logical concepts and forms of reasoning. Unconvinced by his own results, Fergola then formulated a research program that took synthetic geometry—which he considered "intuitive"—as the ultimate foundation of all mathematics, including calculus.

This program was further articulated by his students. Flauti, who emerged as the leader of the synthetic school as Fergola's health declined during the 1810s, exacerbated his school's polemics against the analytics. His works are emblematic of the later phase of the synthetic school, when emphasis shifted from original research to textbook production, systematization, and institutionalization. Flauti's reorientation took place in a context in which synthetic mathematicians were mobilized to shape key educational reforms. Some of them still carried out original research, but mostly in the history of mathematics, focusing on issues of lineages of concepts and methods. Although Fergola's works on pedagogy and the history of geometrical problem solving were reprinted, his epistemological considerations faded into the background. One Fergolian leitmotif remained undoubtedly prominent: the anxiety of preserving the purity of mathematical knowledge against any attempt to pollute it with extraneous considerations and principles.

This mathematical culture's essential objective was the restoration of an alleged original purity. It was about returning to a millenary tradition that had been corrupted during the eighteenth century and the revolutionary storm, and about recognizing the authority of that tradition. In practice, achieving this objective meant expunging all that was not the immediate product of intellectual intuition from the legitimate mathematical realm. Like their maestro, though less subtly, Fergola's students targeted the misuse of mathematical knowledge and its illegitimate applications in the world of human experience. Above all, they denied that mathematics could offer reliable guidance in social, economic, and political matters.

The Geometrical Turn

Fergola began arguing for the need to rediscover the art of Greek geometry during the 1780s. His Euclidean program concerned not only geometry but the entire edifice of mathematics. While other philosophers were questioning the legitimacy of using mathematics as a reliable cognitive tool in the natural and moral sciences, Fergola aimed to oppose the analytic spirit *within mathematics itself.*

According to his biographers, Fergola lived an exemplary life.[1] The son of a civil servant, Fergola was born in 1753 and studied at a Jesuit college, where he excelled in Greek and Latin, as well as in extracurricular activities such as fencing, music, and singing. Then this lively boy underwent what he later called a religious conversion, which put an end to his "worthless days" and "religious indifference." Religion and science became his exclusive interests. Fergola, we are told, restored the study of mathematics in Naples after decades of decline and neglect by bringing back the pure and elegant mathematical style of the ancients. An "angelic, heavenly purity" characterized both Fergola's religious faith and his mathematical work. A bachelor who found the company of women extremely unpleasant, Fergola chose his religious models in the baroque devotion of the seventeenth and eighteenth centuries. He was close to the sentimental religiosity of Alfonso Maria de' Liguori (1696–1787), an influential Neapolitan bishop, canonized in 1839, who championed new forms of Marian enthusiasm—in open contrast to contemporary rationalist and antibaroque trends within Catholicism.[2]

In mathematics, Fergola saw himself as the direct heir of the Greek geometers as well as the Christian mathematicians of the seventeenth century. After the expulsion of Jesuits from the kingdom in 1767, Fergola continued to study mathematics with Giuseppe Marzucco and Marcello Cecere at the Collegio del Salvatore. Here, according to his biographers, Fergola experienced firsthand the inadequacy of contemporary mathematical education. Only Cecere's lectures on Euclidean geometry touched him profoundly—as in a religious illumination: "The light of geometrical truth hit the young man . . . and he suddenly saw the art, the art through which the great Geometer [Euclid] received his *Elements*, and thus he grasped the supreme principles of geometrical invention."[3]

Biographers insist that Fergola was mostly self-taught, emphasizing his genius and his radical break with Neapolitan mathematical life. The image of Fergola as a solitary genius in a mathematical wasteland is hardly defensible, though. Some of his teachers were competent in differential and integral calculus and interested in its application to mechanics.[4] Also, in the 1770s he frequented the salons of the Neapolitan aristocracy, and—as tutor to the son of Giovanni Domenico Berio, Marquis of Salza—he had access to the magnificent family's library, which was rich in modern scientific texts and contained collections of the acts of the main European scientific academies. Here Fergola became familiar with the works of Newton, the Bernoullis, Euler, Brook Taylor, the de Martinos, and many other eighteenth-century mathematicians. He also met the protagonists of the Neapolitan Enlightenment, among whom were cosmopolitan figures like Ferdinando Galiani, who had long frequented the Parisian salons. While privately studying mathematics, Fergola was awarded a university degree in law, not an uncommon combination. His studies included philosophy and metaphysics, and he was particularly fond of the works of Ralph Cudworth, Christian Wolff, and Antonio Genovesi.[5]

Overall, Fergola's educational path and early publications are similar to those of other Neapolitan mathematicians of his generation. The young Fergola knew and appreciated men like Nicola Pacifico, and he entered the Royal Academy of Sciences in 1779 thanks to the support of Vito Caravelli, as future Jacobin Vincenzo de Filippis did in the same year. Around 1779, he took a teaching position at the Collegio del Salvatore, a former Jesuit college, and presented his first paper at the Royal Academy of Sciences. This paper—now lost—provided the solution for two geometrical problems through the application of integral calculus.[6] That year, Fergola edited a textbook of natural philosophy based on Genovesi's lectures on experimental physics.[7] In it Fergola praised the utility of the study of physics for agriculture, medicine, and many other practical purposes and argued that physics should provide the foundation for the morals.[8] Other papers from the early 1780s show that Fergola was interested in using a mixture of geometry and calculus in optical and architectural applications.[9]

There are, however, some distinctive traits in Fergola's early productions. His reading of Genovesi, for example: Fergola emphasizes the link between physics and theology and a teleological and providential vision of the physical universe. He endorses Genovesi's phenomenal empiricism, the irreducibility of consciousness to mechanical motion and, against Locke, the impossibility of distinguishing consciousness from the (incognizable) spiritual soul. Fergola's account of the history of physics was reminiscent of the doctrine of the *philosophia perennis*, linking in one single scientific tradition the Greeks, scholastic philosophy, seventeenth-century rationalism, and Christian Wolff.[10] Indeed he situated himself in the tradition of post-Wolffian philosophy, then entrenched at the University of Naples. The study of law, in particular, was embedded in this rationalist metaphysics, which was a main critical target of reformers like Filangieri and Pagano. They had learned a very different lesson from Genovesi: the battle against metaphysics for a "civil philosophy" whose ultimate goal was public happiness.[11]

The first publication in which Fergola stated openly that he wanted to revive ancient geometrical methods dates to 1786. He noted that it was despicable that the purely geometrical methods of the ancients had been almost completely neglected during the eighteenth century, so that "now they are uncultivated and derelict." Fergola only approved of those uses of analysis that could be reduced in principle to pure geometrical reasoning; he described the recent transformation of continental mathematics as a dangerous "diversion from geometry." Fergola showed how a series of problems "of site and position" could be easily solved through a purely geometrical method of reduction (of one problem into another equivalent problem) that he called the "principle of conversion." His paper aimed for a high level of generality: it argued that all problems of site and position can be grouped into three main classes, then provided a specific problem-solving method for each. Synthetic methods could return to the center of geometrical practice, Fergola argued, once they were properly generalized.[12]

Fergola's interest shifted markedly from practical applications to the study of families of geometrical problems that had a primarily historical interest and were considered difficult to solve through pure analysis. In 1787, presenting the solutions to another family of problems, Fergola described his method as based on "a reflection on the nature of [each specific] geometrical problem," which yields truths that "would be hardly visible within the web of an analytic calculus, no matter how well executed."[13] Fergola also informed the Royal Academy that his student Annibale Giordano had found a simple and elegant solution for a problem of plane geometry known as Cramer's problem. In his paper, Giordano used lemmas he had found in Pappus's collection of ancient problems, as well as Fergola's principle of conversion. The paper was published in the acts of the main Italian academy of sciences and became exemplary for those who wished to revive the geometrical analysis of the ancients. It was later hailed by Michel Chasles and Lazare Carnot as a model of pure—"Greek-like"—synthesis.[14]

In 1786 Fergola applied for a position of pensionary member at the Royal Academy, presenting himself as the one who "first among the Neapolitan mathematicians, applied the integral calculus, so sublime in its manipulation, to the Science of Nature."[15] In a clear sign that Fergola did not enjoy the favor of Royal Academy leaders or of their political referents, his application was rejected.[16] At that point, Fergola's geometrical approach to mathematics was perceived as old-fashioned and inadequate. When he presented a paper on optics at the Royal Academy, for example, a member of the audience challenged him, arguing that such difficult questions were best addressed with purely analytic methods.[17] A biographer tells us that this critic "had been living in Turin, London, and Paris" and was a personal friend of many French philosophes. He could be no other than Domenico Caracciolo, the future prime minister.[18] On his return from Paris in 1780, Caracciolo found that Naples was on the wrong track in its scientific and mathematical development. The following year, he invited Lagrange to move to Naples and lead the Royal Academy, but to no avail. When he died in 1789, Caracciolo was planning a radical reform of the university.[19]

Shortly after Caracciolo's death—a few days after the storming of the Bastille—the political climate began to change in Naples. Chairs were redistributed at the Collegio del Salvatore, and Fergola was appointed to his first mathematical position, chair of Analytic Mathematics and Mathematical Physics; his salary increased from 36 to 240 *ducati* per annum. A new season had begun for him and the synthetic school.[20] Giordano was nominated a member of the Royal Academy and obtained a chair of mathematics at the Military Academy. Carlo Lauberg, who was running for the same position, was excluded from the final selection; Fergola had vetoed him. "When I suggested to our Military Academy D[on] Annibale Giordano," he wrote later, "I let France and northern Italy know that a great geometer was teaching among us. And I freed the Academy from D[on] C[arlo] L[auberg], a great villain, by excluding him from the competition."[21]

Fergola's first book, published anonymously in 1791 and adopted in colleges across the kingdom, contained his lectures on conic sections.[22] The book had a clear polemical intent, as he was convinced that contemporary geometry textbooks were gravely inadequate. Mostly, they used too many "external" (nongeometrical) devices, including "tortuous" pieces of algebraic analysis and propositions from mechanics or optics.[23] Felice Giannattasio, one of Fergola's students, prepared the drawings and a historical introduction.

The presentation of the geometrical properties of the curves was modeled on Apollonius, a paradigm of Fergola's "Euclidean rigor." In a later book on conic sections, Fergola showed the main properties of these curves through Cartesian analysis, "so that the young, while learning these truths, get used to reason correctly and elegantly with both methods."[24] If Apollonius was Fergola's model for pure geometrical methods, Descartes was his guide when it came to the "proper analytic method," a mixed method wherein geometrical reasoning was partially replaced by algebraic calculations, produced via the choice of a convenient system of coordinates and, crucially, transformable back into geometrical terms through the use of specific techniques, such as the construction of equations.

Fergola showed again his fondness for the Cartesian method in a 1818 collection of his geometrical lectures.[25] The book also included the study of what he called the "heuristic art," the method for discovering new geometrical truths. It was a text that had long circulated in manuscript among the synthetics' inner circle, as had most of Fergola's writings.[26] His main goal was to reconstruct the lost heuristic method of the Greeks, the canons of their geometrical analysis. The basic steps of this method were, according to Fergola, the supposition of the fact; its consequences; the reduction of the problem to another problem whose solution is known; and the geometrical composition, which includes construction and final proof. If the first three steps are replaced by algebraic analysis, the fourth step must include a conversion of algebraic expressions into geometrical magnitudes and of equations into geometrical proportions. Fergola also presented new techniques that would support geometrical analysis, such as the "principle of conversion" and the "transfer principle," essentially ways to reduce geometrical problems to a few basic types that could be solved through the application of standard methods. Finally, he articulated aesthetic canons—like the simplicity, clarity, and concision of proofs—which he considered emblematic of "geometrical wisdom."

That the leader of a combative school devoted to pure geometry should take such an interest in Cartesian analysis might be puzzling. Even more so, the fact that his published and—above all—unpublished lectures and treatises reveal that Fergola was an attentive reader of late-eighteenth-century calculus and analytical mechanics up to, at least, Lagrange's 1788 landmark mechanics. And not just a reader: in his manuscripts one finds attempts to systematize research and teaching on topics like series, variation calculus (1795–1798), differential and integral calculus (1788–1810), and algebra (1800–1807).[27] In reality this evidence is consistent with the essentially foundational aims of Fergola's synthetic program. When examined closely, his geometrical turn was far from a mere return to the past. "The geometers who discovered infinitesimal analysis," he wrote, "and those who continued their research, did not attend to the clarification of its principles as they should have done in order to make of it a truly fruitful science." Fergola aimed to provide calculus with the same kind of *rigor* enjoyed by classical geometry, avoiding any reference to induction, empirical truths, or practical applications, and thus making it pure and well ordered.[28]

Manuscript evidence shows that Fergola tried two main strategies to achieve such purification. On the one hand, he tried to give a "Euclidean form" to the algebraic infinitesimal calculus of Euler and Lagrange. Fergola thought of it as a "method" for the manipulation of symbolic expressions that stand for discrete magnitudes. To give this symbolic calculus a Euclidean, logical form, Fergola elaborated a set of axioms and definitions, and then tried to introduce laws through which its theorems could be proved *rigorously*, thus transforming it into an "ordered system." In this way, Fergola was hoping to turn calculus from a set of useful and loosely related techniques into a unitary and well-ordered branch of pure mathematics. As an alternative foundational strategy, he also tried to ground the practice of calculus on geometrical intuition—hence his interest in reviving Newton's theory of fluxions and his struggle to provide a geometrical definition of the infinitesimal. Nothing from this body of work, however, was published.

The 1790s saw Fergola's rapid rise in popularity and prestige. Already charged with numerous positions and official duties and still running his own private school, in 1792 Fergola declined the offer of yet another chair at the Military Academy. In those "happy days" many students arrived from the provinces to study with him. These were "brilliant years for the mathematical sciences," wrote a former pupil and biographer, noting that many of Fergola's students went on to successful academic, political, and military careers.[29]

Not everyone was on board, though. Annibale Giordano, Fergola's favorite student, soon abandoned his maestro's way. In 1789 Giordano had entered the Military Academy as a mathematics lecturer and there met Lauberg, a chemistry lecturer. The next year, as he collaborated with Fergola on a textbook of mechanics, Giordano joined Lauberg's private school as a teaching assistant. Giordano was twenty-one and Lauberg twenty-seven when they began teaching mathematics and chemistry together at Lauberg's house in Vico dei Giganti. In 1792, they opened a private studio on Piazza Santa Caterina, and by that time Giordano had apparently cut ties with Fergola and joined Lauberg's analytic program. On that occasion, a biographer wrote, Naples suffered "a great misfortune, as young and highly gifted men did not trust and follow with docility the voices of the wisest people." The loss of Giordano, however, was "more than compensated" by the doctrine and behavior of many other pupils, for whom "the government had always the highest esteem." These students focused their attention on pure, Greek-like geometry, abandoning other concerns of their maestro. Fergola's quest for generality was also side-lined in their research and teaching, as they concentrated their attention on particular problems with significant historical pedigrees. In essence, a new generation of synthetics left behind Fergola's foundationalist program in calculus and emphasized instead his geometrical radicalism: for them analysis was nothing less than "a curse."[30]

The 1799 revolution had important consequences for the institutionaliza-tion of the synthetic school. "The noble school was gloriously advancing on his path," noted a biographer, "when the devastating storm that had formed in Paris in 1789 finally reached Naples in 1799." Fergola continued to lecture at the College of the Salvatore until it was closed down and transformed into a military hospital. "Not agreeing with the new political aspirations," he then retired to a country house on the hill of Capodimonte, a "solitary and sunny place," far from "the noise of the city."[31] It's a description that echoes that of Giacinto Gigante's favorite working place: distant from the city and politi-cal upheaval. In November 1799, the restored Bourbon government reopened the University of Naples, which had been sacked by the *Santafede*. Eighteen professors were missing: seven had died for the republican cause, and eleven were under arrest. New professors had been "opportunely chosen" by the king himself. The chairs of feudal law and of economy and trade (which had been Genovesi's and the Jacobin Troiano Odazi's) were given to members of the government. The government's aim, spelled out in a dispatch in the sum-mer of 1799, was to promote "true and sound knowledge" and "extirpate the dangerous doctrines that have caused so much ruin and destruction." Mon-signor Agostino Gervasio (1730–1806), the new principal of the university, considered the university "very much infected" and prepared a reform for the entire public education system. It called for the abolition of many university chairs and the closure of primary and secondary schools, forcing children to enter religious schools instead. But the immediate priority was to find reliable professors.[32]

In March 1800, Fergola was immersed in "profound speculations" when he received a royal dispatch offering him the prestigious chair of sublime mathematics. Fergola accepted it but rarely visited the city, as he preferred to work in his quiet country house; a substitute, appointed by Monsignor Ger-vasio, lectured in his place.[33] Meanwhile, Giannattasio and Flauti had taken over Fergola's private studio. They too were lavished with official appoint-ments. In 1803, Giannattasio began teaching sublime mathematics in the Mil-itary School, which had replaced the now suppressed Military Academy—an

institution that had been too close to the revolutionary cause. Giuseppe Sangro (1775–?), another student of Fergola, was also given a chair at the Military School. Vincenzo Flauti, age twenty-one, became professor of synthetic
mathematics at the university; the royal dispatch he received read: "The King
wanted to set an example, rewarding in this young man both knowledge and
perfect morality."[34]

In this period, Fergola was reflecting on the metaphysical and theological dimensions of his mathematical work. In 1804, he exchanged letters with
Francesco Colangelo, an erudite Oratorian priest who had asked him for
mathematical demonstrations of the existence of God to support the great
battle against atheism and social rebellion. In the same year, Fergola was invited by Pope Pius VII to join the Academy of Catholic Religion, which had
been created in 1801 to fight the secularization of knowledge and society.[35]
Meanwhile, Fergola's fervent faith and ascetic lifestyle had become proverbial.
His biographers report numerous anecdotes that emphasize his moral and
religious virtues: he followed a strictly vegetarian diet; he walked, each Sat-
urday, to a distant sanctuary dedicated to the Holy Virgin; he could be seen
frequently among the most humble people, participating in popular processions and listening to charismatic predicators; he wore cilice and chains; he
practiced lifelong chastity; and he never entered the Royal Gallery of Art, as
he was repelled by the nudity of the statues.[36] Fergola's public persona and
his epistemic and moral virtues would form an influential model for devout
Catholic scientists of the Restoration age. His was a model of purity: purity of
faith, body, desires, intellect, and science. Fergola rejected the impure mathematics of the moderns as he rejected their disastrous politics. At their roots
was one and the same moral corruption.

History Matters

Representative of the synthetic-school publications was a collection of papers
edited by Flauti and Giannattasio in 1811. It was intended for training and as
a model for future publications. The goal of mathematics, one reads in the
introduction, is "to guide the spirit in the sublime art of inventing, which
demonstrates the celestial origin of human reason." The Neapolitan synthetics saw their practice as directly derived from that of the Pythagorean school,
which had been active in southern Italy during the fourth century BCE. Fergola, in their view, had revived the heritage of this pre-Platonic tradition for
the glory of the Neapolitan nation.[37]

The collection opened with a generalized version of Giordano's solution to Cramer's problem. This paradigmatic solution had succeeded where

"algebraic analysis was impotent."[38] The editors contrasted its elegance to Lagrange's "complex" analytic solution for the same problem, arguing that only the proper integration of the two methods could produce significant results. Furthermore, Giordano's solution carried the stamp of "Greek genius," whereas other synthetic solutions were "heavier, and less elegant," because they required more lemmas and constructions. Lagrange's solution contained some "sublime analytic flights," but produced equations extremely difficult to construct: "this was a shortcoming of the [analytic] art, not of the great man."[39] The synthetics often associated purely algebraic analytic reasoning to images of spectacular flights, groundless rapidity, and an unbound imagination, as opposed to the slower but trustworthy and elegant procedures of geometrical reasoning.[40]

To strengthen these points, the 1811 collection also contained excerpts from Fergola's own manuscript on the heuristic art, including comparisons of synthetic, purely analytic, and Cartesian solutions for one problem.[41] The difficult and "uncertain" construction of the final equation is presented as the crucial deficiency of the purely analytic method. When geometers leave the path of geometrical reasoning, results can only be "monstrous." In these pages Fergola insisted on the purely intellectual nature of geometrical reasoning: "The operations prescribed in geometry have to be executed by the intellect, not by the hand," he wrote; geometrical instruments such as compass and rule, too limited and empirical, must not be taken as criteria for reasoning. The "Euclidean spirit demands that reasoning on magnitudes be carried out without interposing any manual operation."[42]

The collection was also intended as a response to those who accused the synthetics of ignoring modern analysis. Hence it included excerpts from Fergola's unpublished lectures on "sublime analysis." These texts show how Fergola aimed to "inform analysis with the spirit of synthesis," thus providing it with unprecedented demonstrative rigor.[43] One excerpt describes the reduction of rational functions to simple fractions. "Among the various topics contained in the summing methods of the modern geometers, no one is more elegant and complete than the integration of rational functions," Fergola argued.[44] To accomplish the integration, the functions must be properly resolved into "convenient" elements, "the most difficult part" of the process. Fergola referred to Euler's treatment of this topic, which he found unsatisfactory: calculations should be more "rigorous," "clear," and "elegant." To this purpose Fergola replaced Euler's "speculative principles" with "intuitive" ones, which are less general but clearer: perspicuity was to Fergola a more desirable goal than generality.[45]

Above all, the collection included exemplary solutions to many famous geometrical problems. Fergola organized his research and teaching around clusters of significant problems and theorems, typically derived from Greek texts. Some ancient solutions had been lost and had to be "divined"; others were replaced with more elegant ones. Reflecting on ancient geometrical questions refined and directed the pupil's intellectual intuition and helped to organize original research, which Fergola understood as an endless articulation of the classical tradition. Among the problems considered were the "problem of the cylindroid," which concerns the properties of the solid obtained by the rotation of a hyperbola around its secondary axis, and which had attracted the attention of ancient and modern geometers.[46]

*

Other problems reworked by Fergola and his students since the 1780s were the so-called problems of contacts, which treated the relations between certain given straight lines and certain given circles. Fergola, following the tracks of Viète and Newton, offered his own divination of the ancient solutions, reducing each of them to a single lemma and then proving it; his students then competed to offer the most elegant Greek-like version of such solutions.[47] In other cases, Fergola rescued ancient theorems from oblivion, as in the case of Ptolemy's theorem, which expresses a metric relation between four points on the same circle. This theorem can be significant in trigonometry if one holds a purely geometric conception of this branch of mathematics. In fact, the gradual shift of trigonometry from the realm of pure geometry to that of analytic geometry during the seventeenth and eighteenth centuries had made this theorem less relevant. Fergola, by contrast, considered it "a good source of geometrical principles" and deduced a number of corollaries from it. Some of them, once "translated into algebraic signs," revealed "new truths relative to the circular functions," which Fergola used in his work on elliptic transcendent functions, where he deployed Cartesian geometrical methods to calculate the differential of an arch of the ellipse.[48]

Finally, the collection also included Fergola's solution to the problems "of the inclinations," which he tackled with the method of the loci. It consisted of 1) reducing the solution of a problem to the determination of a single point; 2) eliminating one condition for this point to determine the locus of the infinite number of positions it can consequently assume; and 3) determining one intersection between two such loci. In this way, a number of different solid and hypersolid problems could be reduced to that of determining the intersections between certain circles and certain straight lines. Only synthesis,

Fergola argued, could shed light on these problems; using analysis would produce "dreadful and impracticable results."[49] One of these problems, Fergola
observed, "is particularly famous among us." In 1807 "a professor from northern Italy" had visited Naples and criticized the school's geometrical approach.
Giuseppe Scorza, one of Fergola's students, challenged him to solve a simple
problem of the inclinations. Once the professor had completed his complex
analytic solution, Scorza proudly revealed his own solution, "executed with
few passages of synthesis, in our own way." The visitor was the prominent
Piedmontese mathematician Giovanni Plana (1781–1864), who had studied
with Lagrange and was looking for employment in French-controlled Naples.
Similar episodes, in which French or Francophile visitors are stunned by the
Greek-like ability of Fergola's students, abound in the apologetic reconstructions of the history of the synthetic school. [50]

Ancient Problems and Modern Careers

Fergola remained active until about 1818. By that time, most of his former
students had shifted their attention to the history of mathematics: above all,
they were interested in revising Jean-Étienne Montucla's authoritative reconstruction. They did so at two main levels: by challenging many of his attri-
butions, especially those regarding the achievements of Greek mathematics,
and—more fundamentally—by rejecting Montucla's narrative of the history of mathematics as an essentially linear progress characterized by ever-
increasing understanding. In Montucla's enlightened reconstruction, the
growth of mathematical knowledge mirrored the development of the human
mind: as in d'Alembert, the history of mathematics was a map of the triumphs
of *l'esprit humain*. Against this view, the Neapolitan synthetics held up one of
fall and human weakness: to them, the history of modern mathematics was
characterized by the decline of a distant and pure knowledge. This knowledge could, at best, be partially recovered; there was no linear progress to be
celebrated, just, at most, cyclical iterations in which exceptional geometers
like Fergola were able to reactivate, temporarily and imperfectly, the ancient
tradition.[51]

The most remarkable investigator of the history of ancient geometry was
Ferdinando de Luca (1783–1869) who, in 1832, read a memoir that claimed
the entirety of ancient geometry to "the Italian school." De Luca contested
Montucla's attribution of a large number of geometrical discoveries to Plato
and argued instead that the invention of geometrical analysis and the study
of conic sections and geometrical loci should be attributed to various Pythagoreans, like Archytas, a native of the southern Italian city of Taranto. This

interpretation was certainly original and closer to the present view. De Luca, however, went on to argue that Pythagoras himself was a native of southern Italy—an opinion he had found in Thomas Aquinas. Hence he labeled him an "Italian-Greek geometer" and turned the Pythagoreans into the creators of a mythical "Italian school of geometry" whose lineage continued uninterruptedly until its last champion, Nicola Fergola.[52]

The belief in an "ancient Italian wisdom" that originated with Pythagorean philosophy was not as peculiar as it might seem. Versions of this myth had emerged as part of the reconstruction of Italian cultural identity during the Renaissance and continued to capture the imagination of erudite humanists in the sixteenth and seventeenth centuries. The political events that followed the French occupation of northern Italy in 1796 breathed new life into this set of beliefs, which were mobilized to support a variety of political programs—revolutionary, liberal, and reactionary. In the early nineteenth century the reflection on an alleged Italian cultural primacy intensified as part of the debate on the contested meaning of the Italian nation. In 1806 Vincenzo Cuoco wrote a historical novel about the transmission of philosophical wisdom from the "Etruscans" who lived in the Italian peninsula to Plato himself. This narrative supported, at once, the cultural primacy and the ethnic unity of Italians. Other authors weaved similar reconstructions emphasizing pre-Greek and pre-Roman elements but insisting mostly on the cultural rather than the ethnical unity of the Italian nation. Building up to the 1848 revolutions and the Risorgimento, the patriotic and nationalist movement appropriated some of these narratives, which played a key role in the construction of a new Italian national identity.[53]

There were other versions of the ancient wisdom argument, though. In Naples, some rejected the theories of Italic indigenous unity and emphasized instead the connections with the "Pelasgians" and therefore the primacy and distinctiveness of southern Italy. The ancient wisdom was, in this perspective, an Italian-Greek one, first developed in the territory of the Kingdom of Naples.[54] This narrative too had clear political implications: it supported the 1816 unification of the crowns of Sicily and Naples, which had been resisted by Sicilian elites, and confuted liberal and nationalist arguments about the ethnic and cultural unity of the peninsula. While not mentioning this broader debate, it's clear that de Luca's history of geometry supported the theory of an original Italian-Greek wisdom—philosophical and mathematical—that was rooted exclusively in the south of the peninsula and thus questioned the liberal nation-building project.[55]

As might be expected, Euclid's *Elements* was key to the synthetics' historical research and pedagogy. Flauti's own critical edition of this book was dedicated to King Ferdinand, and had been adopted in schools across the kingdom. It earned Flauti fame as well as fortune—the eleventh edition was printed in Flauti's "private printing office." The last edition, the twenty-second, was published in 1857. Flauti claimed he felt compelled to prepare his edition of the *Elements* because this text "had been largely abandoned by the schools, even in Italy, in favor of much less geometrically-oriented textbooks." It was 1810, and French analytic textbooks of geometry were flooding the peninsula. Flauti described the work of Greek, Arab, and European glossers of Euclid; the Renaissance editions; and the British editions, up to the much-appreciated 1756 edition by Robert Simpson, "great practitioner and promoter of the geometry of the ancients." By contrast, Flauti disliked recent French editions, as their authors had been unable to discriminate between authentic and spurious parts of the original. Flauti aimed at—in his own words—"purifying" the *Elements*, according to the view that its weaknesses must be attributed not to Euclid, but rather to "some ancient editor, misguided by the arrogant desire to innovate." Such was Flauti's eagerness to clean up Euclid's blemishes that in the sixth edition he even supplied a proof for the fifth postulate, which he later retracted.[56]

Scorza was not fully satisfied by Flauti's work and published his own edition of Euclid's *Elements* in 1828, under the title *Vindication of Euclid*. He aimed to relieve all the book's well-known errors and deficiencies—which he too believed should be attributed to later interpolations—and thereby to return the book to its original integrity. Although some of his arguments have been considered convincing by later historians, his attack against anyone—ancient or modern—who disagreed with Euclid's definitions and methods reached extremes.[57] As for the fifth postulate, Scorza could not believe that the "very accurate Euclid" assumed this proposition as a postulate, given that he had previously provided a proof for a very similar proposition (proposition 28); Euclid must have considered it a lemma whose proof had not survived. Scorza thus presented two possible proofs, which in fact shift rather than solve the problem.[58]

Scorza's *Euclid* was also a pedagogical intervention. Euclid's "admirable order" (the "natural" progression from simple to complex: point, line, surface, solid) is taken as paradigmatic and opposed to modern geometrical textbooks that move from the general to the particular. Scorza argued that such inversion derives from a confusion of mathematics with the natural sciences: "[Finding the correct order] is not a question of discovering definitions by analyzing natural bodies," but rather one of "arrang[ing] truths in the most natural way." To say that a geometric solid is more easily conceivable than points or lines is to mistake mathematics with empirical reality: a geometric solid is not a physical body, and our senses are irrelevant in the study of its properties; otherwise, mathematics would "lose all value." Like all synthetics, Scorza defended the purely intellectual nature of mathematical knowledge, which explains why "its principles are stable, and its proofs are certain and evident." Geometrical axioms, for example, are self-evident propositions whose truth is immediately apprehended "with the light of the intellect." Hence, all "mechanical" considerations, which are independent from "the evidence of reason," should be banned.[59]

A bright provincial student, Scorza had arrived in Naples to study medicine; he obtained his degree, but he never exercised the profession. Instead, he studied mathematics with Fergola and soon began teaching for him.[60] He had been trained in higher geometry, algebraic analysis, and general mechanics, but his favorite topic was the divination of the lost methods of geometrical analysis. His most relevant work, *Divination of the Geometrical Analysis of the Ancients* (1823), included many reconstructions of the lost solutions to classical problems. Significantly, Scorza also included a problem relative to triangular pyramids, a response to Lagrange's famous paper of 1773.[61] Apparently he was the one student who responded best to Fergola's religious guidance: "in the middle of so many vicissitudes [the French occupation], he always conducted a most exemplary life, dividing up his day equally between Christian practices, teaching, and research." Scorza was a "man of God, and of Mathematics."[62]

Providing synthetic solutions to problems treated by Lagrange in his 1773 paper was clearly a meaningful task within the synthetic school. After Flauti and Scorza, Francesco Bruno (1790–1862), professor at the Royal Naval Academy, offered a new synthetic solution for one of those problems in 1825.[63] But Bruno's masterpiece, according to his admirers, were his solutions to a group of plane-geometry problems: they were considered particularly elegant because he used "loci whose description and determination is the simplest possible one" and which could be directly extracted from the data, so that no reference was made to geometrical truths external to the problem, thus preserving its true unpolluted nature—a local nature. In this sense Bruno's solution is "more natural" than other possible geometrical solutions. Bruno then moves on to the composition of the problem: his construction is so pertinent to the nature of the problem that it does not even require tracing new figures: a masterful demonstration of the "exactitude and rigor of Greek geometrizing." [64]

Also renowned within the school was Felice Giannattasio, Fergola's first recorded student. He had arrived in Naples to become a priest, had studied philosophy and mathematics, and in 1778 was already collaborating with

Fergola, proving himself instrumental to the publication of his maestro's lectures on conic sections (1791) and mechanics (1792–1793). The two would have published more, Flauti argues, "if the political circumstances of those times" had not "perturbed our happiness, and all kinds of improvement that were taking place."[65] Giannattasio did not publish much, but he was considered a formidable teacher: he became professor of sublime mathematics at the Military Academy in 1802 and member of the Royal Academy of Sciences in 1811. Shortly afterward, he succeeded Fergola as the chair of sublime synthesis at the university. He was "a most respectable and exemplary ecclesiastic" who "always avoided politics and parties." Giannattasio's frugality was legendary; although he was a well-off art collector, his austere table was graced more often than not by boiled cabbage. Like Fergola, he was a vegetarian, and followed dietary restrictions clearly inspired by both Christian ascetics and Pythagorean doctrines.[66]

It was Flauti, however, who took over leadership of the synthetic school as Fergola's health declined in the 1810s. During his sixty-year career, Flauti cumulated innumerable assignments, positions, and honors, and, during the Restoration, he had a remarkable degree of control over academic life and public education. Only the collapse of the Neapolitan state in 1861, and the annexation of its territory to the newly created Kingdom of Italy, forced him out of scientific life. Flauti's career had been precocious and fast: he began teaching in Fergola's private school around 1798; in 1801 he was already directing the school with Giannattasio. In 1803 he held the university chair of synthetic mathematics, then in 1806 moved to algebra and descriptive geometry, and in 1812 replaced Fergola as professor of sublime analysis. In 1818 he was nominated "first professor of mathematics" at the Royal College of Naples, where he taught conic sections, sublime calculus, and mechanics, while acting routinely as external examiner for colleges and academies across the kingdom. In 1822 he became the director of the University Library. In 1829 he was nominated to the Committee of Public Education, presided over by Francesco Colangelo, and was knighted by the king. A member of the Royal Academy of Sciences since 1808 and its secretary from 1817 to 1860, Flauti gradually came to personify this institution.[67]

Unlike Fergola, Flauti showed little interest in foundational issues. He translated and annotated Étienne Bézout's textbook of differential calculus, which he used in his private teaching, and later wrote a textbook of algebra,[68] but his vast production was made up mostly of geometry textbooks. Some of them included parts of Euclid's *Elements* and teaching materials by Fergola and other synthetics—one such multivolume course of geometry was reprinted at least twenty times.[69] His contributions to the pioneering field of descriptive geometry demonstrate well how Flauti conceived the future of mathematics as an emanation of its mythical past, and mathematical innovation as the endless articulation of an original wisdom.

Descriptive geometry is a branch of geometry in which geometrical problems in three dimensions are represented and solved in the plane. The graphical methods in question have a long history and respond to practical problems in the construction of buildings and artifacts. They were systemized and generalized by Gaspar Monge (1746–1818), who published a first textbook in 1799. The Neapolitan synthetics were very receptive to Monge's textbook: Flauti was teaching it already in 1801, and in 1807 he published the first Italian textbook of this discipline.[70] Flauti infused Monge's basic structure with Greek-like grace, which meant providing more simple and elegant proofs for the theorems and showing more clearly "the geometrical nexus that links all the truths presented to the reader." Flauti's enthusiasm for this branch of mathematics derived from the fact that he considered it an extension of pure geometry. The "uncertain and difficult techniques" devised by artists and architects in the previous centuries, he declared, were eventually given generality and certainty through solid geometry. For Flauti descriptive geometry illustrated clearly how the long-neglected synthetic methods were essential for the advancement of mathematics. Descriptive geometry as reorganized by Monge was a new discipline, but it could be "entirely subsumed under the geometry of the ancients." Flauti acknowledged the affinity between modern analysis and descriptive geometry and the possibility of translating its problems into analytical terms; nevertheless he was convinced that "the analysis of the ancients is more proper than that of the moderns to solve these locus problems."[71]

In 1815 Flauti published a second book of descriptive geometry: *Plane and Solid Locus Geometry*. This book was more original—many said more obscure—than its 1807 predecessor, diverging significantly from contemporary Italian and French treatises on the subject. It presented some peculiar innovations, both in form and content, that made it quite difficult to read for those less familiar with Greek-like synthetic geometry. In fact, Flauti wrote this book as he thought a Greek geometer would have written it.[72] Flauti presented the methods of descriptive geometry, defined as the science that determines figures in space, after having presented the "analogous" methods used to determine figures on the plane. These methods were then applied to a number of problems of solid geometry. The last five chapters provided a compendium of Fergola's *Geometrical Invention*, chiefly the methods of conversion and transfer and their application to particular problems. Descriptive geometry was thus conceptualized as a natural complement to ancient solid geometry, providing practical methods by which to construct the solutions to its problems. Problems favored by the synthetics (e.g., Wallis's cylindroid; triangular pyramids) figured prominently. Nineteenth-century historians of geometry praised, in particular, his construction of the solution for the problem of Archytas. More generally, they noticed how Flauti's book, rich as it was in geometrical applications of descriptive geometry, offered an interesting alternative to the more familiar problems in applied mathematics that filled standard textbooks.[73] The overall form of Flauti's book is also Greek-like, from the terminology (he referred to ruled surfaces as *superfici plectoidi*, rather than the customary *superfici rigate*) to the form of the problems themselves. Commonly, a problem was stated in the form of a proposition asking for the construction of a figure satisfying certain conditions, but Flauti put them in the form of Greek "data," propositions wherein the position of one or more entities described was undetermined. A datum was "proved" when one showed that the determination was in fact implicit in the original proposition, of which it was a necessary consequence.

The case of descriptive geometry shows the re-elaboration of contemporary mathematical practice carried out by the synthetics. Flauti had encountered the new discipline in French books; he understood it as a source of techniques for constructing the solutions to geometrical problems and as a rare opportunity for synthetic geometry to reclaim its long-lost priority in contemporary mathematical research. The new French-Neapolitan government had introduced descriptive geometry to the curricula of universities and technical schools for architects and engineers. The space for collaboration with Flauti was thus open. His 1807 book was the result of a political and mathematical compromise: its language and structure are those of Monge (and therefore easily legible to outsiders), but it also embodied the synthetic school's ambitious program. The 1807 book's gesture toward the complete reduction of descriptive geometry to synthetic geometry is fully realized in the 1815 book. After the Bourbon restoration, a compromise with the practically and analytically oriented French-inspired curriculum was no longer necessary. At that point, Flauti felt free to treat the matter in a completely Greek-like, synthetic way.

Almost fifty years later, at the moment of the national unification of Italy, a career move in the opposite direction would mark the effective end of the synthetic school. Nicola Trudi (1811–1884), the synthetic champion in the challenge of 1839 and Flauti's successor at the University of Naples, had become the leading figure of the school's third generation. In 1860–1861 he faced an existential choice: the new regime was eager to co-opt scientific talent but only those who embraced their new national mission could be considered.

Trudi became a professor of infinitesimal analysis at the University of Naples and a member of the new Academy of Sciences. Eighty-year-old Flauti was excluded from both. Their ways had finally parted. As effectively put by Giovanni Ferraro: "Trudi became an Italian mathematician, Flauti chose to die a Neapolitan." Again, looking at the mathematical dimension of this crucial existential and political passage is revealing. A reading of Trudi's 1862 book on matrices and determinants shows that he was by that time situating his research and teaching squarely within the Italian and European context, avoiding any explicit reference to Fergola, even where the connection is obvious. The applications of these functions to geometry are minimal. And yet, Trudi's quest for "rigid demonstrations," perspicuous and rigorous at the same time, and his mistrust of demonstrations that rely solely on the manipulation of concise "symbolic forms" show that, at a deeper level, Fergola's project of rigorization lived on.[74]

Synthetic Pedagogy

Pivotal to the Jacobin experience and to later modernizing efforts, the pedagogy of mathematics was by far the synthetics' main concern. Flauti summarized synthetic pedagogy in his *Project for the Reform of Public Education in the Kingdom of Naples* (1820). Pedagogy, he argued, must be grounded on concrete experience rather than on "vain speculations"—a reference to sensationalist philosophy and the analytic method. Flauti was hostile to the very idea of state-controlled public education, which, he claimed, "was unknown to the world until fifty years ago." Only "the very modern men," the philosophes and their followers, could think of legislating the entirety of human knowledge. When Flauti referred to "the recent progresses of the human spirit," he did so ironically: "Enlightenment and modern civilization produced many educated people, but very few wise ones; the extension of knowledge is indeed inversely proportional to its profundity, and we will finally reach an epoch where everyone will equally know nothing."[75] Flauti planned to restore traditional curricula and methods of teaching, and he proudly recognized the untimely nature of his resistance against "the ideas of the century." Like the authors of the political reaction, he was largely reinventing the tradition he was defending, which he characterized as an idyllic pre-revolutionary world evoked by the phrase "fifty years ago." Flauti's broad definition of education is the process through which the young become "honest and good citizens." Because the primary goal of public education is moral, he held, it is meaningless to separate the transmission of scientific knowledge from that of religion and the morals—and indeed the greatest obstacle to the good functioning of public education is "the moral corruption of our days."[76]

Flauti considered the current education system badly designed from the beginning. Children ought to remain longer with their parents, at least until the age of eight; family is indeed "the first origin of virtue," he claimed, and the original model for the attachment of a citizen to monarchy and divinity. The parental example is the first and most natural form of education: moral and religious truths are taught in vain if this example is inadequate. The state can intervene to reduce moral corruption, and religious authorities should use "every possible means" to supervise parental education. After the age of eight, children could be taught in elementary schools. They should learn reading and writing together with catechism and the precepts of the most perfect morals. Teachers should be selected by the local authorities, not by the government—a clear call to reverse the centralizing policies of the French period.[77]

Secondary education, Flauti continued, was not necessary for the good of the state, as "being ignorant does not cause moral corruption." Twelve-year-old children, however, should be offered the chance to enter college (*liceo*) for a general education. Religious schools should be protected by the state. Eighteen-year-old students willing to continue their study could then enter the university, a military academy, or a "special school" of law, medicine, or engineering. Mathematics courses for civil engineers and military officers should be centered on geometry and its practical applications. Again, Flauti fought all attempts to centralize control and competences: local authorities, for example, should be entitled to train their own civil engineers and land surveyors, without sending them to a special school in Naples.[78]

The University of Naples topped the system of public education, and its main mission, according to Flauti, should be training teachers for colleges and special schools. At the top of the pyramid were, in Flauti's project, three theological chairs, then four chairs of law, seven of practical medicine, surgery, and anatomy, two of pure mathematics (sublime analysis and sublime synthesis), one of mechanics, one of experimental physics, and one each of chemistry, botany, zoology, mineralogy, and astronomy: an order that reflected Flauti's notion of the natural hierarchical structure of knowledge and one that all professors should constantly emphasize in their teaching, mainly through a historical framing of their discipline.[79]

Finally, Flauti suggested restructuring academic careers and scientific curricula. He loathed the bureaucratization of the academic profession, which had turned professors into civil servants. One of his main concerns was that new faculty lacked the experience and ability he thought were typical of the old system. Only those who were "excellent," he claimed, could become professors, and typically these were already distinguished teachers when they were appointed, as in the case of Fergola. Currently, he continued, one need only study a textbook to present oneself as a learned scholar. Most of the young candidates who gathered each time an academic position opened, Flauti lamented, had no teaching experience whatsoever.[80]

After the revolutionary events of 1820–1821, as he was growing influential, Flauti further articulated this critique in a series of polemical memoirs delivered at the Academy of Sciences. His main targets were the officers of the Corps of Engineers, who knew "this science only through those few applications needed in the trivial exercise of their profession." Flauti was convinced that their superficiality was just a symptom of the long-term decline of the discipline. Newton, the modern hero of the synthetic school, had derived calculus from Greek methods of approximation: he was still a member of the "Greek School." In the eighteenth century, however, the discipline had been corrupted: it had been made accessible to an unprecedented number of students who had learned from it just a few useful techniques; for the first time in history, mathematics was not the exclusive preserve of those who pursued it only "for passion."[81]

Changing pedagogy was key to halting the decline. Learning mathematics had been made easy and mechanical through new methods, and the new and superficial practitioners did not realize that each branch of mathematics had its own specific goals and methods. These days, Flauti lamented, "no one remains within the boundaries of that part of knowledge that they cultivate." As a consequence, mathematics seemed to have lost its "nature": it had been reduced to "a science of words." A science of words can only produce discursive knowledge, which is necessarily the result of reasoning, not intellectual intuition. The confusion between these two different kinds of knowledge made some believe they could solve problems "just because they can write a few formulas on a piece of paper."[82]

The Greeks, Flauti argued, used geometrical analysis to discover the "essence" of geometrical truths and synthesis to prove such truths. Most importantly, they never stopped at the analysis of the problem, a "monstrosity" that characterizes modern textbooks of analytic geometry.[83] The fact that the pedagogy of mathematics had never been a controversial issue before the seventeenth century was to Flauti a clear sign that ancient and Renaissance geometers belonged to one and the same tradition. He also noted how the term "method," which had previously meant a system of scientific principles that yielded mathematical discoveries, had come to signify the mere "manipulation and application of formulas."

Modern textbooks, Flauti lamented, were invariably characterized by the "the deepest abstraction" and by an "absence of light." In them, geometry had been reduced "to a simple bunch of letters, symbols, and formulas." The new mathematical pedagogy had "destroyed the good tradition and replaced it with false systems and bad textbooks."[84]

In response, Flauti argued for a return to Greek texts—Euclid, above all. The core of mathematical education consisted in fostering the intuitive apprehension of mathematical truths. Students must learn to see mathematical truths "at a glance," survey chains of deductions without missing any intermediate passage, and acquire a taste for mathematical elegance. All these ends could be achieved without new textbooks: "for twenty-two centuries the geometrical part of mathematics has had only one perfect institution: Euclid's *Elements*."[85] Once accustomed to Euclid's precision and logical rigor, students should learn how the fundamental geometrical truths can be generalized and applied through the study of conic sections. They could then be trained in problem solving and move on to study Euclid's *Data*, Archimedes's and Pappus's passages on geometrical loci, and the works of the modern cultivators of pure geometry, including Fergola's *Geometrical Invention*. The teacher should also introduce Cartesian analysis and show in which cases it could be useful. Lagrangian analysis should also be taught, but the teacher should show clearly that its use is limited.[86]

As for the other branches of mathematics, Flauti warned teachers not to use textbooks where the theory of the calculus was presented without the necessary rigor and sound proofs, while mechanics should not be taught as a set of abstract formulas. Mechanics was not an abstract science, and its practice did not consist in "calculations," but rather in wisely using geometry to reveal the true laws of the world. His polemical target was, once again, Lagrange, whose mechanics Flauti asserted to be so abstract that "descending from its sublime considerations down to the facts" was simply impossible. It was only after the study of calculus and mechanics that students should take special courses in applied mathematics to be trained for the professions—especially engineering and architecture. In this way science would not be "damaged," having preserved the logical rigor of proofs, the intuitive elegance of solutions, and the evident relations between mathematical propositions. One could not modify or eliminate any component of the synthetic curriculum without "breaking the entire chain [of truths], and destroying the method itself." "Strange and meaningless research," Flauti concluded, "is that which aims to prepare special textbooks of mathematics, as if there was more than one way to properly learn it, and as if the principles of a science could be learned differently according to personal taste."[87]

Flauti saw boundaries falling apart, sciences and morals becoming corrupted, and true knowledge disappearing—all at the hands of a crowd of arrogant pseudo-philosophers and pseudo-mathematicians. To him the restoration of ancient mathematics was not only desirable in itself; it was the first step of a veritable scientific counterrevolution.[88]

A Scientific Counterrevolution

In Naples, between the 1790s and the 1830s, mathematics became a key political battleground. Synthetic practice and pedagogy were shaped by, and in turn supported, a broader, reactionary cultural and political agenda. Not only can the rhetorical apparatus of these mathematical texts be labeled reactionary but so can their technical content—*the way this mathematics works*. The aberration of revolutionary mathematics and its destructive effects called for the urgent mobilization of true mathematicians, and for the elaboration of a mathematics that was not distorted by political partisanship: a pure mathematics. Achieving this pure mathematics, however, was only the first step. Restoring order in mathematics would make it possible to rebuild the entirety of the modern sciences on sounder foundations. One thing was clear: there would be no return to the natural order in politics without a return to natural order in scientific knowledge first. The political counterrevolution required a *scientific* counterrevolution.

This chapter traces the contours of this scientific counterrevolution as it was imagined and experienced in Naples. It emphasizes how early-nineteenth-century reactionary culture provided practical orientation to many mathematicians and natural philosophers, while emphasizing a set of moral and spiritual values—embodied in new scholarly and scientific personae—and defending specific material interests. Essentially, this culture opposed the rationalistic view that social relationships were grounded in universal laws, rooting them instead in the concrete experience of specific communities. Not only one can trace a social genealogy of the political beliefs proper to this culture, but also a genealogy of its distinctive scientific concepts and practices.[1]

The scientific work of devout and legitimist mathematicians and natural philosophers in Naples was permeated by the urgency to return the sciences to their alleged natural order. They shared, I argue, a reactionary worldview that shaped their scientific concepts and methods, and their very definition of reason. At the core of this reactionary science was the mathematical culture of the synthetic school. But reactionary priorities also informed scientific research and teaching in the natural sciences and in medicine. I call this widespread perspective "apologetic empiricism." With this term, I do not refer simply to the fact that certain scholars mobilized scientific knowledge to support religious beliefs and values. Rather, I refer to a distinctive, more specific phenomenon: the use of scientific knowledge to undermine a particular image of individual reason as *active* and *autonomous* and, conversely, to reinforce an image of individual reason as essentially passive and dependent on external sources of validation. Apologetic empiricism was therefore not simply apologetic; it was—strictly speaking—reactionary.

Restoration Philosophy

The reactionary turn of the Neapolitan government after 1789 and its alliance with the anti-French coalition were accompanied by a radical reorientation of cultural and educational policies. After decades of anticurial battles, the intransigent component of the Neapolitan Church was given full control over education. Many private schools and academies were closed down during the anti-Jacobin trials. By 1800, one historian of philosophy noted, "the old Neapolitan cultural environment, that of Genovesi, Grimaldi, Filangieri, Pagano, and Galanti, of the *Lessons on Trade*, and of the *Science of Legislation*, had become completely unrecognizable."[2] The government saw religion as the best antidote against political subversion and planned to place it at the core of the education system. In 1797, Monsignor Agostino Gervasio, archbishop of Capua and former professor of theology at the University of Vienna, became the principal of the University of Naples. A learned and intransigent ecclesiastic, Gervasio had long argued that religion should be the compass of the sciences. After 1799, he was given the explicit directive "to eradicate those pernicious maxims that have caused so many disorders and disasters." In 1804, the Jesuits were readmitted in Naples to take up important teaching duties. Education, and the moral conduct of teachers and professors, were also high on the political agenda during the restorations of 1815 and 1821. Tellingly, in 1816 the university chair of ideology was suppressed and that of truth of Christian religion was established.[3]

Neapolitan philosophical life was transformed. Many new textbooks revitalized the late scholastic tradition and the rationalistic systematizations it had inspired. Christian Wolff was seen as an important resource, as he had shown how to integrate the results of the modern sciences into a structure that was still essentially scholastic. Wolff never questioned the priority of metaphysics over the rest of knowledge, an orientation that is reflected in the traditional organization of his works, which typically move from logic to metaphysics, ontology, cosmology, empirical psychology, rational psychology, and natural theology, and from there to practical problems in moral philosophy, natural law, ethics, and politics. The unity of the system was guaranteed by a few metaphysical principles that operated (circularly, one could argue) across the divides of theology and politics, metaphysics and physics. The works of Wolffians like Friedrich Baumaister and the Jesuits Paul Mako and Sigismund Storchenau constituted the backbone of the new philosophical and scientific curricula in Naples.[4]

In his inaugural speech of 1804, Giuseppe Capocasale (1754–1828), a newly appointed professor of logic and metaphysics, sketched the tenets of this Restoration philosophy. Like his maestro Fergola, Capocasale ate frugally and dressed modestly; he was celibate and, at the age of forty-eight, he became a priest. He was, to his admirers, an exemplar of purity (*illibatezza*). Capocasale explained that no scientific investigation is possible without the support of metaphysics; moving from the principle of noncontradiction, which holds for any necessary and contingent truth, he provided an ontological demonstration that reality is structured according to the most harmonic and teleological order. In fact, the sciences reveal the cosmic chain of being, the universal and uninterrupted connection—indeed the *order* (Lat. *ordo*)—of all created entities. Such order takes us necessarily to an intelligent causal principle.[5]

The same basic argument can be found in innumerable philosophical texts of the Restoration age.[6] The primacy of theology and metaphysics did not imply a rejection of the empirical study of reality. Rather, this literature abounds with eulogies of Galileo, Newton, Locke, Condillac, and Genovesi as models for the study of nature; Capocasale himself authored textbooks of geometry and Newtonian physics. The all-encompassing philosophical systems of the Restoration presented empirical research as harmonically integrated with their broader metaphysical structure. Thus, for instance, one could investigate empirically the origin of ideas, but this investigation could not bear upon the metaphysical question of the nature of the mind. In this way, much of the ideological production that had characterized the reformist and French periods could be appropriated and reinterpreted.[7]

Some doctrines, however, were considered decidedly erroneous. The active nature of matter, for example: in Restoration philosophy, matter is invariably inert and passive, and theories of life and organization emphasize their transcendence rather than immanence.[8] Also, although human reason can grasp the formal ontological principles of reality, it was deemed unable to penetrate the mysteries of nature. In metaphysics as elsewhere, the human intellect should never cross its "proper limits." As a result, the constitutive principles of reality—essences, forces, causes, and substances—are simply beyond human understanding. Humans may know only the phenomenal properties of beings, not their real essences. Consequently, certain themes, such as the relationship between body and soul, lay outside the field of legitimate empirical investigation. This awareness did not rule out the possibility of formal proofs of the existence of God or the immateriality of the soul, which abound in these texts. Behind them is the assumption that, while human reason cannot penetrate metaphysical and religious truths, it can ascertain them: human reason can *recognize* fundamental truths and deduce other truths from them. Restoration philosophy, in other words, sought to ground the claim that human reason has no autonomous cognitive capacities.[9]

Another feature of this Restoration philosophy is that it linked metaphysics to the practical disciplines of ethics, law, economy, pedagogy, and politics. Every possible practical question had a clear and necessary answer. From the principle of noncontradiction it followed that in ethics, as in physics, contradiction is evil—it "cripples" the spirit. Such principle is "the first line of the natural code, which is necessarily true, certain, constant, immutable, evident, and adequate; from it all duties and rights of men and citizens must be deduced." Morals and politics are necessarily part of the system of laws that rules the universe as an organic and ordered whole. In their treatises of ethics and practical philosophy—at least since Capocasale's *Eternal Code* (1792–1793)—these philosophers derived a series of duties from the particular position of man in the cosmos. The outcome is an ethics characterized by deference to the existing moral and political order, as any deviation from this necessary state of affairs would put man "in contradiction with himself." The practical sphere of human life is thus deprived of autonomy and taken away from the control of the individual. Ethics, law, pedagogy, politics, and economics are mere articulations of a single fundamental body of metaphysically founded norms, and each individual enters society burdened with natural duties.[10]

Restoration philosophy was an endless, elaborate exercise in the self-limitation of individual reason. The "real philosopher," one reads, is "the man who does not investigate inconsiderately with the senses what must be investigated with reason; does not discuss with reason what must be discussed with testimony; he is able to ignore what is not given to know; he knows how to limit his research, to doubt where doubt is reasonable, to accept the known truth, to submit to infallible authority: he is the man who prefers to follow truth with the common people, rather than falsity with the so-called philosophers."[11] Commoners and tradition are portrayed as repositories of an original and intact wisdom. Political events are complex and even mysterious; we ought, one author writes, to "leave our destinies in the hands of He who controls the fate of men."[12] In the same spirit, Capocasale demonstrated the reasonability of political intolerance, grounding it on a paternalistic and authoritarian conception of power.[13] Ultimately there cannot be social order without religious dogma, as "religion is like a strict chain that anywhere connects one man to the others, one citizen to the others, the subject to his prince, and thus makes us respect the law."[14]

Anti-Modern Science

The makers of these Restoration philosophies did not reject science tout court but reacted to a recent perversion of science. The work of an influential champion of reactionary Catholicism, the Oratorian priest Francesco Colangelo (1769–1836), whose dual career as a bishop and as the Minister of Public Education (1824–1831) marked profoundly the Restoration age in Naples, shows this clearly.

Colangelo, who entered the Oratory in 1783, had studied philosophy and theology and was regarded by many as a man of profound learning—others, however, described him as a "detestable prelate," someone who could have done well as a "hangman or an assassin." Colangelo's early literary career illustrates well the transformation of eighteenth-century intransigent Catholicism into forms of reactionary apologetics. His dark reputation was fostered by his account of the Neapolitan Revolution, dedicated to the queen and published while Jacobin prisoners were still being executed. In Colangelo's diagnosis, those "disorders and crimes" were the outcome of Rousseau's "pernicious system," brought to Naples by the likes of Genovesi, Filangieri, and Pagano. They were directly responsible for a revolution that "assaulted the altar, desecrated the throne, upset nature, and massacred humanity."[15]

Colangelo was no ordinary ecclesiastic scholar: he had a penchant for prohibited books, could work in several languages, and published extensively on literature, philosophy, history, and the sciences.[16] In *The Irreligious Freedom of Thought, Enemy of Scientific Progress* (1804), Colangelo reversed the enlightened and Jacobin perspective that the autonomy of the sciences from metaphysics was essential to their progress; rather, he thought, at the core of all good science is the "science of God." Terms like "philosophy" and "wisdom," he wrote, have always referred to the knowledge that moves from the contemplation of nature to the contemplation of the creator, and thus to the understanding of one's duties toward both God and society. This precept was the core of Platonism and, more generally, of ancient wisdom. The science of God has always been the noblest and the most fundamental of all sciences, the others being merely instrumental to its enhancement.[17]

It is therefore ironic, Colangelo continued, that modern philosophers should abuse the term "philosophy" to refer to the "horrible project" of banishing God from the universe and reducing all virtues to pleasure and utility. These fundamental philosophical errors have produced endless violence and destruction, and they are rooted, ultimately, in moral corruption: modern atheist philosophers aimed at breaking free from all natural ties and limitations. Colangelo argued that religious dogma and reason should coincide, given that revelation and nature are simply two different manifestations of God. If they do not coincide, the corrupting effects of self-interest must be at work. "If men had an interest in doing so," Colangelo quips, "they would even doubt Euclid's *Elements*."[18] Colangelo also accused modern philosophy of being disappointingly unoriginal. The moderns were simply rehearsing the arguments that ancient philosophers—skeptics, cynics, and Epicureans—had raised against the divinely inspired systems of Plato and Aristotle. And if Cicero had deemed Lucretius's atomistic system a "childish tale," the same was true of the godless universe of eighteenth-century philosophers. Ultimately, they had not been able to produce a truly new philosophical system, or even a convincing argument to bring God "down from His throne."[19]

Colangelo was a champion of the harmony of science and religion. Christianity, in his view, was a "friend of the sciences" and shared with them its basic principles. After all, that God exists and has absolute dominion over nature had been stated by the likes of Plato, Aristotle, Descartes, Leibniz, and Newton. In fact, there was no philosophical system that "submits the machine of the universe to the power of God in such a perfect way as the Newtonian one." Colangelo had many reasons to praise *Principia Mathematica*, including the fact that it proved the existence of a "non-mechanical force" that pervades the entire universe.[20] Mathematics, he concluded, "do not favor unbelief, as libertines maintain, or as some not-too-wise men fear."[21]

Colangelo described the development of modern science up to Newton as consonant with religion. The mathematical harmony of the Newtonian universe inherently supports religion, as do other modern discoveries, like the nutrition system of plants or the physical constitution of the "animal machines." All these advancements leave the "philosopher-contemplator" in awe of creation. The contemplation of nature involves "the purest and more spiritual part of the intellect." Colangelo described it in mystical terms: philosophers experience a "scientific *ecstasy*" and the "genial enthusiasm of [their] reason," as they "measure the oceans and count the stars."[22] Further, religion actively promotes and defends the sciences, while atheism and skepticism tend to destroy them. The Christian faith prevents philosophers from wasting time with "old and childish speculations," such as materialistic cosmogonies; it instills humility and restrains passions, which obstruct the acquisition of knowledge. Colangelo wrote an entire book to criticize Giambattista Vico's *New Science*, which he read as a Lucretian fantasy that denied the divine origin of language and society. Vico's "portentous fantasy" certainly matched, for Colangelo, the "corporeal fantasy" of his *bestioni*. As for the argument that religion originated from the phenomenon of thunder, Colangelo wondered whether Vico's masterpiece might be, after all, an erudite divertissement.[23]

Religion is also beneficial to the sciences because it counteracts the inclination to ignore experience and produce abstract metaphysical systems. This opposition between empirical, inductive procedures on the one hand and abstract reasoning and the spirit of system on the other is the basis of the natural "friendship" between Christianity and modern science. Not surprisingly, Colangelo considered Francis Bacon a giant of modern science, and many Baconian themes run through his remarks on science.[24] In Colangelo's narrative, the artificial world systems of French philosophes were not only scientifically erroneous and religiously heretical (the two coincide), they were merely ghosts of past philosophies. What happened in the eighteenth century was nothing but a pale reproduction of the original battle between the defenders of a divinely inspired wisdom and the philosophical sects. Nothing new under the sun: just the age-old struggle of pious men against atheists.[25]

Restoration philosophy argued for a few all-encompassing principles as well as for the locality of epistemological criteria. Colangelo acknowledged that there are many heuristic methods and "kinds of certainty"; the important thing is to understand which one is relevant to a given inquiry—that is, its legitimate jurisdiction. In this sense, religion and science are part of two "completely different orders of truths and arguments."[26] The sciences themselves are far from uniform, as shown by the profound divide between mathematics and the rest. Mathematics deals with intellectual intuition and demonstrable truths—hence its superior kind of certainty. But mathematical certainty is not required in all matters, commented Colangelo, echoing Aquinas.[27] Moral and legal issues, for example, do not lend themselves to mathematical study. Much of the present philosophical confusion about mathematics and its legitimate applications, Colangelo argued, derives from confusing the form of mathematics with its content. A proof's essence does not coincide with its form, which is mere "bark covering mathematical truths." Certainty in mathematics derives from the *nature* of mathematical truths, *not from the specific method employed to reach them.* In fact, mathematical methods are designed for specific purposes, and they change over time.[28] During the eighteenth century, many "sophists disguised as geometers" exploited this confusion and, through a misleading apparatus of analytic formulas, introduced mathematics into fields like politics that, due to the nature of their objects, "could not receive it."[29]

Colangelo's fight against the spirit of system was an important component of a position I call "apologetic empiricism." The apologetic interpretation of experimental philosophy was no novelty. In Colangelo and some of his contemporaries, however, it assumed distinctive traits, such as a pressing sense of urgency, an eminently polemical orientation, a focus on a restricted subset of Christian values, and—above all—an understanding of individual reason as invariably weak and dependent.[30] Apologetic empiricism is clearly articulated in Colangelo's 1815 *Galileo as a Guide for the Young Student*, dedicated to Nicola Fergola—"another Galileo." For Colangelo, Galileo was exemplary because he was both a devout Catholic and a sagacious investigator of nature. He showed that it was possible "to be a sublime philosopher without being deceived by the insanity of the so-called free thinkers." In fact, if properly understood, science should teach philosophers modesty and temper their arrogance: few truths can be proved, and most human knowledge is only probable.[31]

Colangelo wondered why the progress of the sciences slowed down so dramatically after the age of Galileo and Newton. Despite a vast array of new machines and mathematical techniques, eighteenth-century philosophers made few fundamental discoveries: many had written wisely on mechanics, he pondered, but could they be called lawgivers like Galileo and Newton? They could not, he concluded. At the root of the problem was their betrayal of empiricism and the fact that "the contemplation of nature has been oppressed by a wealth of analytic formulas that for the great part are extraneous [to it]."[32] Returning to Galileo meant returning to the source of modern science and to the methodological principles that guided it into the "sanctuary of nature," without using the mysterious arcana of analysis. In particular, Colangelo was interested in those principles through which Galileo "regulated [his] mind" during his investigations, thus showing students what it means to have "the rare and sublime prerogative of the contemplator." Galileo practiced a "wise and careful freedom of philosophizing," refusing to take for granted any abstract system of the world. It's the Newtonian *hypotheses non fingo*, the prelude to a strictly empirical study of nature. Galileo's investigation of nature was informed, according to Colangelo, by a few "canons": Do not construct fanciful systems of the universe. Test all opinions empirically, and be ready to abandon your own beliefs if experience or a clear argument proves them false. Be aware that men will never know the ultimate essence of things; they will never fully comprehend the order of the universe. Colangelo also identified more-specific methodological principles, such as the "supposition of the fact," whereby one takes for granted a certain reasonable assumption and verifies whether its implications are consistent with the observations—the empirical version of one of Fergola's favorite geometrical methods.[33]

Colangelo was especially fond of principles that assert the priority of observation and experience over calculation: it is only with great caution that one should impose geometrical rigor on phenomena, he argued. Granted, the great book of nature is written in geometrical language, but Galileo was well aware of the boundary between physics and mathematics. He knew that the "very pure mathematical demonstrations" should not be "contaminated" by nature's imperfections. Therefore he "kept his eyes fixed upon nature" while trying to graft geometry onto it, and "he did this in such a way that makes it almost impossible to discern whether it is the geometer or the observer who is speaking." This way of doing science does not "make violence to nature" but "follows its inclinations." Following nature rather than suffocating it with mathematical formulas also enables rigorous, clear, and elegant arguments. Many recent texts of natural philosophy were obscure, according to Colangelo, precisely because their authors were trying to force natural phenomena into the straitjacket of abstract mathematics.[34]

Colangelo's eulogy of Galileo appeared right after the Bourbon restoration, in 1815, when the *Dialogue* was still on the index of prohibited books—it was removed in 1822. Clearly, the formal condemnation had not prevented apologetic and even openly anti-modern readings of the Galilean legacy, a motif with a long tradition in intransigent Catholicism. Colangelo failed to mention Galileo's trial, suggesting that Galileo had not been properly rewarded in his own days mainly because contemporary peripatetic philosophers opposed his genuinely empirical stance.[35] Colangelo invited his students to keep their faith pure and follow the luminous path of Galileo's empiricism. He was convinced that in science, as in politics, the time had come to return to order: "Political states need, every now and then, to be brought back to their original principles; and the same holds for the sciences in the different nations, given that every product of man carries the fatal mark of fragility and decadence, which is proper to its author." Return to political order had just been achieved, more or less satisfactorily, with the Congress of Vienna. As for science, Colangelo believed that the examples of the past's great natural philosophers should now be used "to call men back to the ancient order."[36]

The Holy Mathematician

Colangelo's *The Irreligious Freedom of Thought* was revised and corrected by Nicola Fergola, who also contributed, as an appendix, a short apologetic text on how the sublime sciences lead to religion.[37] It was 1804, and recent upheavals had made it very clear who was on the side of throne and altar. Colangelo referred to Fergola as an example of religious devotion, a friend, and a maestro. Indeed, many of his arguments were taken verbatim from Fergola's texts, showing how the anti-encyclopedic battle of the erudite clergymen at the Neapolitan Oratory was connected with the most renowned school of mathematics in the city. Why had Fergola become such an important referent for Neapolitan reactionary Catholics?

Let's begin at the end. Fergola died in 1824, "turning his eyes to the beloved image of the Virgin." In a significant exchange of glances, Fergola, almost blind, looked at Mary, whose concerned gaze had long been fixed upon the abominations of the revolutionary age. His devotion to Mary was well known, and until his very last earthly moment Fergola kept together the faith of the learned—Christocentric, focused on the attributes of God's power and the image of the Pantocrator—with the faith of the poor, the marginal, the women, the feared urban crowds, and the recalcitrant people of the highlands who would revolt in the name of a wronged Madonna.[38] Fergola, the famous man of science, did not disdain baroque or sentimental devotions: he knew well the point at which reason should bow before the mysteries of creation.

Fergola's religious trajectory effectively captures the transformation of the Catholic Church and the reconfigurations of its relations to power and the faithful through the revolutionary age. In Italy, the "war against God" waged by Jacobin revolutionaries first and then by the empire had been, as Michael Broers puts it, "an epic struggle."[39] Its outcome was complex, but some features of the new relations between religion and politics are clear. While weakened in its socioeconomic structures, the Catholic Church emerged from the confrontation as a powerful interlocutor of the modern state, one with a firm grip on popular piety; firmer, in fact, than ever before. The eighteenth-century alliance of local episcopates and lay power, which we have seen so clearly in the case of Naples, was a thing of the past. Instead, the Church had forged an alliance with popular piety—whose manifestations it had often fought—showing an impressive capacity to reinvent itself, certainly superior to that of the ancien régime states. In Restoration Italy, France, and Germany, the Catholic Church appropriated and sanitized forms of popular religion, mostly in opposition to the direct heir of the Napoleonic state: the liberal state. The Church controlled a large and disciplined faithful, and it was a force the emerging nations recognized and feared. Gramsci didn't fail to note this crucial misalignment of Church and state at the opening of the nineteenth century, and concluded that its outcome was a fundamental incompatibility of the liberal state and the Church, a problem that would mar the Italian Risorgimento. Free from any dangerous solidarity with legitimism, the Church emerged from the struggle with the upper hand, as a vital source of social control.[40]

Fergola, who came from a relatively affluent urban background, went through a religious transformation that culminated in the crucial decade of the 1790s. Moving from a religiosity that was in no way outstanding, and that he later judged superficial, he became absorbed with that world of southern baroque devotion that had repelled generations of enlightened commentators, including many ecclesiastics. While exemplarily modest and retired, Fergola did not shy away from exhibiting his fondness for practices that Jansenists, enlightened reformers, Jacobins, and French officials had scorned as ignorant and superstitious. His devotion, it was noted, was that of a "simple woman." He venerated relics and holy images, went on pilgrimages to visit sanctuaries dedicated to the Virgin, and was often seen among the humblest people, whom he joined in popular processions and hours-long prayers. At his house, guests could not miss his blood-soaked scourge and cilice, and he made a point of wearing a conspicuous rosary before people whose morality and religiosity he deemed questionable—especially during the republican and French periods.[41]

Emblematic of these popular baroque devotions was the miracle of Saint January. While not officially recognized by the Catholic Church, the miracle of the ancient blood that comes back to life offers—as argued by Francesco de Ceglia—a trove of meanings and a fascinating historical-anthropological object, indissolubly related to the cultural and political history of Naples. In the eighteenth century, the periodical liquefaction of the saint's dried blood had been scorned by local reformers as well as prominent philosophes like Voltaire. Montesquieu had suggested that the blood might function as a sort of thermometer, turning liquid as the external temperature rose. In 1794–1795, as the Jacobin crisis ravaged the Neapolitan cultural and political worlds, Fergola decided to study the question scientifically. For about a year he measured the temperature around the blood as it changed state, reaching the conclusion that the liquefaction takes place under very different physical conditions, and thereby ruling out all natural explanations based on factors like temperature or pressure. Fergola would later integrate this scientific report with familiar anti-skeptical arguments: when dealing with empirical truths, one can only be skeptical up to a point. Otherwise, "all moral certainties would be suspended, and as a result society would be destroyed." Those who look for absolute necessity and mathematical certainty where they cannot be found are a danger to society. Questions and doubts must end at some point for social life to be possible; authority and tradition tell us where that point lies.[42]

Fergola was not engaging with fading remnants of past devotions, and his was not an antiquarian curiosity. The popular religion he encountered was dynamic and highly responsive, as were the provincial communities and urban social groups that supported it. Whether phenomena like the wronged-Madonnas revolts should be considered as straightforwardly counterrevolutionary is open to debate. Some specialists have coined the term "anti-revolution" to capture their essential feature: a resistance to revolutionary social transformation that is not necessarily articulated in an explicitly counterrevolutionary discourse. These acts of resistance can however, be reasonably described as politicized acts. Through them, communities resisted the destruction of ways of living based on kinship, clientage, corporatism, and shared rituals. Popular religion, with its innumerable concrete, sensuous manifestations, was a key component of this political culture. In a similar vein, René Dupuy has read a significant level of politicization in the *Chouannerie*, which he interprets as an attempt to recover lost power; and Donald Sutherland, while emphasizing the inarticulate and unconscious elements of the same movement, sees it as "one of the great moral struggles of the revolutionary era."[43]

Fergola embraced popular religion when it was under pressure, when a spate of crying and bleeding Madonnas signaled the crisis of a peripheral world that would soon explode in the 1799 anarchy. The last decade of the eighteenth century was a time of millenarianism across Italy, both Jacobin and Catholic.[44] In this climate, Fergola took a side. He embraced not simply popular religion, but its most politicized devotions—such as the cult of the Virgin Mary. "Under the French," writes Michael Broers about the Italian situation, "when a Madonna cried, moved, fell over or worked a miracle, French officials and local *giacobini* reached for their guns." Embracing popular religion in 1790s meant embracing it as resistance. In general terms, it constituted a resistance against "the modern state" that originated from the Great Revolution, and against that state's secularized and rational organizing principles. In this sense, Fergola's conspicuous rosary and bloodied scourge were politicized acts of resistance. In this sense, I argue, his mathematics too was a politicized act of resistance.

This is not to say that Fergola's faith was not sincere and intensely experienced. There is plenty of evidence to the contrary. While his school prospered in the aftermath of the 1799 Restoration, Fergola's health was failing, strained by innumerable political and scientific duties. In all these positions he was fighting, anxiously, the degeneration of learning in the kingdom, and the "sacrilegious horde" of Freemasons, Jacobins, and liberal conspirators, whose numbers and influence, he was convinced, kept growing. A member of the powerful Committee for Public Education since 1815, Fergola played a main role in reshaping the kingdom's education system. But even there, among his legitimist colleagues, Fergola reported he had discovered religious hetero-doxy and political subversion.[45]

Meanwhile, he suffered numerous and often inexplicable "organic" and "moral" ailments that progressively hampered his activity. Yet, we are told, he gazed serenely at his sore body as if its flesh was not his own: his entire life, scientific and moral, could be recounted as a triumph of spirit over flesh and matter. Through his letters, though, we glimpse the existential struggle of a tormented man, one exhausted by mysterious "convulsions" and prostrated by horrific demonic visions. His faith was constantly put to the test. On one occasion, while praying in one of his favorite churches, Fergola saw the im-mense abyss of hell opening under his feet and the entire church collapsing into it. There, on the brink, he felt he had been abandoned by God, and even Mary, the powerful advocate of those who are lost or forgotten, seemed un-able to rescue him. Fergola became convinced that he was experiencing a condition known in theology as "mystic hell," whereby the most perfect souls are subjected to a last and extreme purification via the absence of God. This absence causes profound spiritual suffering, of which earlier ascetic practices and material deprivations can only be a pale anticipation. Eventually, such rare form of purification opens the door to beatific vision: the intuitive ap-prehension of God. Fergola lived his final tribulations and suffering as the culminating point of a life that had been entirely devoted to purity.[46]

Fergola's body was transported from his house facing the church of the Oratorians to the church of the Theatines, where he was buried. The funeral was sumptuous; the entire university faculty was in attendance. The Theatine priest Gioacchino Ventura, a renowned and passionate speaker, gave the eu-logy. In his speech, Ventura aimed to defend those "principles of order" that, he believed, were "the only hope for a society threatened with total dissolu-tion." Among these principles was the secret affinity between religion and mathematics, their relations, and the reciprocal advantages of their coopera-tion. While the postlapsarian human condition is characterized by the im-placable conflict between celestial and mundane wisdom, their harmonious integration is proved possible by certain great figures—like Nicola Fergola, "the holy mathematician."[47]

All attempts to secularize science, Ventura argued, result in its corruption, as they strip from it all that is "solid, noble, and advantageous." When it's not "consecrated" by religion, science is lost. Moral corruption weakens reason and "materializes" it, as what corrupts the affections "obscures" reason, and what deprives the heart of its virtues distracts the spirit from science. By contrast, Fergola showed that the religious virtue of purity (*castitas*) is necessary for the acquisition of sound scientific knowledge. After what he liked to call his conversion, Fergola realized that prayer and self-purification were the ultimate way to knowledge, so he took a vow of chastity and "closed his heart to the seduction of the passions." He embraced a life of severe asceticism, characterized by continuous fasting and physical mortification; it was also peculiarly secluded, away from most relationships and from the riots of the "noisy capital." Fergola thus created the conditions for extraordinary scientific achievements; his reason, "almost completely freed from the weight of the senses," was uplifted and ennobled. Thanks to this spiritual and cognitive purification, Ventura remarked, "mathematics began to shine with a new light among us."[48]

Ventura framed his speech within the narrative of the great philosophical conspiracy. The tone was apocalyptic: "In these last days of the world," he thundered, the evil use of science "has proved fatal." If with the fall science had lost man, "now it has lost society." In particular, it was mathematics that took the "most erroneous and fatal path." Its key role in the philosophical plot against throne and altar had gone largely undetected because of its obscurity. But mathematic transgressions "spread error and foment revolt" most effectively. Indeed, "set square and compass became deadly weapons in the hands of impiety and pride"; they "broke any restraints, unchained all passions"; and in this way they "eroded the foundations of religion and order."[49] How could all this have come to pass? Mathematicians like d'Alembert, Condorcet, and Laplace had tried, illegitimately, to turn mathematics into a "*universal science*, the key and the foundation of all human knowledge." Thus they rejected the dogmas of religion simply because such dogmas could not be proved mathematically. All branches of knowledge, even emotions, had been "algebraized." Fergola, by contrast, was aware that there are different "kinds [*ordini*] of truth" and different ways to acquire knowledge. He followed the path of the great philosophers of the seventeenth century and considered human reason simply as an instrument to investigate earthly matters, knowing that one needs a different guide when it comes to heavenly truths. In fact, the more his mathematical knowledge grew, the humbler and more docile he became in religious life.[50]

The point, continued Ventura, is that seventeenth-century mathematics was a truly "intellectual" and "spiritual" science. During the following century,

however, moral corruption turned mathematics into a "material science" practiced by "cold, arrogant, resolute, *algebraists.*" These men's reason had been "degraded and sterilized by atheism"; they had become "geometrical machines,"
satisfied with the mere execution of complicated calculations. If Plato, Leibniz,
and Newton were "great mathematicians," their eighteenth-century epigones
were mere "calculators." And calculators do not recognize God's powerful
hand behind their "cold formulas." All they see is matter, and "behind matter,
they found nothingness"; there was "neither truth left in their minds, nor virtue in their hearts."[51] Ventura was skillfully deploying arguments that were becoming familiar to reactionary authors across Europe: that sciences could be
material or spiritual, that the materialistic century had destroyed the speculative and intellectual sciences, and that mathematics had been key to the diffusion of "error and insurrection" and had eventually turned itself into a new
religion. Ventura concluded his speech by exhorting the professors assembled
in the church to continue fighting those who "bring anarchy and disorder"
into science and into the life of the states. His words echoed Fergola's own
belief that synthetic mathematics produced evidence, certainty, and "political
tranquility."[52]

No doubt Fergola offered powerful intellectual resources to the champions of reaction. Especially influential were his lectures on natural philosophy, published between 1792 and 1793 as *Lectures on the Mathematical Principles of the Natural Philosophy of Sir Isaac Newton.* This book—a survey of
eighteenth-century mechanics and machine theory—was praised for its clarity and substantive historical commentaries. Fergola introduced each question through its historical genesis, and he accurately presented and weighed
all contributions in the style of the influential Geneva edition of Newton's
Principia (1739). The starting point was usually a passage from the *Principia,* followed by references to relevant research up to Lagrange's *Mechanics*
(1788).[53]

Among the *Lectures's* distinctive features were its detailed historical apparatus and the geometrical form of its demonstrations. As for the contents, the
key notion in Fergola's mechanics is "force." In his 1789 essay on mechanics,
Lauberg had rejected the study of the nature of forces as metaphysical and
therefore meaningless: he was interested only in their measurable effects. Fergola, by contrast, wrote of "the aggregation of forces" as "the soul that informs
the immense mass of the universe, and gives life to it." To him the study of
forces "is not a vain or despicable effort, but truly a way to essay the laws of
the Universe, and the deep wisdom of He who rules and sustains it." Granted,
philosophers cannot penetrate the intimate nature of the forces that pervade
the universe, but they can glimpse enough of their reciprocal relationships

to work out the basic principles of mechanics. The real essence of forces and the modes of their transmission and composition will always remain mysterious. Yet we see enough to know that at the end of the chain of "moving powers"—such as muscles, elasticity, and gravitation—there can only be "the Hand of the Living God," the ultimate moving power. Through the study of mechanics, Fergola glimpsed God as the free agent who gave the universe its contingent laws, choosing those laws that would maintain natural motions with the least action.[54] Toward the end of the book, Fergola presented his four "canons" to guide the geometer in the contemplation of nature. First, geometry and analysis have to be the handmaids of nature, not their masters. Second, "the contemplation of Nature must not be oppressed by a set of analytic formulas that are mostly extraneous to Nature itself." Third, final causes in physics must be restored consistently with Fergola's providential view of the universe. The fourth and final recommendation is that geometry be blended with observation and induction in the search for natural laws; Fergola pointed to the second book of Newton's *Principia* as a luminous example of this process.[55]

In 1804 Colangelo asked Fergola to join him in the literary battle against impiety and offer the public a scientific proof of the existence of God. Fergola obliged with a letter in which he rehearses arguments from natural theology.[56] He then expanded these reflections into a peculiar manuscript published posthumously as *Theory of Miracles*, which also included materials Fergola had promised to Ventura for his *Ecclesiastical Encyclopedia*. For Fergola, miracles are phenomena that cannot be explained naturally: they support the revelation and are therefore essential to the Christian faith. Human reason will never be able to grasp how miracles work, our knowledge of nature being only "slightly superior to absolute ignorance." As shown by the mystery of the gravitational force, the real workings of nature are inscrutable to human reason. In fact, to the wise contemplator, nature in its entirety is "a continuous miracle."[57] Miracles break not a logical but a physical necessity, which is the result of our ability to make inductive predictions about natural phenomena. In other words, all natural truths are contingent. This contingency is even more obvious in the case of the knowledge of society. In fields like politics, knowledge cannot be acquired through the application of logic and mathematics, but through the study of history and local traditions and the acceptance of the principle of authority, which—as Restoration philosophy would make abundantly clear—can be justified on metaphysical grounds. A teleological perspective and a few basic metaphysical principles frame the entirety of human knowledge. At the same time, knowledge is epistemologically stratified and hierarchically structured—it is *ordered*. In fact, and this was a common assumption in continental rationalism, the *ordo cognoscendi* maps onto the *ordo essendi*, which means that there is only one fundamental and universal order.[58]

Well before the official rediscovery of Thomism and scholastic philosophy by Catholic theologians, Fergola was clearly appropriating and mobilizing components of Aquinas's epistemology. Not that he was interested in an overall revival of Aquinas's rationalistic perspective. Fergola was fiercely antimechanist and believed that not all variations in the forces are reducible to some kind of contact. To him, the universe was pervaded by innumerable "hidden forces," from gravitational forces in the solar system to "muscular forces" in animals to electrical and magnetic forces. He was also convinced that the overall quantity of force in the universe was decreasing, meaning "nature cannot sustain itself," and therefore that the stability of the universe implied a continuous infusion of new forces *ex nihilo*. In this respect, he sided with Newton's theological voluntarism against Wolff's rationalism. Fergola was especially fond of Newton's definition of God as *imperator universalis*: "The Lord rules the heavens and nature like a sovereign. He didn't write the destinies of things like constitutional laws of the universe, to which He Himself is subject."[59] The nature of God's dominion was of the essence to reactionary Catholics. Ventura, in his eulogy of Pius VII, attacked the "erroneous philosophy" according to which kings—through constitutions—are subject to the will of the multitude and thus become "temporary representatives of the people." Similarly, he believed that this "constitutional philosophy" was being used to reduce the power of God and banish him from the universe. It is not coincidental that Fergola chose the topic of miracles to answer Colangelo's call to support with science the political restoration of the Bourbons. Fergola's mechanics supported the claim that God's power is absolute, and if God himself did not rule the universe "as a constitutional king," then the restored Bourbon king could legitimately abolish—as he did—the 1812 constitution of Sicily.[60]

Fergola's theory of knowledge shared much of its conceptual scaffolding with Restoration philosophy. He was receptive, in particular, to Christian Wolff's appropriation of Thomistic epistemology and to his enthusiasm for the methods of the new mathematical and physical sciences. Wolff's remarkable success in Catholic Europe can be understood, above all, in structural terms: the Protestant philosopher provided what was still essentially a scholastic framework for a capacious encyclopedia within which both faith and reason were recognized as fully legitimated and whose jurisdictions extended over neatly separated domains. Thus the system afforded the possibility of up-to-date versions of the Catholic conciliation of reason and revelation.

Catholic theologians and natural philosophers were swift in using its potential. The most influential reinterpretations were crafted in Vienna, in the
mid-eighteenth century, by Jesuit mathematicians and natural philosophers
gathered around Roger Boscovich, who in 1758 had authored an original synthesis of Newton's physics and Leibniz's theory of monads. Fergola certainly
knew Wolff through his maestro Genovesi, who in his early Latin works had
deployed the philosopher's geometrical style and referred to his renewed ontology. More important, though, was Fergola's reading of these Jesuit scholars,
mainly the logical and metaphysical compendia by Paul Mako and Sigismund
Storchenau. Fergola's was a Wolffianism that had reached Naples by way of
Vienna, so to speak.

   This Jesuit-Wolffian tradition played a key role in shaping mid- and late-
eighteenth-century scientific culture in Catholic Europe and deserves more
attention than it has received so far. For our present purposes I'll simply note
that cosmology has a privileged place in these compendia, and that logical
and ontological questions relative to the principle of sufficient reason are
prominent and frame the entire system.[61] Also, compared to Wolff's works,
these Catholic compendia strengthen the apologetic dimension and expand
the sections on metaphysics and pneumatology. Interestingly, the traditional
theoretical demonstrations of the immortality of the soul are followed by
other ethical and teleological arguments, such as the need to preserve society, the infinite wisdom of God, and the appeal to "common sense," a notion
derived from classical antiquity but also from a contemporary ethnographic
literature that would have been familiar to members of the Society of Jesus.
Concerning the relationship between soul and body, pre-established harmony and occasionalism are rejected in favor of a real influence (*influxus
physicus*) of the spiritual substance over the physical body. While influxionism had a long tradition, this version is distinctive of late forms of Catholic
Enlightenment, and its distance from scholastic hylomorphism is obvious: it's
a modern, Cartesian borrowing (note that the brain is described as the site
of the soul) that introduces a rigid dualism into the system while opening up
the possibility of integrating into it recent physiological theories and Locke's
sensationalism and theory of ideas.[62]

   From this literature Fergola derived a stratified epistemology according to
which knowledge can be acquired through faith and tradition, the faculties of
reason, and the senses. Correspondingly, the ordered hierarchical structure
of knowledge moves downward from theology to metaphysics, mathematics,
natural sciences, and the various articulations of practical philosophy. Each
region has its own ontology and its own epistemic criteria, and to it pertains
a specific kind of attainable certainty. In theology, for example, the ultimate

source of knowledge is the authority of the Church; reason and senses can be used to argue for the existence of God and other theological truths, but they cannot go far without the support of faith and authority. Instead, the fundamental metaphysical axioms and mathematics are the area in which human reason can reach its highest intellectual achievements. The intellect is the highest cognitive power of the soul, and its nature is spiritual, meaning that it's not intrinsically dependent on a bodily organ as sensation is. In mathematics the intellect can be deployed at its purest.[63]

If faith is "much nobler" than intellect, the achievements of the latter are immensely superior to those of the senses. The most spiritual faculty of human reason works at its best when dealing with pure metaphysical or geometrical entities. But when it comes to empirical matters, reason must deal with objects whose components are indefinite and whose properties are largely unknowable. No wonder geometrical reasoning is inadequate to deal with these objects; other forms of reasoning, such as induction or probabilistic considerations, should be adopted instead. Different degrees of intellectual purity translate into different degrees of certainty. The physical sciences are partially mathematizable as their objects are relatively simple and abstract; the moral sciences are much less so; and the science of society is completely nonmathematical.

For Fergola, therefore, geometry falls within the reach of pure intellect—at least the human version of it. He defined the art of geometrical discovery as "the most beautiful intellectual virtue," one through which the evidence of formal relations forces human reason to assent to geometrical statements. Geometrical ideas are the paradigm of evidence and clarity: one *sees* them, and this knowledge is not mediated by any symbolic language. The acquisition of geometrical knowledge is invariably associated with visual metaphors to signify its immediate intuitive nature. Solving a problem is seeing how to analyze it, how to reduce it to simpler geometrical constructions. Proving a theorem is seeing that certain relations among the figures hold. The resulting knowledge is an immediate apprehension of reality in its concrete existence. The intellect's abstracting capacity removes the outer envelope of the concrete to reach the core of reality. The intuitive evidence of geometrical truths is what makes them essentially different from the outcomes of algebraic *reasoning*. While praising the power of algebra and calculus as heuristic methods, Fergola insisted that their discoveries need always be confirmed by geometrical intuition.[64]

Fergola's epistemology and foundational quest were thus grounded on a specific image of human reason, one assembled out of the scholastic tradition and modern philosophy. While recognizing that the intellect can be an active agent, as when it directs its attention to or extracts intelligible forms from concrete objects, Fergola emphasized its essentially passive nature. Intellectual cognitive processes involve, above all, contemplation, mirroring, abstraction, and recognition. Reason discovers geometrical truths in the sense that it recognizes them intuitively. The wealth of visual metaphors only emphasizes this passive condition. By contrast, when calculating the values of a certain algebraic function, reason is not mirroring anything: algebra is artificial, a human creation. This artificiality is both the root of its power and its essential weakness. Its foundation does not lie in the intuitive apprehension of a real form, but in reason itself. But how could reason create true knowledge out of nothing? Ventura, in his eulogy, summed this critique up effectively: Fergola saw God behind triangles and circles, whereas the arrogant algebraists saw nothing behind their formulas. Fergola's geometrical knowledge was grounded on vision, intuition, and the recognition of the divine order of the cosmos. The algebraic knowledge of the analytics was an artificial creation of their own minds, it was blind, and it was *imaginary*.

Fergola was wary of imagination. In synthetic narratives imagination is usually diseased, rampant, unbridled, and distorted by passions. The good mathematician has a robust intellect and a controlled imagination. The problem with the faculty of imagination, which has both retentive and creative functions, is that it's already too distant from the spiritual intellect and too close to the material and the corruptible. It's not just distant from but heterogeneous to the intellect. Images and imaginary activity are essential components of intellectual knowledge; yet everything that is intellectual transcends them. Or, to put it differently, the imagination is necessary for the construction of the intellectual (and therefore ontological) order, but it does not—*and cannot*—constitute that order by itself.

Purification in this context is a distancing and ordering strategy, one that separates the spiritual intellect, capable of recognizing and enforcing a transcendent order, from the lower functions of human reason, which are entangled with imagination, senses, and matter, and which need to be controlled and guided. Fergola had a dramatic view of the present human condition, characterized by the conflict of spiritual and material forces on the battleground of human reason. A healthy reason is one wherein the two components cooperate and respect their reciprocal limits, but this equilibrium had—he felt—been broken. Some modern philosophers had materialized reason, reducing it to its inferior, sensuous components, and concealing the fundamental truth that the substratum of all faculties is the spiritual soul, the thinking substance.[65] Through purification the intellect recognizes most clearly a transcendent order in the universe, a system of laws imposed by a

FIGURE 7. Cristoforo Russo and Francesco Malerba, *The Triumph of Faith*, first half of the eighteenth century. Oil on canvas, 1800 × 850 cm. Courtesy of Biblioteca e Complesso Monumentale dei Girolamini, Naples.

superior will that human reason must accept dogmatically, one based on an internal sentiment of subjection and obligation and on the awareness that not doing so would compromise "the order."[66]

If the structure of knowledge that framed the Jacobin discourse of science and politics can be illustrated by Cestari's epistemological-classificatory table, the most powerful representation of the reactionary Catholic theory of knowledge is—quite appropriately—nondiscursive: it's an image. I saw it as I raised my eyes while working in the reading room of the Neapolitan Oratory's library. It's a large rococo painting dating to the first half of the eighteenth century, when the library was restored and began hosting anti-encyclopedic activities. At first sight, it's just another triumph of faith, modeled after a famous fresco by Francesco Solimena in the church of San Domenico Maggiore. At a closer inspection, however, it becomes clear that this image is not a triumph of faith against heresy but of faith over scientific error.

Surrounded by the portraits of illustrious Oratorian scholars is an oval whose content is organized through a Rococo spiral movement. Let us follow it, starting from the top right. Behind a cloud is the throne of the unknowable, invisible God. Angels testify to his presence: we know of him through authority and tradition—not through reason. Below, on the left, faith is represented as a woman with a severe expression; on her left shoulder is a wooden cross, in her right hand a golden chalice. She is stretching her left arm down toward a group of contemplating figures, and her open hand releases crowns of roses and anemones to them as prizes. The figures, a group of seven women, are making offering gestures, and from their six visible symbols it is clear that they are the sciences (golden circle, golden lamp, mirror) and the virtues (silver shield, water, scale). In the lower part of the oval are four colorful figures, the continents, signifying creation. In the midst of them, a blonde angel strikes some giants—the erroneous doctrines—with a scourge, sending them beyond an arch, toward a dark cave. One of the giants is trying not to fall by leaning on a book, but the book sticks out in the void: it will certainly take the giant down. Therefore: faith receives its truth from God himself; in turn, it enlightens the sciences and the virtues, which have no value otherwise. The whole world receives civilization and progress from the sciences when they are joined to the virtues. Truth—in the form of an angel of God—attacks erroneous doctrines and puts them to flight. They are illusory, without foundations.[67]

Apologetic Empiricism

Fergola and his students were not the only prominent men of science who supported the scientific counterrevolution. Many academic philosophers published textbooks of Newtonian physics between the 1790s and the 1830s, although their texts were simply rehearsing Fergola's with less technical competence. Leaving them aside, let's focus instead on two prominent scholars who were known well beyond Naples and are not normally associated with reactionary politics. I argue that their works were informed by priorities similar to Fergola's and can be understood as sophisticated versions of apologetic empiricism.

The natural philosopher Giuseppe Saverio Poli (1746–1825) had studied medicine at Padua and practiced in his native town of Molfetta before devoting himself to academic life. He taught medicine and experimental physics at the University of Naples, became the president of the Military Academy, and was a founding member of the Royal Academy of Sciences. Poli enjoyed the trust and patronage of King Ferdinand and his son—the future Francesco I—whom he tutored. He received lavish support for his European travels and scientific instruments and was fascinated, above all, by magnetic and electrical phenomena. Particularly significant were his contacts with British and French physicist-electricians, and with Benjamin Franklin, whose theory of electricity he favored. He was nominated a fellow of the Royal Society of London (1779), and joined numerous other prominent scientific academies.

He contributed to the foundation of the Botanical Garden of Naples, and assembled impressive natural history collections, which included items he had bought directly from James Cook.[68]

Poli's main publication is a study of Mediterranean mollusks: two magnificent folio volumes enriched with more than sixty hand-colored copper plates (1791–1795), the outcome of a massive collective effort that involved dozens of collaborators—including fishermen, anatomists, and artists. Credited with a number of discoveries, Poli presented the first complete classification of this branch of zoology, articulating and modifying Linnaeus's scheme. Through his meticulous observations and new histological methods, he was able to describe the morphology and function of these animals' organs. He also shifted the description and classification of mollusks from one based on the superficial features of their shells to one based on their anatomy and physiology. His research project, however, came to a sudden halt during the 1799 revolution, and then again in 1806 as Poli abandoned the capital twice to follow the royal family into Sicilian exile. The publication of the third volume was suspended indefinitely, and Poli's naturalistic collections and scientific instruments were plundered or, as in the case of a set of finely crafted wax models of mollusks' internal organs, seized by the French army.[69]

An experienced teacher, in 1781 Poli had published a textbook of experimental physics that would be widely adopted across the kingdom and in other Italian states: by 1825 it counted more than twenty editions and reprints. The book was organized as a series of lectures around topics like matter, motion, cosmography, mechanics, hydraulics, air, gases, sound, water, caloric, and light. More than three hundred pages were devoted to electricity, including the phenomena of magnetism and Galvanism. For each phenomenon, Poli offered a brief account of competing theoretical explanations, weighed their respective strengths and weaknesses, and described the experiments that revealed its properties.[70] Poli reconstructed the scientific debate from the Nollet-Franklin-Symmer controversy to the Galvani-Volta controversy over animal electricity. He favored Franklin's theory of electricity on experimental grounds; and, in general, he cautiously defended a view according to which the electrical fluid produced by electrical machines was a phenomenon analogous to magnetic and Galvanic fluids. While dismissing Mesmer as a "charlatan," he defended the use of electrical machines to treat pathologies related to the difficult circulation of "internal bodily fluids." Poli replicated Galvani's experiments on frogs, and in the textbook he offers a detailed description of each one, including the provenance and price of the necessary equipment.[71]

Poli's scholarly persona was that of an enthusiastic and skilled experimenter, immersed in observation and Baconian-style data collection, reluctant to make any theoretical commitments, and with no interest whatsoever in the turbulent political life that surrounded him. And yet Poli's empirical approach was complementary to Fergola's view of mathematics, and its apologetic dimension is similarly pervasive. His discoveries in mollusk anatomy, for example, depended essentially on his unshakable belief that nature ascended continuously from the simplest forms of life to the most complex: the result of the constant and uniform action of the creator. Hence he searched for anatomical and physiological continuities between species and was able to identify intermediate organs and functions that had until then escaped detection. In a celebratory engraving that illustrates Poli's teleological view of nature, against the background of the Gulf of Naples, a pyramid rises from a terrain covered with mollusks, and reaches for the clouds—we know who's behind them. On the ground are emblems that remind us of Poli's studies: a book, a globe, a parchment, a scythe, shells, vases, and ancient coins (he had assembled a formidable numismatic collection). Allegorical figures—Atlas and Janus—stand for human empirical knowledge. The distich under Poli's bust states that he discovered many hidden natural phenomena that clearly point at the creator. The study of mollusks was the basis of a teleological pyramid of knowledge, within which all his diverse interests found their ultimate meaning.[72]

Poli's electrical studies and physics lectures contained explicit arguments from natural theology, as when he celebrated "the prodigious simplicity and economy through which the general system of the universe is ruled," and the many sure signs of "an infinite and inscrutable Wisdom." More subtly, he told students that true wisdom consists of a balanced mixture of knowledge and ignorance: being aware of the *limits* of human reason is the first step toward sound science. This awareness requires one to stick to a strictly phenomenal approach and to avoid at-best useless metaphysical debates. Consistent with this approach, Poli constantly drew boundaries between the certain and the hypothetical, emphasizing that most natural questions are surrounded by "the darkness of uncertainty."[73] At the basis of his metaphysical modesty was the belief that sound empirical science should aim to observe, discover, and classify the properties of phenomena, without trying to explain their alleged true essences. Natural philosophers should stick to facts: understanding ultimate components and essential properties is, quite simply, beyond the capacities of the human intellect. As often noted by Fergola, philosophers are completely ignorant about, say, the real nature of forces in physics; and those who try to unveil it cannot but produce imaginary claims and deceptive systems.[74] Still, contemplators of nature can recognize that the universe is a unitary, organic, and harmonic structure, a "chain of created beings depending on each other." They can also detect various "affinities and attractions" acting in the universe and emanating from "a sovereign wisdom."[75] Poli described the creator as a "free God," the only true agent in the universe, who acts on an invariably inert and passive matter. To him, this statement derived from experience: it is part of our "concrete" knowledge of the world—as opposed to abstract mathematical knowledge. Poli admonished his students to stick to "true, clear, indubitable facts," "leave aside the abstract dimension," and not be seduced by flights of the imagination.[76]

Poli constantly opposed two constellations of terms: "individual imagination," "abstraction," "mathematics," and "fiction" against "common sense," "experience," "physics," "facts," and "trustworthy descriptions." Strikingly, his celebrated textbook of experimental physics does not contain a single mathematical formula. This unusual choice, which his admirers hailed as "revolutionary," served a pedagogical function: the book was addressed to all students, including those studying medicine. But it was, above all, a reflection of Poli's conviction that algebraic formulas tend to hide nature's behavior rather than reveal it.[77] The universe that Poli explored and described was very much the universe of Fergola's mechanics: one pervaded and animated by mysterious, incalculable forces whose workings cannot be submitted to the yoke of analysis without being misunderstood. No wonder that, in physics, Poli favored the study of electricity and magnetism, which he saw as essentially irreducible to mathematical formalism. Like Fergola, in the 1780s and 1790s Poli was fighting against the rapid expansion of the empire of analysis. And with Fergola he shared a boundary-drawing strategy that emphasized the limits of individual reason and the ontological and epistemological stratification of reality. Poli's empiricism was apologetic not simply because it turned easily into natural theology but because it sustained an image of human reason as essentially limited and nonautonomous.

To Restoration philosophers, Poli was the exemplary naturalist and physicist. Their renderings of his life are strikingly similar to the idealized reconstructions of the lives of synthetic mathematicians. Poli, we are told, joined the study of nature with a fervent study of theology, convinced that only their conjunction could reveal the real design of creation, and the "destiny of man." He had shown an "uncommon docility" toward his superiors since he was a child; his mind was "serious" and "quiet" rather than "fast," "brilliant," or "impetuous." Modest and frugal, in full control of his imagination and his passions, Poli had resisted the flatteries of Naples as a young man; in fact, like Fergola and Capocasale, he remained celibate—"in peace." He also disdained honors and wealth and was among the few who did not become "dishonorable accomplices of the secular corruption." Poli, we are told, fled the noise and disorder of the capital, finding solace only near his king and amid his collections, his instruments, and his "scientific ecstasies."[78]

In reality, Poli had important institutional responsibilities, including the vice-presidency of the committee that advised the king on matters of government and legislation. Like Fergola, he played a major role in redesigning public education after the restorations of 1799 and 1815. In his comprehensive reform for the University of Naples, approved in 1805 but blocked by the French occupation, the didactic backbone ("primary chairs") of the new university was formed by three theological chairs, two chairs of law (one of feudal law), three of medicine (anatomy, Hippocratic medicine, and practical medicine), one of experimental physics, and one of mathematics. He advised that the contents of all lectures be published in advance and approved by the university prefect. He also advised that the Jesuits, whom the king had recalled in 1804, should guarantee that "the good doctrines" were taught in royal colleges across the kingdom.[79]

Poli abandoned his customarily detached, objective tone only in his poetry. His occasional verses mostly mark important events in the life of the royal family and sometimes introduce scientific concepts and experiences for the amusement and edification of his public. Poli's most ambitious poem celebrated the divine harmony of the universe and the restoration of the Bourbons on the throne of Naples after the repression of the 1799 Republic. The poet, guided by Urania, muse of astronomy, explores the heavens in a physical and allegorical ascension toward the creator. Poli slipped much astronomical knowledge into his verses, but ultimately the journey humbles human reason. Astronomers, in their attempt to comprehend the universe, keep stumbling into unsurpassable obstacles, such as the limits of their finest instruments and of their "most admirable calculations," which cannot help them to make sense of the immense distance between earth and the stars. Thus the poem celebrated the admirable but incalculable order of the universe, as well as the recently restored King Ferdinand, "great and clement monarch."[80]

*

We can conclude our considerations on apologetic empiricism with a brief survey of the publications and scholarly persona of the most celebrated physician in late-eighteenth-century Naples, Domenico Cotugno (1736–1822). Cotugno enjoyed the trust of the royal family and would become the chancellor of the University of Naples as well as the president of the Academy of Sciences. He too contributed to reshaping public education in Naples after the two restorations of 1799 and 1815. Like Poli, he was a talented anatomist, and among his works was a 1761 study in the physiology of the auditory organs:

Cotugno offered a detailed description of the vestibule, semicircular canals, and cochlea of the osseous labyrinth of the inner ear, and argued that the labyrinth was filled with fluid, contrary to the predominant assumption that it was filled with air, as fluid was considered incapable of transmitting sound waves. Cotugno's morphological observations thus suggested that the phenomena of resonance and hearing had not then yet been correctly described.[81]

Cotugno had little patience for philosophical controversies: medicine needs facts, he protested, not words. A sworn enemy of systems and metaphysical assumptions, he argued that medical knowledge can be grounded only on observation and experience. In 1772, in a lecture "on the spirit of medicine" at a large Neapolitan hospital, Cotugno described medicine not as a science but as an art, and a very difficult one. Medicine consists of "mere empirical knowledge," which can be acquired only by observing and following nature with the utmost attention. Nature should be contemplated rather than interpreted, "known" rather than "understood," and physicians should approach it with "pure eyes" and without prejudice. If, as Cotugno argued, the goal of medicine is to treat diseases rather than understand the infinitely complex human machine, then medical practice is best guided by the heuristics of ancient medicine, based on intuition, analogy, and trial and error. Physicians should never think they are pursuing ultimate truth, nor can they extend the application of their therapies based on reasoning alone, as "nature is so free and masterful" that each one of its endless productions must be studied on its own terms. Generalization and systematization are always highly problematic.[82]

Cotugno believed that, in recent times, the age-old practice of medicine had been perturbed by reason's "despotic rule." The self-declared priests of reason had brought causal explanations and theoretical disputes into medicine, turning it into an obscure and useless knowledge. Against the grain of this "unhappy epoch," Cotugno mobilized Genovesi's phenomenal stance to argue for a return to a Hippocratic medicine that valued diligent observation over philosophical subtleties. Certainly reason could play an important role in perfecting medical cognitions, but "it should know its own limits." Reason can "examine, compare, and calculate physical effects," but it cannot take an observer to the study of first causes. The "spirit" of Cotugno's medicine is therefore that of matter-of-fact knowledge based on anatomy and oriented toward the well-being of the patient.[83]

Medical research, like any other scientific investigation, requires a specific spiritual preparation, Cotugno argued. In his rigidly dualist view, the mind is a celestial element imprisoned in the human machine, like a light immersed in fog. In those conditions, it can reason properly only with great difficulty, and only if able to focus its attention. The capacity of attention can be strengthened through meditation, which Cotugno recommended constantly to his students. Meditation strengthens the intellect by increasing self-control, and its effects on the brain are material as well, as it stimulates those brain fibers that enlighten the ideas stored in the brain. By contrast, when the mind is not properly controlled, passions and imagination can overwhelm the intellect, with nefarious consequences. The mind can become the slave of ambition and pride, failing to recognize the existence and attributes of the author of the universe. Cotugno was convinced that humanity had reached such a "state of depravation" that the body, created to obey, contrasted instead the rule of the spirit, often reducing it into a "shaming slavery." Medicine, however, could help to restore the dominion of the spiritual component by fostering meditation and self-control. Education, the cornerstone of a stable and well-ordered society, needed to integrate morals, religion, and science: "if they do not proceed together, everything is doomed."[84]

Cotugno's concern for the disorder of both mind and society helps us thus to understand his keen interest in the study of brain activity, sensory organs, and the nervous system. After numerous anatomical observations, he concluded that the blood flow to the brain diminishes during the state of profound absorption that he called meditation, as does the excitability of semifluid cerebral fibers, whose vibrations become weaker. This physiological state corresponds, in his view, to the superior spiritual component's full control of the brain.[85]

While most of Cotugno's scientific contributions predate the political turmoil of the 1790s, the role of the "Neapolitan Hippocrates" in shaping scientific culture continued to be significant well into the Restoration age. The hegemony of his anti-theoretical medicine contained the diffusion the theory of excitability promoted by the Scottish physician John Brown (1735–1788), whose scientific-mathematical nature and revolutionary potential were much appreciated in Jacobin circles.[86] Always a devout man respectful of authority "and public order," in his later years Cotugno gave his fellow citizens a most edifying spectacle. Retired and frugal, he offered his services to the poor and toured daily the churches of the city, practicing penitence and long hours of contemplation of the celestial mysteries. It was during one of these meditation sessions in a church that, in 1818, he suffered what is described as an epileptic seizure. As his health declined rapidly, Cotugno—like Fergola—showed a supreme detachment from his tormented body and faced his destiny with tranquility. He behaved like someone who "had never belonged to the Earth," thus reinforcing what many considered his most important lesson: that one can be an excellent physician while preserving "the most austere probity."[87]

Cotugno's work and scholarly persona can be interestingly contrasted with those of Domenico Cirillo, a key figure of the Neapolitan Enlightenment and the 1799 Revolution. A reformist and a Freemason, Cirillo was the teacher of both Pagano and Filangieri, and a correspondent of Voltaire, d'Alembert, Diderot, and Rousseau. He introduced the works of Linnaeus and Lavoisier in Naples and studied plant physiology through the conceptual framework of Lavoisier's chemistry. He claimed that it was possible to ground morals upon human physiology experimentally and was an outspoken advocate of the expulsion of religion from the discourses of science and politics. Cotugno and Cirillo both studied in 1750s Naples with Genovesi, began teaching at the university in the 1760s, and worked at the main hospital of the city. They were renowned for their experimental ability in the fields of, respectively, human anatomy and biology. While Cirillo's last public duty as a member of the Legislative Committee of the Neapolitan Republic cost him his life, Cotugno showed no interest in politics. A recent biographer notes that he "remained always extraneous to active politics, and he professed a neutral and non-philosophical conception of scientific knowledge."[88]

Cotugno's work is often described as contiguous with that of his colleague Cirillo—the fact that the latter ended up hanged as a Jacobin in 1799 while Cotugno was in Palermo with the royal family and died a wealthy man in 1822 being a mere contingency. Earlier reconstructions have indeed grouped these two physicians together under capacious categories such as "neonaturalism"—which can also accommodate Fergola's perspective as well as that of Mario Pagano—or the "Catholic Enlightenment." What I'm emphasizing, instead, is a rupture that runs across Neapolitan science, a rupture that is at once epistemological and political. Rather than engaging with the misleading question of whether science in late-eighteenth-century Naples was declining or not, I have highlighted how different forms of scientific life emerged that could be mobilized to support concrete political action. Men of science like Fergola, Cotugno, and Poli were not enlightened Catholics as they did not champion a renewal of Catholicism, whether liturgical or cultural; their devotion was fiercely baroque, popular, and Jesuitical—that is to say, at odds with the religious sensibility of contemporary Catholic reform movements. Their defense of noncommittal empiricism was also profoundly apologetic and, in fact, could easily coexist with the restoration of an up-to-date scholastic metaphysics. Similarly, labeling them as neonaturalists hides more than it reveals; it hides, in particular, their distinctive and urgent efforts to craft the image of a nonautonomous human reason: a reactionary reason.[89]

A Reactionary Reason

We have followed the thread of an emerging reactionary reason through politics, technology, art, mathematics, the natural sciences, and medicine. It is now time to focus on reactionary reason itself, and therefore on those who articulated and mobilized it most explicitly. I'm interested, in particular, in arguments against the autonomy of individual reason, and in their trajectories from theology to the sphere of the technical. This chapter characterizes these arguments as reactions against the impious century and as attempts to restore an allegedly lost political and cognitive order. In this perspective, reactionary anti-modernism did not translate necessarily into anti-science or anti-reason positions. Rather, many reactionary authors aimed to *redefine* reason, its operations, and its legitimate scope, rejecting the Enlightenment's and French Revolution's celebration of the rational independent subject. The moral and political errors that brought upon the world ruinous doctrines and social unrest, they argued, originated in a misunderstanding of the workings of reason, and in the mistaken assumption that individual reason could operate autonomously.[1]

We begin with the critique of the autonomy of reason within the intransigent Catholic literature of the second half of the eighteenth century. I refer to groups and publications from Rome and the Italian peninsula whose main target was the new secularized philosophy and its emblematic production: the *Encyclopédie*. The organization of explicitly anti-modern theological and philosophical academies in Naples began in the 1740s, with campaigns that targeted the new philosophy as well as cultural and political reforms. By the 1790s, reactionary Catholicism had condensed these earlier experiences into a repertoire of a few simplified and obsessively repeated themes, among them the inherent passivity of human reason and the necessity that reason operate in harmony with authority and tradition.

Pre-revolutionary expressions of intransigent Catholicism were typically associated with ecclesiastical circles and institutions. That state of affairs changed dramatically after 1789, when legitimist governments reoriented their cultural policies toward explicitly reactionary aims. The fortune of French traditionalism derives precisely from its capacity to mobilize arguments from previous theological critiques to address urgent social and political problems. Traditionalists built on the historical apologetics of the late eighteenth century and its attempt to ground religious truths in historical facts—we have seen an example of this approach in Colangelo's work. On this basis, traditionalists elaborated theories of knowledge that historicized moral and political truths. The authors of the aristocratic reaction bolstered traditionalism while crafting new languages that proved popular among new and larger publics. By the mid-nineteenth century, Neo-Scholasticism was grounding a comprehensive anti-modern perspective on a sound philosophical foundation, in alleged continuity with the rationalist tradition of medieval scholasticism.

As diverse as they were, all these movements reacted against a common enemy, the subordinate's exercise of agency. They considered agency the prerogative of traditional elites, and challenges to this order were invariably cast as the monstrous fruits of secularization; only the reinstatement of the principle of authority could reverse them. In religious matters, this stance affirmed the Church as the repository of absolute truth, with God-given authority that must not be challenged. Within the Church, the authority of the pontiff was supreme and ultimate. Hence traditionalists emphasized the institutional, hierarchical, and missionary dimensions of the religious experience, while largely ignoring more individual dimensions such as faith by grace.

In political matters, they argued for the unity and absolute autonomy of sovereignty: any constitutional concession would necessarily result in political chaos. They believed the modern state to be essentially unstable because it is founded on contingent, transient interests. What these reactionaries strove for was a political order grounded upon the values of tradition: the transcendental and immutable values of revealed religion and its historical incarnations, be they institutions, language, or common sense—a central notion to modern political life.[2] The idealized past that guided their vision of the future could not conceive of separating church and state. Reactionaries sought to prove the principles of the French Revolution unnatural and irrational, nothing more than a historical aberration: a true revolution was, ultimately, a logical impossibility.[3]

As for reason itself, its secularization had to be reversed. The root of all modern political and religious upheavals was, according to the reactionary literature, the rebellion of individual reason. The Protestant Reformation was the first visible and highly destructive product of this rebellion, the fruit of a morally corrupt reason, which had turned itself into the ultimate source of universal order. A series of calamitous events had, since then, constellated the modern age, culminating in the Great Revolution. This immoral escalation was written off as reason's hubristic delirium, best incarnate in the godless mathematicians, who mistook the products of their febrile imagination for reality. Reinstating the principle of authority meant revising an understanding of what reason really was, its scope, and the necessity of its integration with tradition. The reason of the godless mathematician, individual, universal and autonomous, needed to be replaced by a reason that was collective, contingent, and dependent—by a reactionary reason.[4]

Against Enlightenment

Historical scholarship has long undermined the image of the Enlightenment as a unified project with a coherent set of values and objectives that was championed and diffused by a few great men. By contrast, the perception of the opposition to the Enlightenment as a coherent and unified front has proved more durable. Many have imagined "the" Counter-Enlightenment as a single historical movement, with its own logic and principles.[5] One problem with this reconstruction is that it assumes the existence of a single and unchanging notion of reason, supported by the philosophes and attacked by their enemies. In reality, definitions of reason varied significantly, even among the friends of Enlightenment. Most anti-Enlightenment arguments were not arguments against reason; rather they were arguments about reason's veritable nature and scope. To many of its enemies, Enlightenment—and, more broadly, modernity—was the direct outcome of a misguided use, and ultimately a *betrayal*, of reason.

Recent research acknowledges the plurality and social differentiation of positions that have defined themselves in opposition to Enlightenment.[6] Earlier on I advocated for recognizing a plurality of "resistances" to revolutionary transformations, introducing the term "reaction" to refer, more specifically, to the conscious articulation of a counterrevolutionary discourse and action. This chapter makes it clear that even the notion of reaction needs to be articulated, and that we should refer, more properly, to a plurality of reactions. The authors I mention have been referred to, in the different national historiographies, as representatives of *cattolicesimo reazionario, traditionalisme,* and *Konservatismus*. Their views and objectives differed significantly—their views of capitalism, for example, were hardly similar. And yet, while acknowledging these divergences, I refer to them simply as representatives of reaction. This usage does not mark them as part of a single coherent movement; it is a methodological move to reveal a distinctive trait of their productions: a certain way of thinking about reason. They did not endorse a coherent shared theory, and they did not share a clear set of common objectives; rather, in pursuing their battles, they mobilized and deployed an image of reason, which I refer to as reactionary. Reactionary reason is best defined by negation: it's noncreative, non-sovereign, nonlegislative and, above all, it's nonautonomous. It cannot direct itself and cannot rely entirely on itself. It is essentially dependent *on something else*. We have been tracking the emergence of this subordinate reason in mathematics and science for the last few chapters. It is now time to consider the elaboration of doctrines that explicitly crafted and celebrated reactionary reason.

"Reaction" is a term that comes with baggage. It enters the modern age with a scholastic pedigree, primarily in mechanics and the life sciences; the semantic shift to the world of politics is sudden and takes place—you guessed it—in the 1790s. As "progress" loses its neutral meaning and becomes associated with the idea of perfectibility, "reaction" turns into that which impedes or opposes progress. For Kant, what opposed progress was still essential to man's very nature; after Thermidor, that opposition took the form of social groups and political forces. The term *réactionnaire*, coined to mirror *révolutionnaire*, was brought into political language in 1797 by the moderate liberal Benjamin Constant, who set the scene for the use of the postrevolutionary concept of reaction. Scientific reason and political freedom, he argued, are closely linked, as reason is naturally at the service of political progress. Mathematical precision can and should be applied to the calculation of "moral forces" and the practices of administration. Those revolutionaries who pursued the rational transformation of institutions through revolution, however, committed a series of "excesses," such as the attack against property. These excesses provoked reactions, guided not by reason but by passion: the partisans of the backward pre-revolutionary order aimed to return those prejudices and injustices that constituted their world. In his rendering of revolution and reaction, Constant opened up the space for a middle ground, a defense of individual freedoms that is fully compatible with the *Directoire* first and with a constitutional monarchy later. It was the space where a new kind of scientific knowledge could be deployed by the technical elites of the modern state.[7]

The meaning of "reaction" that became standard in the nineteenth century emphasizes its anti-modernity, backwardness, and incapacity to imagine the future. For Jules Michelet, what followed the Revolution was "the blind reaction"; for Edgar Quinet, reaction is "passive and enraged"; for Victor Hugo, reaction is the lashing out of an agony—the aggressive face of that which has no future.[8] By contrast, the reading I offer of reactionary authors does not interpret their ambition as a straightforward return to the past. For one thing, the past that they evoked was either long gone or, more likely, had never existed. Thus their exhibited nostalgia, their anxiety in the face of accelerating transformations, and their pervasive sense of historical betrayal should be handled with care. Far from signifying inertia and passivity, or being blind reflexes, they take us straight to the core of the modern age. "Political nostalgia," notes Mark Lilla, "settled like a cloud on European thought after the French Revolution and has never fully lifted."[9] The very militancy of their nostalgia is what makes reactionaries distinctly modern figures. Even more significantly, reactionary productions such as those mentioned in this chapter offered new and profoundly historical ways of understanding society, politics, and knowledge. Like Fergola, these authors looked at the past to surpass their contemporaries: they strived for an alternative to enlightened modernity—one built on reactionary reason.

Hence the radicalism of many reactionary arguments and the sheer modernity of some of their insights. Reactionaries are repelled but also fascinated by revolutionary violence and methods, and they cultivate assiduously the literary and argumentative styles of the philosophes and the Jacobins. They are no less radical than the revolutionaries, which is why there is an unresolved and palpable tension, in their productions, between claims about returning to a past that has been lost and claims about making a new beginning, establishing a new order—or rather, a new old order. In his novel *The Leopard*, Tomasi di Lampedusa captures this distinctive trait of reaction in a sentence uttered by the scion of a family of Sicilian feudatories, the Prince of Salina: "If we want things to stay as they are, things will have to change."[10] The relation between reaction and modernity is not as straightforward as it might seem at first. In his literary analysis of "the anti-moderns"—which moves from Maistre, Chateaubriand, and Baudelaire to reach Roland Barthes—Antoine Compagnon closes the circle in an illuminating and paradoxical way: "the veritable anti-moderns are also, at the same time, modern"; they are in fact "the salt of modernity."[11]

*

In order to trace the contours of reactionary reason, let's focus on the cultural resources that were immediately available in Naples between the 1790s and the 1830s: those of the Catholic reaction. Let me briefly unpack this term.

"Catholic" refers to the key role of religion for these authors and to the significant ecclesiastical component in their ranks. Also, more subtly, it refers to the continuity of their political and philosophical arguments, as well as their rhetorical strategies, with the theological arguments of eighteenth-century intransigent Catholicism. "Reaction" captures the militancy of these arguments and the fact that they are invariably framed as necessary and urgent responses against some recent aberration, be it the impiety of the encyclopedic movement, the horrors of the Revolution, or the absurdity of constitutionalism and political liberalism. All these aberrations derive from the abuse of reason that, according to the reactionary imagination, characterizes modernity. They take the form, invariably, of challenges from below, such as democratic doctrines, calls for juridical equality, or universal education. Reactionaries envision instead a return to order that can only take the form of a return to hierarchy—order produced by subordination. Reactionary doctrines are thus activist doctrines for an activist time, and their arguments are invariably "forged in battle": it's only by contrast with an abhorred revolutionary reason that the invocation of ancient forms of wisdom can captivate the modern mind.[12]

The anti-modern battle had begun around the mid-eighteenth century, with the shift of the Catholic Church toward the intransigent critique of contemporary culture. The year 1758 marks an important step in this transition, with the passage from the pontificate of the "pope philosopher" Benedict XIV to that of Clemens XIII, who was convinced that the fight against the philosophes should be the common aim of both the Church and the monarchies. In an early version of the throne-and-altar argument, the Roman Curia identified the corruption of fundamental moral and religious values as the roots of the philosophes' doctrines. This corruption, which stemmed from the Protestant Reformation, seeded a crisis in European culture and society. From this perspective, Christianity could only escape self-destruction by returning to a mythical medieval theocratic structure, wherein the pontiff would play the role of mediator between divine authority, the monarchies, and the people.[13]

The Church, however, was forced to scale back further its temporal pretensions during the pontificate of Clemens XIV. Following the intense anti-Jesuit campaign of most European monarchies, the pontiff even decided to abolish the Society of Jesus (1773). The intransigent components of the Catholic Church reacted to these traumatic events with a flourishing of anti-modern publications, just as a wave of supernatural phenomena was being registered across the European lands. Portents and mystic visions were interpreted as signs of an impending divine intervention to punish the impious actions of the pontiff and the monarchs, who were stripping the Church of its juridical and economic privileges. The accounts of these supernatural phenomena, the reactivation of baroque practices and devotions, and the propagandic activity of the clandestine "Christian Friendships" network prepared the terrain for reactionary Catholicism.[14]

Pius VI's aptly titled encyclical letter *Inscrutabili divinae sapientiae* (1775) openly condemned modern philosophy. The encyclical argued that the spread of atheism was the main cause for contemporary social disorders and emphasized the role of the Church in maintaining social stability. The Church, however, was far from a monolithic bloc. Those who argued for a renewal of its structure and a redefinition of its social function contested the anti-modern turn. In the late eighteenth century, Jansenism was the most influential among these internal movements, its rigorous moral connotations providing theological support to national governments fighting Rome's economic and cultural pretensions. The crisis of the temporal power of the Church was taken by Jansenist ecclesiastics as an opportunity to return to the original spirit of the evangelical message. This internal debate came to a sudden halt with the anti-Christian policy of the newborn French Republic. The Catholic reaction to the "great insurrection" had two main components: an idealized representation of medieval political life in which the pontiff's authority guaranteed the stability of the *res publica christiana* and a scathing critique of the culture of the eighteenth century. Philosophers, Freemasons, Jansenists, and Jacobins were, in this perspective, the authors of a massive conspiracy against the Church and therefore against society and the natural order.[15]

After the Napoleonic normalization, the debate inside the Church polarized between the supporters of a pragmatic neo-Constantinism and the reactionary wing that had rallied around the theocratic project. In the same years, the number and relevance of the religious popular missions increased. Initially organized and directed by the reconstituted Society of Jesus, the missions intended to regroup believers around their parishes and, more specifically, to spread baroque devotions—such as the Sacred Heart and the Sacred Blood of Jesus—that were at odds with Jansenist rigor and individualism.

Among the most representative publications of reactionary Catholicism were the 1791 book *The Rights of Man* by Nicola Spedalieri (1740–1795) and the 1794 *Theological-Political Letters* by Pietro Tamburini (1737–1827). The first book attempted to combine a theocratic conception of power with the principle of popular sovereignty, and presented Christianity as the necessary foundation of society and the only defense against the excesses of a popular government. It argued that the French Revolution had suppressed the "real" rights of man, which are those found in the New Testament. Any attempt

to build a society based on natural or deistic principles is doomed to failure; atheism, religious reformism, Jansenism, and the French Revolution are,
ultimately, manifestations of one and the same pestiferous phenomenon.
Spedalieri implored the monarchies to renounce their reformist policies and
reinstate the Church as center of social life. Tamburini's book responded to
Spedalieri with a rejection of the association of Jansenism and revolution.
Himself a counterrevolutionary Jansenist, Tamburini argued that Jansenists
cannot be confused with revolutionaries because they acknowledge the divine right of kings and the absolute power of the sovereign. In fact, he argues,
only the moral rigor of Jansenism can turn the religious man into a faithful
subject.[16] By the time these books were published, key themes of reactionary Catholicism, such as the great philosophical conspiracy and the opposition between abstraction and experience, could be found well beyond the
theological discourse. In his "Thoughts on the French Revolution," Pietro
Verri, a prominent representative of the Milanese Enlightenment, described
in dismay the widespread hostility of Italian elites toward the French Revolution and the doctrines of the French philosophes, which were scorned as "abstract" and "artificial." "The most essential and concrete principles about government, human rights, and the nature of monarchy," he commented wryly,
"are called, among us, metaphysical principles."[17]

Traditionalism

From the 1790s onwards, the arguments put forward by the enemies of Enlightenment acquired unprecedented visibility. Particularly significant are
the productions of traditionalism, a large body of literature that includes the
works of Joseph de Maistre, Louis de Bonald, and Félicité de Lamennais. Traditionalist writings have been described by Catholic historians as part of a
"spiritual spring" of renewal and reinvigoration of Catholic thought characterized by the preeminence of apologetics and new demonstrations of the
existence of God, the immateriality of the soul, and the necessity and social
utility of religion.[18] However, seeing their contributions as simply internal to
the theological discourse would be reductive: these authors were in fact crafting new anti-modern theories of history and society.

Among the most popular traditionalist texts were Maistre's *Considerations
on France*, Bonald's *Theory of Power*—both published in 1796—and Lamennais's *Reflections on the State of the Church*, published in 1808. These texts responded to the exceptional event of the Great Revolution: every available cultural resource, from the fathers of the Church to Bossuet and Rousseau, was
mobilized against eighteenth-century culture and its sociopolitical outcomes.

The traditionalists' body of work is expansive and reading through it can be a tedious experience, as every argument reinforces one or another of traditionalism's few basic tenets, above all that social crisis originates from the arrogance of individual reason, for which the only remedy is the restoration of the principle of authority within both church and state. The skills and historical sensibility of some of these writers are, nevertheless, remarkable.[19]

Maistre, a politician and diplomat at the service of the King of Sardinia-Piedmont, was undeniably a gifted writer. He artfully ridiculed the ideals of the Revolution, the principles of Enlightenment, and the pride of modern science—demolishing with cynical irony the philosophes' naivete and progressivism. It's a straw-man argument, to be sure, but an interesting one that reveals the profound and long-lasting connection—rooted in intransigent Catholicism—between pessimism and political reaction.[20] Modern philosophy and constitutional forms of government were to Maistre artificial and paradoxical: the traditional order had been perturbed and must be restored, in knowledge as well as in society. Individual reason must recognize its own limits: "the masterpiece of reasoning is the discovery of the point where one must stop reasoning."[21] And yet, Maistre cannot be easily written off as an irrationalist or a fideist. His hero of reason was none other than Descartes, and he found inspiration in Cartesian philosophers of nature and mathematicians like Nicholas Malebranche, who understood intellectual activity and scientific inquiry as the site of the encounter with the divine. Reason, wrote Maistre evocatively, is like the "flickering flame" of a candle in the midst of darkness, unable to guide human action without the support of tradition. The peoples of Europe, he stated, did not need "unfounded systems based only on what they call *reason* and instead is simply *reasoning*"; they needed "prejudices, practical rules, and sensible, material, palpable ideas."[22]

Paradigmatic of the abstract and misleading nature of many notions of modern philosophy is the universal concept of man. There never existed any such thing, Maistre argued; what exist are only particular individuals. Similarly, there never existed such a thing as the state of nature. All forms of contractualism are unsound, and the original covenant is only a myth: society is not the sum of individuals, and its origin is necessarily transcendent. Similarly, "to talk of nature as opposed to society is to talk nonsense," as they are both created by God, the only legitimate legislator, and the only source of sovereignty. Indeed, "one of the greatest errors of this century was to believe that political constitutions were a human creation; and that a constitution can be made as a clock is." For this gravest of mistakes one had only to thank modern philosophy, that is, the calamitous result of individual reason *acting on its own*.[23]

If God is its only legitimate source, sovereignty must be indivisible and unaffected by mundane limitations. Here the distance between Maistre and the nostalgic champions of the ancien régime becomes obvious. Whatever form a state takes, absolute power must be wielded by a single actor. Monarchy can therefore be an advisable form of government, but not necessarily; its advisability depends on specific local conditions. In fact, "the art of reforming governments does not consist in subverting and re-building them according to some new theories, but rather in bringing them back to those internal and hidden principles discovered in the ancient times."[24] Maistre is referring here to an intuitive knowledge that precedes, logically and chronologically, written culture. This knowledge is essentially nondiscursive, and any attempt to conceptualize and systematize it—any *reasoning*—ends up compromising its truth: "writing is always a sign of weakness, of ignorance, of danger; the more perfect an institution is, the less it writes."[25]

Skeptical about the outcome of the Congress of Vienna, Maistre presented his ambitious theocratic vision in *The Pope* (1819). In this vision, Europe could be saved only by the "Catholic principle," according to which truth and authority are one and the same. This metaphysical principle of the unity of authority finds its best expression in the conjunction of the pontiff's spiritual infallibility and temporal sovereignty. The pontiff's authority is of the same kind as that of kings, but much superior by true universality. The authority of the pontiff is therefore the necessary basis of any temporal authority. Take this basis away, and the whole society collapses: there is no society without government, no government without sovereignty, no sovereignty without infallibility. Maistre's considerations on power are, so to speak, structural. He did not deny that the exercise of power by certain monarchs can be tyrannical, but such contingent considerations did not affect the argument for the necessity of the integrity and infallibility of power. Europe had to find its way between the "two abysses" of tyranny and anarchy: only the supreme authority of the pontiff—as universal mediator between peoples and temporal sovereigns—could save it from self-destruction.[26]

As Maistre worked on his *Considerations*, Bonald, independently, was completing his massive *Theory of Power*, in which he tried to ground his radical opposition to the idea that societies are a human construction and that men can truly be legislators in civil and religious matters. The first step in the restoration of the principle of authority must be the confutation of eighteenth-century individualistic philosophy. The focus of philosophical reflection must shift from the single individual to society as an organic whole. Rather than study the abstract notion of man, one should try to understand how men are shaped by society, as "man only exists in society." In fact, society preexists man, and its origin is divine. If God gave both society and the church their "natural constitution," then the alliance between absolute monarchy and the Catholic Church has a sound metaphysical justification. For Maistre, providence drives human history, often mysteriously; for Bonald, the emergence of a specific form of organization is due to metaphysical necessity: such form is not the result of *calculation* but is the incarnation of the divine constitution of society—constitution here meaning the set of necessary relations existing among society's members. Ill-judged human action can temporarily skew trajectories, but in the end the natural constitution will always impose itself.[27] Like Maistre, Bonald conceived of power as absolute and indivisible; its fragmentation can only yield the disintegration of society, as—in the absence of a true general power—everyone exercises their own private power. Equally necessary for the maintenance of social order are religious dogmas like the existence of God or the immortality of the soul. *And therefore* they are true: "All that is useful for the conservation of society is necessary; all that is necessary is true." Conversely, "all that is dangerous for man and society is an error."[28]

Bonald's doctrine of the logical and chronological priority of society over the individual found an effective instantiation in his theory of language. Language is necessary to articulate the simplest thoughts, so it cannot be a human invention: "man cannot invent without thinking, and he cannot think without signs." The origin of language can only be transcendental. Its basic elements are a divine gift and inherently contain all fundamental truths about nature, religion, and society. These truths manifest themselves through the use of language. The result of this process is the creation of a cultural and social tradition (the Judeo-Christian one) that must be held as the only guide in political action. Moving away from Maistre's Platonic innatism, Bonald offers instead a form of social innatism. The theological resonance of his argument is clear: as God is known through his word, so the laws of society are known through language. Language, which is conceived as coextensive with thought, contains the idea of God, the basis for all other moral, social, and political ideas. Truth manifests itself in tradition through language; it cannot be reached individually. The ideological investigations inspired by Locke are therefore useless exercises where the human spirit "extenuates, consumes, and desiccates itself in a sterile self-contemplation." Instead, the foundations of human knowledge and the criteria for discerning truth from error are external but not material, objective but not sensible. They are in the words that constitute individuals as well as society—not surprisingly, Bonald's thinking will fascinate late-nineteenth-century French sociologists.[29]

Lamennais, a man of the Church, defended similar principles well into the age of Restoration. He wasn't keen on medieval mythology, and his theocratic project was based rather on a spiritual renewal of the Catholic Church. His influence on the debate within the Church was profound—he soon became the most read of the traditionalists. His 1817 *Essay on Indifference in Matters of Religion* sold around forty thousand copies.[30] He, too, insisted on the organic relation between social and religious principles, described the Reformation as the original rebellion of individual reason against the divine order, and condemned the subversive philosophical and theological doctrines of the eighteenth century, whose outcome had been the deification of reason itself. A distinctive claim of Lamennais's polemic is that European monarchies' longstanding support of secularization and reformism were themselves responsible for recent social disorders. Interestingly, though, the main target of his *Essay* is not erroneous doctrines but the modern "spirit of indifference": the attitude of those philosophers who have renounced the search for ultimate truth and are in a state of "spiritual sleep." Under the label of "religious indifference," Lamennais grouped a number of different positions whose common aspect was the nonrecognition of the supreme authority of the Catholic Church and of the pontiff. Following a typical argumentative form of the traditionalist literature, Lamennais moved from the alleged effects of ideas to their theoretical evaluation. To him, the effects of religious indifference were, quite simply, devastating.[31]

Lamennais insisted on the essential limits of individual reason, emphasizing the distinction between the capacity to "perceive" truths and the operation of reasoning (*raisonner*), which is "the spiritual operation by which one discovers relations among known truths and deduces consequences from them." The apprehension of truth—a perfectly clear intuition—excludes reasoning, which by contrast can be said to be weak, fallible, and not self-sufficient. If individual reasoning is deceptive and unreliable, one can only ground epistemologically sound arguments in "common sense," the universal consensus. All true knowledge, starting from the existence of God, is warranted by the authority of universal consensus.[32] Lamennais espoused Maistre's radical historicist perspectives on truth and the ways of expressing it and, like Bonald, provided a social alternative to individual reason as the source of certain knowledge in both science and religion. Believing in the autonomous and self-sufficient nature of individual reason was modernity's great error, he held—an error that yields only social chaos and philosophical skepticism. The political equivalent of the primacy of individual reason over authority and common sense, Lamennais argued, is the democratic doctrine (*democraticisme*). In a democracy nothing is stable; everything changes following passions and opinions: democracy inevitably drags the people and their leaders toward self-destruction.[33]

Reactionary Catholicism in Naples

By 1820, a well-organized network of individuals, institutions, periodicals, and publishers was promoting the tenets of reactionary Catholicism. In Italy most authors were ecclesiastics supported by the restored monarchies. The key sites for the elaboration of reactionary Catholicism were Turin, Modena, and Naples. These cities had been hosting institutions at the forefront of intransigent Catholicism since the mid-eighteenth century, and they were the capitals of the most reactionary regimes of the peninsula. The first "Christian Friendships" were created in Turin in the 1780s; they constituted a network inspired by the Masonic experience and devoted to the diffusion of the "good press" and the anti-encyclopedic campaign. By the early nineteenth century they promoted authors like Gerdil, Bossut, Marchetti, Haller, Maistre, Bonald, and Lamennais. The Friendships had a distinctive aristocratic and antimodern character, in clear continuity with the work of Turin-based Maistre. They enjoyed the patronage of the restored monarch, who had abolished the Napoleonic code and every law issued after 1800. He had also handed control of education to the reconstituted Society of Jesus, restored guilds and corporations, limited imports and exports, and removed all university professors who had cooperated with the French during the occupation. In 1821 the first reactionary Catholic periodical was founded in Naples. Its title was programmatic: *Ecclesiastic Encyclopedia*. In 1822 another periodical, *The Friend of Italy*, was established in Turin. In the same year the *Memoirs of Religion, Morals, and Literature* appeared in Modena.[34]

These periodicals shared a similar ideological framework, but their tone and quality varied. The Neapolitan periodical was immersed in the local political struggle, its articles violently polemical and their content very specific. The Turin-based periodical, weaker in terms of content, pursued a straightforward throne-and-altar campaign. Political life was relatively quiet in Modena, and there the academic component was more relevant; indeed, the *Memoirs* aimed at renewing Italian culture, denouncing the dissemination of "evil books" and modern philosophers' and Jacobins' "abuses" of scientific knowledge to legitimate their subversive political action. Also noteworthy was the critical response to the liberal conception of *patria* (fatherland), which referred to the entire Italian peninsula and was the basis for the national unitary movement. Contributors to the *Memoirs* accepted the classical value of *patria*, but in its traditional meaning of the native city, the family, and the religion of the ancestors.[35]

Naples had seen an early reaction against the philosophes and the *Encyclopédie* in correspondence with the first wave of reforms promoted by

Genovesi and his school. The context of this reaction was the controversy over the influence of the Roman Church in the kingdom. Even at the peak of the reformist season, resistance had been significant, as shown by the fact that the university reform of 1777 had not eliminated a chair of canon law (*cattedra delle decretali*), which Genovesi disparaged as the bastion of Roman pretensions. This resistance anticipated many themes that were to emerge after the Revolution, such as the critique of secularization and individualism and the need to reposition the Church at the center of social and cultural life. What makes this debate particularly interesting is that, unlike in Modena or Turin, Neapolitan reactionary culture had to face a significant and well-organized opposition even after 1815, as the technocratic elites of the Napoleonic period continued to support liberal and centralizing policies, in line with the general orientation of France and Austria.

An important site for the elaboration of reactionary Catholicism in Naples was the Congregation of the Oratory. A typical counter-reformist institution, its main goals were to renew the apostolic mission and educate the young poor. By the early eighteenth century, it had become a philosophical powerhouse with a renowned library strong in patristic literature, dogmatic and moral theology, ecclesiastical history, and biblical chronology but also well stocked with texts on Protestant theology, anti-Jesuit literature, modern natural philosophy, and mathematics. Giambattista Vico was among its most assiduous frequenters. In 1741 the priests of the Oratory asked permission to open an academy of ecclesiastic sciences in their residence, under the patronage of the Archbishop of Naples. The thirty academicians were mostly Oratorians, but Theatines, Franciscans, and Dominicans were also represented. In 1747 Genovesi was invited to the academy but found its program at odds with his efforts to integrate Christianity and modern philosophy. The academy's explicit goal was to train the local clergy in the systematic confutation of heresies and to coordinate the defense of Catholic truths. Academicians were likened to "warriors" whose capacity to fight increases enormously when acting as a unit.[36]

After the appearance of the first volume of the *Encyclopédie* in 1751, deism and "religious indifference" became the main polemical targets of the dissertations read at the academy. The academicians were encouraged to refine their philological and exegetic skills but always within a rigidly predefined perspective, which excluded doubts and interpretative variations. Among the themes treated were the Immaculate Conception, miracles, transubstantiation, the primacy of the pontiff, and the infallibility of the church. To maximize its effect, each year the academy focused on a particular "enemy of the Catholic faith." The "pure and spotless truths" of religion were thus defended against both Protestant scholars and enlightened Catholics. Any expression of religious tolerance was condemned as religious indifference, a flexible concept that was also used to attack prominent local figures like Raimondo di Sangro, the midcentury leader of the Neapolitan Anglo-Dutch Freemasonry. The fight against Masonic "indifference," "pure naturalism," and "Spinozism" characterized this early phase of the academy's anti-modern campaign.[37]

The academy was active only intermittently until 1780, when it returned forcefully to the cultural scene, thanks to the commitment of Archbishop Serafino Filangieri, who was interested in the modern sciences and had briefly taught physics at the University of Naples. He divided the academy into two branches, Moral Sciences and Theological Sciences. Among its members was the well-known apologist Giovanni Camillo Rossi, a sworn enemy of deism. His work shows how the style and goals of apologetics were changing in the 1780s and 1790s: the new apologetics had explicit political implications and aimed to prove—above all—the divine nature of political and religious authority.[38] This apologetics provided cultural legitimization to the unprecedented political alliance between the Neapolitan government and Rome, which had major repercussions on Neapolitan cultural life. Accepting the Roman interpretation of the Revolution as a great philosophical rebellion, the Neapolitan government renounced its traditional anticurial policies, while the Neapolitan church marginalized its Jansenist and Gallican components. The supremacy of the pontiff was no longer up for discussion, and the crusading spirit of the intransigent clergy was fully supported by the Bourbon police and censors, especially after the discovery of the Jacobin conspiracy of 1794.

That very year saw the creation of the Royal Arcadia, a new literary society whose frenzied activity characterized reactionary Catholicism in Naples. The Royal Arcadia aimed "to fight the atheistic and heretical errors of the eighteenth century," and its influence spread rapidly. Its statute was a manifesto for the mobilization of culture in defense of throne and altar. The Arcadia held its meetings in the main churches of Naples and distributed free pamphlets and books to the public—its printing office had the exceptional authorization to print texts without waiting for approval. Around one-third of its members were ecclesiastics, followed by representatives of the lower provincial aristocracy and lawyers, many of whom had invested in land that they administered as absentee landlords. Among its most prominent members was the Dominican Vincenzo Gregorio Lavazzoli, who emerged, between 1794 and 1796, as one of the champions of the campaign against modern philosophy. In his publications and speeches, the defense of religion was invariably coupled with arguments for a theocratic organization of the state.

To counteract the false Jacobin regeneration, one must indeed recover "the anchor of public safety": subordination to religious and political authority.[39] Another Dominican friar, the Sicilian Domenico Crocenti, published, in that same 1794, a three-volume denunciation of the "anarchic Jacobin system of liberty and equality." This system, he argued, is against both experience and reason, and his arguments are filled with geometrical axioms and experimental evidence. The principles of liberty (which he interpreted as being isolated and antisocial) and equality (an abstract, mathematical equality) are proven artificial and, ultimately, irrational, the outcome of an unbridled philosophical enthusiasm. Interestingly, Crocenti envisioned Jacobin leaders enforcing equality by calculating it algebraically (*con calcolo algebratico*).[40]

During the Republic of 1799, the Arcadia converted itself into a clandestine network, and its capillary structure allowed the organization of an effective counterrevolutionary action in support of Ruffo's *Santafede*. Ruffo himself was an Arcadian, as were the bishop-officers who assisted him on the battlefield. The Arcadia was placed under the protection of the Holy Trinity, one of the symbols of reactionary Catholicism—and the patron of the Holy Alliance of Russia, Austria, and Prussia. "The Very Holy and Indivisible Trinity" was considered the basis of all the "mysteries" of Catholicism, the anti-deistic dogma par excellence—the one requiring the most obvious submission of individual reason to dogma and authority.[41]

The Aristocratic Reaction

Among the books given away by the Arcadians during their public gatherings was *The Trinity*, published in 1795 by the Arcadian Isocrate Larisso, an alias for Antonio Capece Minutolo (1768–1838), who would become the Prince of Canosa. Canosa was a leader of the Neapolitan aristocracy in 1799, a position that warranted his arrest on the return of the king. His political life was difficult and discontinuous, as his uncompromising reactionary vision clashed with liberal views as well as with the moderate policies inspired by Metternich's plan for a new European order. Metternich backed seasoned politicians like Luigi de' Medici, who enjoyed the trust of the capital's financial oligarchy and was considered essential to the health of the kingdom's economy. By comparison, Canosa's ideas and methods seemed not just old-fashioned—as did the black seventeenth-century suit that had earned him the ironic nickname "Black Prince"—but thoroughly dangerous as well. To Metternich, Canosa was an unreliable hothead. Yet Canosa—the hammer of Freemasonry and Jacobinism—was repeatedly summoned by the Bourbons as a counselor, an ambassador, and even a minister. His writings enjoyed a remarkable success in Naples and abroad. The eccentric prince expressed the ideas of many, including King Ferdinand, who, giving in to Vienna's pressure, reluctantly let him go in 1821.[42]

Canosa came from a family that had served the rulers of Naples since the thirteenth century and had the privilege of a chapel in the city cathedral. The family suffered from the progressive erosion of feudal privileges: typical of those among the ancient aristocracy who were unwilling to turn themselves into courtiers and unable to face the agricultural crisis. The French occupation and the anti-feudal laws of 1806 precipitated its decline. With the loss of the feudal rights upon the town of Canosa, economic ruin became unavoidable: by 1816 Canosa was living on a state pension.[43] Historians have described him as a naive idealist, a disinterested paladin of the ancien régime who sacrificed his fortune to an anachronistic cause. In fact, when Canosa decided to enter politics his fortune had already vanished, and the persistence of this image has much to do with his own self-fashioning as an idealistic hero. Benedetto Croce seemed to accept Canosa's idealized persona when he commented on the values that informed his education and the way in which "he had sucked Catholic religion with his mother's milk." In fact, Canosa declared that he "became a Catholic" in the early 1790s, as he began studying theology pressed by political events. Before that, Canosa had studied natural philosophy at the Collegio Nazareno in Rome and in 1787 had begun administering the family's feudal lands. "I entered the world," he wrote, "precisely when the disorders were beginning in France, thanks to the sects and to that perverse philosophy that brought so many calamities to that very rich kingdom and the entire Europe." He dated his decision to abandon his previous "Pyrrhonism" and "semi-atheism" to the crucial year 1794.[44]

Canosa chose as his guides two champions of reactionary Catholicism whom he had met in Rome in 1795: Nicola Spedalieri, the theorist of theocratic society, and Cardinal Stefano Borgia (1731–1804). Borgia was a leading intransigent ecclesiastic who had been fighting Jansenism through the 1780s and had recently penned a few counterrevolutionary pamphlets. Canosa also read Tamburini, whose argument about the empirical and nonmathematical nature of the science of society he would often cite. To William Hamilton (1730–1803), the British ambassador in Naples, Canosa expressed his admiration for Edmund Burke's idea that the moral and political sciences must be based on experience and not on abstract speculations. He also praised the 1797 *Memoirs* of Augustin Barruel, who had developed Spedalieri's claim that the French Revolution could be explained only as the outcome of a hidden conspiracy organized by philosophes, Jansenists, and Freemasons. This conspiracy theory proved to be extremely popular with the Bourbon court, which promoted the Italian translation of Barruel's book in 1803. Through his entire life Canosa remained convinced that history has a hidden dimension, a space where obscure forces act, and where the truly important events take place—invisible to most. He also embraced Barruel's contention that the conspirators' moral corruption was rooted in their belief that individual reason could be taken as a supreme criterion for discriminating between good and evil, truth and falsehood.[45]

Canosa was hardly an elegant or original author, but his works are emblematic of the cultural and political transformation of Neapolitan life in the 1790s. They also addressed a much larger audience than earlier reactionary texts. In his 1795 pamphlet on the dogma of trinity, Canosa confuted deism in its many variants, arguing that any attempt to rationalize religion ends up destroying its very essence. He then used a sketchy version of the doctrine of *philosophia perennis* to suggest that the dogma was known to ancient Jewish and Indian wise men and was secretly passed on until it was finally codified by the Catholic Church. Canosa defended the infallibility of the pontiff—which was not dogma yet—as the necessary basis for the stability of both church and society: religious insubordination cannot but lead to political insubordination.[46]

In *The Utility of Monarchy in the Civil State*, also published in 1795, Canosa argued the impossibility of maintaining social stability under a democratic government and articulated a defense of those intermediate institutions that made up the feudal-communal society. Any attempt to dismantle these institutions undermines the monarchy itself, he contended, as monarchical power is based on the value system of aristocracy. It is impossible to establish durable republican regimes in the modern world because the virtues that supported the ancient ones have disappeared, replaced by ignorance, corruption, and violence. A democracy, a "government of the multitude," would inevitably self-destruct. Canosa's notion of monarchy was of an idealized feudal monarchy—an "aristocratic monarchy"—whose power is sustained by an array of feudal and religious institutions.[47] Coherently with this vision, Canosa criticized the current government for abusing its power and introducing new taxes and obligations without consulting the institutions that represented the traditional orders. This passionate defense of the feudal system vis-à-vis absolute monarchy earned Canosa visibility in the world of reactionary Catholicism, and his book was reviewed positively in the *Ecclesiastic Journal of Rome*.[48]

But Canosa was not just a man of letters. In 1798, as General Championnet led the French army toward the Neapolitan border, Canosa recruited soldiers and joined the military campaign. When the government and the royal family left Naples, the feudal parliament of the capital elected him as a member of an "extraordinary deputation for good government and internal tranquility."

He refused to recognize the authority of the king's representative and argued that, in the absence of the legitimate king, the City of Naples was called to lead the entire nation. When it became clear that the king's representative had signed an armistice with the French, Canosa helped to organize the anti-French resistance, calling for the Neapolitan urban masses to rise and defend their city and religion from the foreign invader.[49]

During the Jacobin Republic, Canosa benefited from the protection of a philo-Jacobin aristocratic family and was able to participate in the heated political discussion about the abolition of feudalism. He supported the moderate faction that believed the republic had inherited the duties of the extinguished monarchy, and therefore that the abolition of feudal rights should include some form of compensation. Canosa argued that the real enemies of feudal rights were those wealthy entrepreneurs who were in control of local administrations and "tyrannize[d] the people and bec[a]me rich at their expenses." The debate was ongoing when Canosa was arrested in connection with a legitimist conspiracy and sentenced to death. However, the republic fell before the execution could be carried out, and Canosa was freed from prison by the *Santafede* only to be sentenced to five years in prison for attempting to establish an aristocratic republic. He was amnestied in 1801 and thenceforth devoted himself wholeheartedly to the fight against atheism and subversion.[50]

His essays from this period are typical of the new apologetic style, with its immediate and indissoluble connection between religious and political considerations. Canosa opposed all reforms of Catholicism, the essence of which he saw enshrined in the immutability of dogma. He also rejected all attempts to separate religion from social and political life, a separation that he believed would destroy the basis of morality and therefore the possibility of a stable social structure. The foundations of morality and politics could only be found in the supreme authority of God, mediated by the temporal structures of the church and the state. Canosa was not keen on philosophical debates, and one would look in vain for theoretical subtleties in his writings. He ventures into this terrain only to satirize the pride and ignorance of those modern philosophers who fashioned themselves as the true interpreters of nature. Antonio Genovesi himself, according to Canosa, was a pernicious philosopher, as he had played in Naples a role similar to that of the philosophes in Paris.[51]

During the French decade, Canosa followed the court to Sicily, from where he coordinated a secret legitimist network while trying to convince the king to rethink his anti-feudal policies. When the Bourbons were restored once again to their throne, Canosa was nominated Minister of Police, but his repressive plans, inspired by the Spain of Ferdinand VII (who had awarded him the Great Cross of the Immaculate Conception), clashed with the moderate conservatism of Prime Minister Luigi de' Medici. Medici and Canosa disagreed on key points such as economics (free market versus protectionism and controlled food prices); the Sicilian question (unification of the two kingdoms and rationalization of the administration versus preserving the two separate identities); and the question of the former employees of the French government (inclusion versus exclusion). Canosa was relieved of his position in 1816, as it turned out that he had created a secret network—clearly this was an irrepressible inclination—to fight the liberal Freemasonry.[52]

Pushed out of active politics, Canosa published a memoir that became his most successful book, with six editions between 1820 and 1831. It originated from a critique of his police methods that had appeared in the *Literary Gazette* of London in 1819. Canosa retorted that legislation and police methods should never be judged in abstract but only in relation to the situation in which they are deployed: what works in Great Britain cannot work in Naples. The local nature of legislation—emphasized in obvious contrast to Filangieri's universal science of legislation—depends on the fact the moral sciences are nonmathematical in nature. In particular, politics is an empirical art, not an exact science. Thus, policy-making cannot be based on universal principles and rational calculations but requires expert judgment instead. The art of politics is difficult and calls for experience as well as profound historical knowledge. The tragedy of the Neapolitan Revolution of 1799 originated precisely from the mistaken belief that a perfect society can be built through the application of scientific and mathematical principles. The Neapolitan Jacobins, Canosa wrote, can be compared to someone who has a perfect understanding of the theory of conic sections, and on that basis only decides to construct real burning mirrors. The result, Canosa concludes, would be a waste of time and good metal.[53]

The pro-constitutional insurrection of 1820–1821 gave Canosa his last stint in government, again as the Minister of Police. In 1822 he left Naples for good, offering his services to various Italian governments as an expert in anti-insurrectional tactics. In 1830 Canosa settled in the Duchy of Modena, where he wrote for the local periodical the *Voice of Truth*. The motto of the periodical—*non commovebitur* (he shall not be moved)—suited well the aging and increasingly anachronistic gentleman.

*

Monaldo Leopardi (1776–1847), from the States of the Church, continued Canosa's anti-modern battle well into the 1830s. He was a sworn enemy of egalitarian doctrines, universal education, and even nurseries, which he believed undermined that unconditioned submission to authority that, alone, could save society. His target was not the by-then-distant revolution but liberal and patriotic movements. To fight them, Leopardi founded the periodical the *Voice of Reason* and published a book: *Short Dialogues on the Matters of 1831*, a series of witty, ironical dialogues in which fictitious characters ridiculed tenets of liberalism such as political constitutions, popular sovereignty, and public education. Lauberg had urged Jacobins to wield the weapon of ridicule against the "Gothic barbarism" being prepared by counterrevolutionary forces. Canosa, like Maistre, employed the same weapon against Jacobins and liberals. But the masterpiece of counterrevolutionary and anti-liberal satire are probably Leopardi's dialogues, which went through six reprints in three months.[54]

Pulcinella's Travel is one of the most famous dialogues. Pulcinella is the traditional Neapolitan mask-character, who represented, in the commedia dell'arte, a simple but resourceful man of the people. In Leopardi's piece, Pulcinella—common sense—and a Doctor—the liberal intellectual—discuss the establishment of a constitutional government in Naples. The Doctor explains to Pulcinella that the king of Naples is an absolute king:

PULCINELLA: What does absolute king mean?
DOCTOR: It means that he rules according to his own will, without depending on anyone else.
P: What a mess! But tell me one thing. The shoemaker rules in his shop, the host rules in his tavern, the head of the house rules in his family, and why should the king not rule in his own kingdom?
D: He can rule, but according to the laws.
P: This is right. I understand. Justice for everyone. But tell me, signor doctor, does not the king of Naples rule exactly like this? He makes the laws and rules according to them. When the laws are not good anymore, he makes new laws, and again he rules according to them.
D: And this is precisely what is wrong.
P: Why?
D: Because the king should not make the laws.
P: Why?
D: Because he is not the sovereign.
P: Hell! And who's the sovereign if not the king?
D: The people.
P: That's the best one. And the people didn't know it?
D: We used to live in ignorance.[55]

Political constitutions are artificial creations, extraneous to the traditions and natural inclinations of the people; they are, at best, useless. When a constitution is imposed from the outside, as in French-occupied Naples, it has the effect of a disease on the organic body of society: hence Pulcinella keeps mistaking the word "constitution" for "constipation." Constitutional systems lead invariably to a paradoxical situation. If sovereignty resides in the throne, the king is subject to his own laws, a state of affairs that is considered absurd. If sovereignty is of the people, then the king is reduced to a mere representative figure and the principle of authority is fragmented, with consequences that can only be disastrous for the stability of society. Instead, as argued by Pulcinella, laws are best understood as instruments in the king's hands.

Leopardi's tirades celebrated a fanciful world of quiet peasants and laborious artisans living around the manor of their good feudal master and under the wing of the Catholic Church—a world that looked much like an idealized version of his own provincial town. Similarly, Canosa, oblivious to the socioeconomic tensions that ravaged the Neapolitan countryside, argued that the French had occupied and ruined "a kingdom that was an earthly paradise."[56] As in a romantic view of the Posillipo school, these narratives represent the countryside as a bucolic landscape replete with natural beauties and populated by peasants wearing traditional clothes and engaging in agricultural and artisanal labor or resting in the shade of Roman and medieval monuments. A student of local history, Leopardi wrote against any innovation that threatened the alleged integrity of his small *patria*, be it a new land register or the railway. During the revolutionary upheavals of 1831, he actively participated in the establishment of an independent government in his town. Like Canosa, he was fascinated by local revolutionary governments because they could revive, in his eyes, medieval self-governing institutions. A biographer remarked that during the revolutionary experience he acted like "a medieval count," arguing for his town's communal prerogatives and against any form of administrative centralization.[57]

The success of Canosa's and Leopardi's publications shows that their evocative and highly idealized narratives captured the imagination of large and socially distant audiences. Theirs was a conscious and reflective effort to recall forms of experience which could "no longer be had in an authentic way."[58] Born in the feudal aristocracy, they excelled at reconstructing the everyday experience of ruling other men and women, "the private life of power" as Corey Robin calls it, now threatened by insubordination.[59] In doing so, they made appalling simplifications, but they demonstrated skill and historical sensitivity when they described traditional sociopolitical institutions as the outcome of centuries-long processes. At the core of their historical reconstructions was a continuous and coherent tradition built around an ancient, God-given wisdom, one that had found its best incarnation in medieval Christian society. Modern history is, essentially, the story of a ruinous decline, as this organic society and its Catholic culture dissolved into a set of artificial—and therefore unstable—political and cultural formations. At the root of this dissolution is a fundamental moral corruption, evident in the hubris of deranged philosophers who mistook their fanciful inventions for the true laws of the universe. While distant from Maistre's and Bonald's structural understanding of power and society, the representatives of the aristocratic reaction partook in crafting powerful narratives of loss, and in taking up arms against the rational and autonomous subject—an artificial construction whose only goal is to subvert natural order.

Neo-Scholasticism

By the mid-1830s the culture of aristocratic reaction, with its colorful gothic imagery, was on the wane. Figures like Canosa might have been useful for hunting down liberal Freemasons and crushing revolutionary conspiracies, but they could hardly fit into a normalized political life. Their literary productions, although vastly influential, were ill-suited to intellectual debate: while suggestive and punchy, they were considered too anecdotal and theoretically shaky. By contrast, a new body of knowledge was emerging that could be deployed to fight the philosophical foundations of liberalism: Neo-Scholastic philosophy.

Neo-Scholasticism was a sophisticated response to the crisis of the modern age, programmatically opposed to modern thought, political and economic liberalism, social reformism, and all democratic doctrines. It originated in Naples, and the Theatine priest Gioacchino Ventura, who had given the eulogy for Fergola, was one of its architects. Ventura played an important role in Restoration Naples as a member of the Commission for Public Education and a royal censor. In 1821 he founded the *Ecclesiastic Encyclopedia*, an influential periodical aimed at refuting the philosophy of the Enlightenment and liberalism, while offering an alternative thought system that would sustain reactionary politics. He defined his periodical as a "counterrevolutionary encyclopedia."[60]

Ventura's tone was dramatic: the walls are being stormed and no compromise is possible; to concede one point means to lose everything. The editorial line of his journal was clearly skewed toward the position of Lamennais, Ventura's personal friend, but he mobilized the entire Italian reactionary culture, from Canosa to the groups of Turin and Modena. The framework of his articles was rather simple: the only two "healthy" social actors left—the people and the legitimate kings—should once again ground society on traditional Catholic values. The enemy was an undifferentiated mass of religious reformism or religious tolerance, secularized philosophy, and liberal elites who advocated for a constitutional monarchy and economic liberalism. These morally corrupt doctrines were advanced by the materialist and profit-driven bourgeoisie, a crowd of "ghastly men who emerged from the mud." Ventura was against the expansion of trade and the full property of land, which he described as the poisonous fruits of human cupidity. The idea of a social contract was equally nefarious, as it too rests on unstable individual interests and ultimately on materialism and atheism. Other favorite targets were mass education and the freedom of the press, which foster rebellion and social envy.[61] Ventura saw only one possible remedy for these ills: restoring the felicitous state of the pre-revolutionary age by acknowledging the supremacy of the pontiff within the Church; the supremacy of religion in every aspect of social and cultural life; and, more generally, the priority of society over the individual—of "the whole over the part." Not surprisingly, Ventura championed Lamennais's theory of common sense as, among other things, it equated modern philosophers who scorn testimony and authority to heretics who refuse "to receive the scripture from the hands of the Church." These modern philosophers failed to understand that truth cannot be reached through individual perception and reasoning alone.[62]

Lamennais had been a useful resource to the Italian reactionary clergy in the aftermath of the revolutionary emergency. However, by the mid-1820s, his writings were no longer philosophically adequate. His marginalization was only accelerated by the pontifical condemnation of 1832, as the French abbé began questioning Rome's authority. The ambitious objective of the Neo-Scholastics was to construct an alternative to modern philosophy: an organic system of knowledge embodied by a Catholic encyclopedia. The restoration of Christian society required the restoration of Christian philosophy. Unlike traditionalists and early reactionary Catholic authors, Neo-Scholastics aimed to provide a rational foundation to their philosophical system in continuity with the bimillennial tradition of the church. Appeals to the principle of authority and to social considerations were insufficient to ground a theocratic theory of society. Hence the enthusiastic "rediscovery" of Aquinas during the first half of the nineteenth century. Some members of the Society of Jesus were especially active in the elaboration of Neo-Scholasticism which, in spite of significant internal resistances, was identified as the official philosophy of the Catholic Church at the First Vatican Council (1870).[63]

Still, the notion of common sense continued to be important to many reactionary Catholics, who believed that "Aristotelian philosophy was common

sense transformed into philosophical method." Ventura himself was convinced that the scholastic tradition could be harmonized with Lamennaisian
common sense. He saw no contradiction between the argument that religion
cannot be founded on reason and the logical demonstrations of religious
dogma. Similarly, in his fight against contractualism, he blended traditionalist themes with arguments from Karl Ludwig von Haller's *Restoration of the
Science of the State*, which was translated and published in Naples in 1826–
1828. Haller argued for the principle of authority by constructing his authoritarian and inegalitarian conception of political life "scientifically," through
empirical observations and rigorous deductions.[64]

Ventura tried to harmonize these dissonances by arguing that in the absence
of hubristic corruption, the scientific method sustains rather than undermines
religion and tradition. Religious indifference is not the natural outcome of
scientific progress but is the consequence of a specific—and misguided—
scientific methodology. In the past, he explained, distinct methods were
deployed to study different kinds of objects: the principle of authority in
theological matters, common sense and tradition in moral and political matters, and the empirical method in the natural sciences. Eighteenth-century
philosophy tried to erase these epistemological boundaries and replace this
plurality of methods with a unique universal method. Thus the traditional hi
erarchical structure of the sciences had been destroyed and mathematical and
empirical methods were illegitimately universalized. But such universality is
absurd for Ventura, as it requires excluding a priori religion and metaphysics from the body of legitimate knowledge. Moreover, while previous natural philosophers acknowledged that their descriptions of the universe only
scratched the surface of reality, the modern ones were convinced they could
reach ultimate truths.[65]

To Ventura, Neo-Scholasticism is the philosophical remedy to this intellectual catastrophe. The principle of scholasticism, he argues, is the "substantial unity" of form and matter in any reality. Thus, against sensationalist
doctrines, a human being is an indissoluble composition of body and soul.
Similarly, human reasoning is the unity of "truth" and "intelligence." Ventura
deploys scholastic categories to make it clear that true knowledge cannot be
the outcome of the individual mind's independent process of discovery. By
analogy, Ventura concludes, unity in the social and political order derives
from the substantial union of the state and the Church. The Christian *recon-
quista* of the modern world needed to start from a reconfiguration of human
reason.[66]

Ventura appropriated scholastic thought to provide solid philosophical
foundations for the counterrevolutionary encyclopedia. Neo-Scholasticism

offered a comprehensive and well-structured system of knowledge fit for the challenges of the modern age. Within its framework certain themes and ideas that had long been part of the reactionary discourse were given rigorous reinterpretations: Bonald's definition of "human being," Maistre's theory of innate ideas, and Lamennais' doctrine of common sense as the ultimate foundation for knowledge. Most importantly, the theme of the weakness of individual human reason—evocatively captured by Maistre's image of the "flickering flame"—was integrated with the Thomistic doctrine of human reason as mirroring external reality, thus emphasizing its passive and merely explicative and deductive nature, as opposed to an inquisitive and creative one. According to Neo-Scholasticism, the assent of reason to "primary truths" is immediate and intuitive, not the product of reasoning. Unlike in early reactionary Catholic thought, this assent is not merely an act of submission but "an act due to nature, which imposes those truths, without any proof, to any healthy intellect, as it lets any healthy eye see daylight without the mediation of another light to make it visible." Appropriately, Ventura wrote that scholasticism was the most effective foundation for the principle of authority, as it offered the definitive confutation that individual reason alone—the "private sense"—could ever produce certain knowledge about anything.[67]

8

Mathematical Purity as Return to Order

Introducing a pioneering historical reconstruction of the social dynamics that characterized nineteenth-century mathematical life, Henk Bos noticed how the opening of the century was a moment of major upheavals in politics, commerce, industry, and the arts, as well as in religious and scientific thinking. In mathematics too, he noted, this period was one of "deep change." In saying so, he meant that what can be observed in that field looks not like incremental development but rather like a major shift in European mathematics, one that would bring about a new view of mathematics as a whole: a new understanding of its foundations, the nature of its fundamental concepts, and its overall educational function. As part of this process, new institutions for the pursuit and teaching of mathematics were created and old ones were radically transformed.[1]

In this book, we have focused on a key aspect of this shift: the urgent need to provide solid foundations for mathematical procedures that had been developed in the eighteenth century and were now perceived as unwarranted and unreliable. One cannot understand nineteenth-century mathematical life in Europe without grasping the pervasiveness and seriousness of this foundational anxiety. Historians of mathematics have often accounted for the foundational projects of the early nineteenth century by endorsing the argument that then-current practices were plagued by unacceptable logical gaps. From this historiographical perspective, the process of rigorization is self-explanatory. Addressing the concerns of a later generation of committed foundationalists, Ludwig Wittgenstein pondered the reasons why logical contradiction should inspire fear. After all, he quibbled famously, it's up to us to decide what follows from a contradiction; logic does not decide for itself. The curious historian should wonder similarly about all foundational quests in mathematics and investigate them as veritable historical problems. Why did practitioners begin to perceive these alleged logical gaps, and why did they consider them *unbearable*?[2]

Steering clear of teleological reconstructions, some scholars have explored the institutional conditions of these conceptual and methodological changes. In particular, they have related the new foundational concerns to changes in European educational systems. The need to teach large classes in the new French schools of engineering, for example, made it desirable to clarify and systematize mathematical theories, providing a pedagogical rationale for rigorization. In Great Britain, debates for or against analysis were typically framed in a pedagogical discourse freighted with moral significance. The mathematics-morality nexus is also visible in the context of the new Prussian research university, where "pure mathematics" emerged as the cornerstone of a self-standing and newly autonomous world of mathematics that helped professionalize the discipline and fashion the persona of the modern mathematician.[3]

Building on this sensitivity to the ecologies of mathematical practice, I have explored the case of an early manifestation of this foundational anxiety and offered a sociohistorical interpretation for it. In order to do so, I have focused on specific groups of actors and on their experience of this anxiety, or lack thereof. It turned out that I could not write a history of their mathematical concerns and achievements without also writing a political history. The anxiety of the synthetics, I argued, was political *as well as* mathematical. I'm not saying that it was political rather than mathematical. Nor am I implying that their mathematics was distorted by their politics. This example is not a pathological case in the history of mathematics, and mine is not a cautionary tale about what can happen to good mathematics when it falls prey to ideology.

Rather, the strange case of the Neapolitan controversy over methods effectively illustrates the social shaping of mathematical knowledge. This case is strange because the social-shaping process is in full sight—not a common occurrence in the history of mathematics. The Neapolitan Revolution was, at once, mathematical and political, and so was the ensuing Restoration. Seeing these two dimensions as mutually constitutive implies the inescapable historicity—and therefore contingency—of mathematical knowledge. Like all knowledge, mathematics is produced, validated, used, and modified by social collectives and through social processes. All mathematics bears the marks of this contingency. It is, in other words, the outcome of choices that could have been different.

How can we construct narratives that effectively reveal this conventional dimension? We need stories in which sameness and universality in mathematics are not a given but the outcomes of historical processes of knowledge-making and circulation. In this book, I wove microhistories of mathematical choices, revealing how they were invariably collective, contingent, and goal oriented. The groups that made these choices shared orientations towards social life, orientations that sustained and benefited from those choices. Synthetics and analytics argued strenuously over mathematical methods, the role of mathematics in science and engineering, the function of higher education, and role of mathematics in political and economic matters. The synthetics' foundational anxiety in mathematics was a dimension of their overall struggle to restore legitimate authority and "natural order" in what they experienced as a deeply dissonant and corrupted world. The analytics, by contrast, did not share this anxiety. The algebraized mathematics they promoted was seen as the finest expression of a universal rationality that informed and legitimated their disruptive social planning: they saw it as the very logic of modernization.

Mathematics enters this book *as* politics then, not because I think mathematics should or can be reduced straightforwardly and invariably to politics, but because the constraints and affordances that characterize the mathematical imagination of these groups were shaped by a social experience overwhelmingly informed by concerns for social order. And, in turn, their political imagination was shaped by their mathematics think, for example, of the analytic structure of the Jacobin secret society in 1794. My answer to the historical question of the transformation of mathematics in the early 1800s thus historicizes the foundational anxiety, which was by no means self-evident. Its emergence depended on a particular and contested perception of the current status of mathematics, a perception channeled and articulated by a specific mathematical culture. Is it better to solve a geometrical problem via analysis or synthesis? The answer to this question could only be technical. But this technical answer was structured (via priorities, assumptions, legitimate procedures) through a process that could only be social. Technical choices, in other words, are inherently rooted in the culture, and therefore social experience, of the relevant collectives.

More specifically, in response to historical questions about the rise of pure mathematics and the transformation of mathematical practice that accompanied it, I have offered an interpretation of pure mathematics as *return to order*. Up until the eighteenth century, pure mathematics was understood as the result of a process of abstraction from empirical reality, with "mixed mathematics" occupying the intermediate space. Pure mathematics as it emerges after 1800 is something different. It is the very core of mathematics, and its separation from empirical reality is not one of degree: it describes *another kind of reality*—one that is purely intellectual, and independent from any empirical state of affairs. In Naples, foundational concerns surrounding the rise of pure mathematics were part of a broader sociotechnical attempt to resist revolutionary upheaval and re-establish social and cognitive order. Reactionaries perceived algebraized mathematics—analysis—as degenerated knowledge, responsible for the moral and political breakdown of Europe. By contrast, they promoted synthesis as the return to an alleged ancient wisdom that would restore and support traditional hierarchies, both social and natural.

These associations were not superficial or generic. Synthesis, for instance, was not favored by reactionaries simply because it was part of an alleged tradition, nor was analysis rejected just because it was seen as new and French. Rather, the specific *technical features* of these approaches were morally and politically charged. For Fergola, the divination of ancient methods and the quest for mathematical purity could restore the dignity of mathematics while shielding metaphysics and religious dogma from the corrosive critique of "arrogant mathematicians." Flauti's geometrical program undermined the credibility of engineering knowledge: the analytic savoir-faire was not—*and could not be*—universally valid. Conversely, for the analytics, the quest for general methods was legitimated by an image of mathematics as universal language—not a game, as it would be for a later generation of mathematicians, but the very essence of human rationality. In sum, the two problem-solving methods and the two competing images of mathematics that shaped them were sustained by distinct sets of resources, assumptions, metaphors, skills, tacit knowledge, and goals, the scope of which reached well beyond the realm of mathematics—that is, by two alternative mathematical cultures.

In turn, these mathematical cultures were shaped by and sustained two alternative images of reason. Hence my use of analytic and reactionary reason as organizing principles of the book. A mathematical technique is necessarily embedded in a specific set of epistemological assumptions. Referring to images of reason is thus a way to capture key aspects of these sets of assumptions and emphasize their systemic traits. Also, doing so makes it easier to see how certain technical choices had political significance as they shared and reinforced epistemological traits of relevant political concepts and practices. I focus primarily on the opposition of analytic and reactionary reason, which I deploy as ideal types, but that cleavage is not the only important one in this book. Within the analytic genealogy, the shift from a revolutionary reason—and therefore a revolutionary mathematics—to a moderate liberal one is of the essence in my story. The moderate liberal reason of the Napoleonic and Restoration state is bound and self-disciplined, and while it legitimates a neutral, instrumental technical dimension, it confines the *truth* of its mathematics to the ethereal realm of pure mathematics.

At various points in this book I have described change in mathematics as a dimension of a broader change that is at once cognitive and social. In this perspective, it would be misleading to understand the Neapolitan case as one in which politics influenced technical solutions. There was no such distance. The technical *was* the political. And the genealogy of mathematical purity that I offered is ipso facto the genealogy of a form of political rationality.

Small Places, Large Issues

The history of science of the last few decades has been keen on using the microscope and rather allergic to the sweeping generalizations of old times. As a consequence, some have lamented, it has become more difficult to make sense of large-scale change and longue-durée processes. A spate of sophisticated microhistories has enriched the field and enthralled scholars, but what about the big picture? In the history of mathematics this problem takes an even sharper form: how can one craft stories that bridge the chasm between carefully reconstructed local experiences and the global transformations of a distinctively universal discipline?

Focusing on a small place at the margins of the enlightened mathematical regime enabled me to explore empirically the interplay of cognition and social interaction, and thus to address the large issue of mathematics as politics. Why do we find such an extraordinary interest in promoting (or opposing) mathematical abstraction and generality in Naples around 1800? One might speculate about the recurrent debates on abstraction that are part and parcel of the history of mathematics itself and be content with identifying different mathematical styles at work in different places and times. A more profound historical understanding can be achieved by emphasizing how a certain spirit of abstraction seems to pervade European mathematics in the late eighteenth century and associating it with new teaching and research institutions and with new professional personae, such as the "modern engineer" or the professional mathematician, and hence with the sets of cognitive and moral values that they embodied.[4] A further step, one I have taken in this book, is to reconstruct what "abstraction" meant concretely within different mathematical cultures and how these meanings were constitutive of competing ideals of reason and rationality that carried immediate political significance.

Where does this reconstruction leave us in terms of the big picture? The Neapolitan case was not typical. If we were to focus on another small place, there is no reason we should expect to find the exact same associations—for example, that of a sophisticated form of Euclidean geometry with theologically and politically reactionary positions. Local stories would likely point in different directions, but this variability should not be seen as a problem, as one of microhistory's lessons is precisely that "similar effects can be due to very different causes; and similar causes can—depending on background conditions—have very different effects."[5] In fact, I do not claim that the mathematical concepts and practices described in this book had a univocal political meaning. On the contrary, it is easy to see that the assumption of a deterministic relation would not stand the test of historical evidence.

One needs only to move up the Italian peninsula to the city of Modena, home to physician and mathematician Paolo Ruffini (1765–1822). Ruffini is credited with the first version (1799) of the proof that polynomial equations of degree five or higher cannot be solved by radicals, later developed in the Abel-Ruffini theorem (1823). This fascinating limitative result contradicted entrenched expectations. Lagrange himself was convinced that quintics could be solved by radicals; it was just a matter of figuring out how. The troubled history of this theorem, including the fact that Ruffini's result was ignored for many years, shows that it met a remarkable resistance. Ruffini's scholarly persona has many similarities to Fergola's: they shared a fervent Catholic devotion, modesty, asceticism, and an alleged lack of interest in political matters, though they both feared and loathed French-inspired moral and social upheavals. They both supported cultural and institutional initiatives that aimed to restore religious and political order. Like Fergola, Ruffini published a defense of the immateriality of the soul (1806), dedicated to Pope Pius VII, in which he engaged in a systematic rebuttal of all forms of scientific materialism. Ruffini structured his critique in a rigorous Euclidean fashion, to prove the "theorem" that any being endowed with cognitive faculties must necessarily be immaterial. Ruffini and Fergola crossed paths often, and the Neapolitan synthetics cited Ruffini's work with appreciation.[6]

One might expect that Ruffini was a staunch defender of synthetic geometry. He wasn't. He was a committed algebraist operating in northern Italy, where algebra had a long and glorious tradition and eighteenth-century analysis was authoritative and institutionally entrenched. After all, Lagrange was born and had taught for many years in Turin before moving to Paris, and Ruffini saw his own work as a continuation of the study of equations carried out by his famed predecessor. The set of cultural resources that Ruffini had at his disposal was different from Fergola's, and different were the assumptions and expectations that constituted the persona of the professional mathematician in northern Italy. Given this set of conditions, a return to an alleged Greek-style pure geometry as the cornerstone of mathematical practice was not a viable option for Ruffini. But he too felt acutely the need to oppose what he saw as the eighteenth-century *abuse* of mathematics. He agreed that the doctrines of sensationalism and materialism had corrupted mathematics, with disastrous moral and political results. Ruffini devoted an entire book to refute Laplace's determinism: he was appalled by the Frenchman's ambition to apply his theory of probability "to the phenomena of Nature, the moral actions, and the most important questions in life." Ruffini rejected the idea that one can discuss voluntary actions in terms of Laplacian probability. Such determinism was for him exemplary of that typically eighteenth-century philosophical aberration: the delusional use of mathematics to understand nature and society in their entirety.[7]

Like Fergola, Ruffini believed that one way to oppose and possibly revert this catastrophic philosophical turn was to defend the boundary between what he would call the moral and the physical. This he did both as a practicing physician and as a professor of mathematics: Ruffini's algebra was disembodied, like the souls that left the bodies of his dying patients. In his hands, and in those of a few other devout north-Italian mathematicians like Gabrio Piola (1794–1850), the techniques of algebra and calculus were explored, developed, and rigorized while being disconnected from their operative dimension. Their significance did not depend primarily on their empirical applications (though of course they could be used to predict and control phenomena, *to some extent*), nor did they offer an infallible and universal method for penetrating the essence of reality. Reality was conceived as ontologically and epistemologically stratified, and its study required the use of multiple cognitive approaches. Piola and Ruffini valued analysis, above all, as a form of metaphysical contemplation.[8]

In this perspective, the priorities that set the mathematician's research agenda were the rigorization of mathematical intuition, the search for the internal foundations and overall consistency of mathematical analysis, and the realization of its limited scope. The relevant similarity between Ruffini and Fergola is not to be found in the specific area of mathematics that they pursued and considered fundamental. There is nothing intrinsic to synthetic geometry that associates it with reactionary orientations. Instead, what they shared was the urge to react against the perceived abuses of algebra and calculus. These disciplines had been hijacked to support a perverted materialistic and deterministic philosophy, which they saw as the direct cause of the political upheavals that plagued their homelands. Their reaction was philosophical, political, *and mathematical*. Mathematics needed to be restored to its proper role as a means for the contemplation of metaphysical truths. The validity of its statements must remain bound to a specific epistemological level. Mathematicians must avoid two major mistakes: they shouldn't focus exclusively on practical applications, as if the meaning of mathematics came from its relation to empirical reality, but they should not indulge in "sterile abstractions and generalizations" either, as the latter might be "founded upon not-too-sound hypotheses." It followed that the most pressing problem was one of *order and discipline*: mathematical practice needed to be "rigorized," grounded on solid logical or intuitive terrain. Eighteenth-century analytic practice, with its unbounded algebraic optimism, was not just morally and philosophically reprehensible: it was *technically unsound*. The return to order needed to begin from mathematics.[9]

Is this interpretation only useful in the context of the Italian peninsula, a land of precocious and profound anti-Enlightenment reactions? There are other debates between supporters of geometrical and analytic methods on the Continent and in Great Britain during the first half of the nineteenth century, but I do not assume that they can be interpreted along the same lines. In any case, concentrating only on the opposition between synthesis and analysis would distract our attention from what seems to be the real leitmotif that links the research of Fergola and Ruffini: the defense of reactionary reason *through* mathematics.

To look for similar mathematical experiences means searching for attempts to rethink mathematics as a self-standing body of knowledge, not defined by its relation to empirical reality, and whose applications can be useful but are limited in scope. Such experiences are often flagged by the appearance and success of the new notion of "pure mathematics," a field that, once properly defined, would be the mathematician's *legitimate* kingdom. The nature of the foundation can vary from geometrical intuition to some form of logical consistency. But the key move is severing once and for all the connection between formal algebraic procedures and the inner structure of empirical reality—that is, the link that had legitimated the rampant mathematization of social and political life, emblemized by the likes of Condorcet and the Neapolitan Jacobins. Emphasis on such a notion of pure mathematics, often framed in terms of resistance against French military and philosophical occupation, figured quite prominently, for example, in the German-speaking lands. But rather than describing yet another case from the margins of the French Empire, let's take one last step toward its very center.

Et in Paris Ego

Arriving in Paris from Naples via Modena sharpens our gaze. Parisian mathematical life lacked the extreme polarization and the institutional tensions that we have found in Naples. Mathematical practice here seems, at first glance, more progressive and continuous. The standard narrative of its transformation does not center on spectacular epistemic ruptures. Mathematicians did not routinely accuse each other of not having a clue of what mathematics was all about or of not understanding the very nature of proofs. And yet, familiar as we are with the Neapolitan debate, we have developed an eye for cracks in seemingly solid mathematical regimes.

Back in 1799, Ruffini believed that his limitative result had momentous implications. He was eager to share it with leading scholars—Lagrange above all. But the French mathematician found Ruffini's proof too obscure to fathom and declined to comment, consigning it to temporary obscurity. However, an acknowledgment of the importance of Ruffini's result arrived in 1821, from Augustin-Louis Cauchy (1789–1857), who wrote that Ruffini's "memoir on the general resolution of equations is a work which has always seemed to me worthy of the attention of mathematicians and which, in my judgment, proves completely the impossibility of solving algebraically equations of higher than the fourth degree." Cauchy had generalized some of Ruffini's results, and he realized that Ruffini's techniques could be turned into useful tools to tackle what he thought was the most urgent mathematical task: the rigorization of analysis.[10]

The transition from Lagrange to Cauchy is usually seen as a key moment in the coming of age of modern mathematics. Its importance lies not only in the fact that Cauchy contributed significantly to the fields of mathematical physics and higher analysis. His research and teaching were framed within a new image of mathematics: he aimed at providing rigorous foundations for—and a reinterpretation of—all existing methods and concepts. The logical rigor, clarity, and precision that came to be associated with Cauchy's work were not, for him, merely desirable features that could be dispensed with if needed. Analysis, in other words, was not just about solving problems through the skilled manipulation of symbols. Eighteenth-century mathematicians had obtained a wealth of useful results by extending the use of algebraic methods from the finite to the infinite domain under a set of unwarranted assumptions, as if formal validity were by itself a warranty of truth. Cauchy aimed at reinterpreting all these results and concepts by situating them within a legitimating logical structure. His final goal was to make clear what can or cannot be done in mathematics: what methods are legitimate and under which conditions they can be deployed. In order to do so, he required much higher standards of rigor than those shared by his predecessors, and the development of new technical tools for the study of mathematical procedures. Historian Judith Grabiner has characterized Cauchy's new analysis—and its development by the likes of Nils Abel, Bernhard Riemann, and Karl Weierstass—as a "true scientific revolution." She uses the term in its Kuhnian sense:

nineteenth-century analysis with its rigorous definitions and proofs had not "grown naturally or automatically out of earlier views." The new "mathematical climate" of precision made it possible to define concepts (such as uniform convergence) that could not be studied or even expressed in the framework of eighteenth-century mathematics.[11]

Cauchy, who had been trained as an engineer and was for a while a professor at the École Polytechnique, did not aim for a straightforward return to Euclidean geometry. Yet he was proud to describe his project as one that would restore the *rigor* of ancient geometry in mathematics. The Greek spirit, he thought, would live on in the rigor of his definitions and in the overall logical structure of his mathematics. In many passages, as argued by Michael Barany, it's his mathematical imagination that is best characterized as geometrical.[12]

Eighteenth-century mathematicians had devised several strategies to address the then-not-very-pressing question of the foundation of calculus, which they understood mainly as justifying the rules for differentiating algebraic quantities. Lagrange had a more systematic view of the question and was convinced that analysis could ultimately be reduced to algebra. Such a reduction would provide the necessary grounding, as algebra was considered a generalized arithmetic for which no further foundation was required. The algebraization of calculus satisfied Lagrange's quest for maximum generality while omitting geometrical and empirical intuition as well as philosophical considerations such as those on the nature of infinitesimals.[13]

Like Ruffini, Cauchy saw his work mainly as building upon Lagrange's achievements. In fact, he turned the foundational program into a reality, though in a way very different from what Lagrange had envisioned. To Cauchy, Lagrange had not been rigorous enough: he had admitted into calculus almost all methods used in the algebra of the infinite, simply because they seemed to work. Lagrange's own plan for reducing calculus to algebra relied on the assumption that certain relations about infinite series hold under any condition. Instead, Cauchy argued, "most formulas hold true only under certain conditions, and for certain values of the quantities they contain." Such conditions need to be determined *with precision*. Cauchy thus used Lagrange's work extensively but transformed its meaning and purpose; his ultimate objective was to infuse algebraic formalism with geometric certainty.[14]

Why did Cauchy have such a heightened sensitivity for the question of foundations? As in the case of Fergola and Ruffini, it's Cauchy's orientations to social and political life that help us reconstruct the meaning of his mathematical anxiety. Lagrange had been the first prominent mathematician to turn the question of the foundation of calculus into a relevant problem. Interestingly, his attention seems to have turned toward foundational questions mostly in response to criticisms raised against the rigor and reliability of calculus by theologians defending religion from the attacks of modern philosophers. By far the most famous was George Berkeley (1665–1753), whose sharp remarks on the paradoxical features of Newton's fluxions and Leibniz's infinitesimals had already stimulated some of the most interesting exchanges on foundations both in Britain and on the Continent. Lagrange also took very seriously the critique by Catholic theologian Giacinto Gerdil (1718–1802) of the way mathematicians handled the concept of infinity. By the close of the century, Lagrange was convinced that the question of foundations required more attention than it had received so far.[15]

As for Cauchy, the foundational question was *the* scandal that shapes his entire work. As argued by his biographer Bruno Belhoste, Cauchy was, first of all, a *mathématicien légitimiste*, a staunch defender of the alliance of throne and altar who looked back in horror at the Great Revolution and its political and cultural aftermath. Throughout his life, he seemed truly anguished by the destiny of Catholicism in France and would commit himself to various attempts to restore the dignity and honor of the "true religion." He sided with the Jesuits in their fight against the monopoly of higher education held by the University of Paris and used his influence in the academic world to demolish any materialist philosophy he encountered. Of course, such behavior came at a cost, as the official world of science, mathematics, and engineering in postrevolutionary France was by and large liberal and anticlerical. In particular, the 1830 revolution would be for him a "national catastrophe" as well as a "personal drama." On that occasion, this "veritable Jesuit in short frock," as Stendhal dubbed him, left the country and his academic positions and spent a few years abroad. The reader will not be too surprised to learn that he traveled to hotbeds of reactionary Catholicism such as Modena and Turin, wrote to reactionary rulers across Europe for financial support, and had an audience with Pope Gregory XVI. On his return to France in 1838 he occupied a marginal position in the academic world and was unable to secure a teaching position until the upheaval of 1848.[16]

In his letter of 1821 to Ruffini, Cauchy wrote that although he owed much to Laplace, he was strongly opposed the latter's use of the theory of probability: "History ought not be investigated from a standpoint of formulas; nor should sanctions be sought for ethics and morals in the theorems of algebra or integral calculus." Cauchy's understanding of mathematics, its epistemological status, and its role in human affairs echoes the tenets of reactionary Catholicism and is well-aligned with those of Fergola, Ruffini, and other devout and legitimist mathematicians. Mathematics and science, for Cauchy, are about the contemplation of truth; in fact, "the pursuit of truth should be the sole aim of any science." Scientific truth is necessarily incomplete, a pale reflection of supernatural truth. But the two are related, as Cauchy argues, using well-known arguments from natural theology or emphasizing the divine nature of the human intellect. Contemplating natural and mathematical truths prepares us for other, far more important contemplations: "The mere contemplation of these heavenly wonders, of their divine beauty, suffices to compensate us for all that we may sacrifice in discovering it, and the joy of heaven itself is no more than the full and complete possession of immortal truth."[17]

Like Fergola and Ruffini, Cauchy argued that there exist different kinds of truth: the dogmas of revealed religion; the principles of sound metaphysics and morals that should inform social life; mathematics; and the knowledge of natural states of affairs. He was acutely aware that there are intrinsic limitations to the knowledge that can be gained through the sciences: they will never let humans discover "the innermost nature of beings, of things, and would never discern the secret mechanisms that underlie the motion of things." "The great crime of the last century," he declared, "was that of wanting to raise nature itself up against its very Author, of desiring to set creatures in a state of permanent revolt against the Creator and even to arm the sciences against God himself." The sciences should be cultivated "with ardor" but "without falling prey to the desire to extend them beyond their proper domains." How absurd, continued Cauchy, to believe that "we can come to grips with history by means of formulas" or "base morality on theorems from algebra and from integral calculus." It was time for the true mathematician to determine with precision the domain of validity for methods and results that the mathematicians of the Enlightenment age had improperly extended. Mathematics should retreat from empirical fields to which it was inapplicable: "Thus," comments Belhoste, "it was that the confident age of indefinite extension was followed, in analysis, by a more rigorous age of voluntary restriction."[18]

This age of voluntary restriction originated the concepts, methods, and domains of research that contrived modern analysis. Cauchy's mathematical practices of self-discipline and self-limitation were expressions of a new kind of reason, a reactionary reason. Grabiner is right in seeing a major break in the history of mathematics where others saw simply a process of self-explanatory rigorization. The term "revolution," however, might not be the most suitable to capture the meaning of Cauchy's newly rigorous and pure mathematics. What he accomplished was instead a mathematical counterrevolution.

Mathematics and Modernity

In the 1820s Cauchy set the tone of European mathematics for decades to follow. In Grabiner's words, he created the "climate" for a profound transformation of the entire edifice of mathematics and for the reinterpretation of all its previous results and methods. The principles that guided this transformation were those of rigor, self-restraint, and purity. A new age had opened—the age of modern mathematics—in which the meaning and use of mathematics was subject to a "voluntary restriction." At the same time, some traits of late eighteenth-century analysis too strike us as distinctively modern: the quest for generality, the emphasis on practical applications, the ambition to mechanize mathematical procedures, and the interest in developing algorithms that would increase our ability to predict and control phenomena. So how should we understand the nexus of mathematics and modernity?

Historians of mathematics use the terms "modern" and "modernization" to refer to different periods and features, including early modern algebraic innovations, early-nineteenth-century rigorization, and, most commonly, the period between 1890 and 1930. The latter was characterized by the acceleration of some of the trends we have encountered in this study: foundationalism, abstraction, axiomatic and formalist approaches, mathematics as a self-standing system of knowledge, structuralism, and the sense that one is at a moment of radical rupture with the past. The purely syntactic notion of a formal system did not belong to the mathematical imagination of Fergola and Cauchy: it emerged fully only in this later wave of foundational research and followed the legitimation of non-Euclidean geometries. It has been often noticed that this new, modern way of understanding mathematics shared some of its features with contemporary visual arts, music, and architecture. Jeremy Gray has given us a detailed overview of the pervasiveness of these modernist traits in logic and mathematics. His assessment, however, falls short of trying to explain the simultaneous fascination with abstraction and structure across socially distant fields in the arts and sciences. Historians of art have explored possible connections, showing how some artists looked at the natural sciences and mathematics to address pressing problems in their own practices. This approach goes beyond the recognition of common features and engages with possible historical explanations for their emergence. The discourse of science is thus described as a resource at the artists' disposal. But was mathematics itself shaped by the broader modernist culture?[19]

The most impressive attempt to answer this question is Herbert Mehrtens's *Moderne-Sprache-Mathematik*, an ambitious and original book, published in 1990, that deploys high-powered semiotic and philosophical tools to study the discourse of the autonomy and meaning of mathematics in Germany between 1890 and 1930. Mehrtens divides up the leading mathematicians of that period into "moderns" and "countermoderns," with David Hilbert and L. E. J. Brouwer as representatives of the two groups. The moderns saw mathematics as an autonomous space of free human creation, logical coherence being its only constraint. Mathematics per se did not capture any essential trait of the empirical world; it was an artificial symbolic language whose axioms and rules were explicit and conventional, like those of the game of chess. The countermoderns, by contrast, emphasized the role of intuition (*Anschauung*) as the basis of mathematics and the field's essentially constructive nature. The truth of mathematics ultimately depended on something that was ontologically independent from the mathematical language in which it was expressed. The two groups also diverged significantly on the role mathematics should play in society. For Mehrtens the foundational crisis of the 1920s was, above all, a crisis of identity of an entire discipline. He did not shy away from trying to pin down the causes of such a crisis: it was, for him, a reaction to the acceleration of the social process of modernization that had invested Europe in the second half of the nineteenth century. This process, the privileged object of study by the newborn discipline of sociology, did not leave mathematical life unaffected. Issues of productivity, labor, and disciplinary identity shook its academic community and shaped its debates, especially the foundational one. Moderns and countermoderns both reacted *mathematically* to the great social problems of their age.[20]

In this book I have clearly followed Mehrtens's invitation to explore the politics of mathematical modernity, and, like him, I believe that historical understanding requires going beyond morphological descriptions and comparisons. Similarities, per se, might not mean much, nor do they point necessarily to any profound connection. Like Mehrtens, I have not shied away from trying to understand historically the emergence of key traits of modern mathematics—purity, above all. My strategy, however, has been different. Rather than relying on discourse analysis, I showed how synthetics and analytics mobilized available materials and symbolic resources to address questions of sense making and problem solving. Moving from their concrete experience, I reconstructed their competing mathematical cultures and their broader social orientations. Consider the relation between Fergola's mathematics and a painting of the Posillipo school. What connects these two historical objects is not simply the fact that they share certain formal traits or are semiotically related. Rather, they offer ways of tackling two concrete and distant field-related problems (What does it mean to prove a geometrical statement? How best to represent visually the Neapolitan landscape?) while pointing toward a common solution to urgent problems of social order (resisting the transformation of Neapolitan social and political life along revolutionary and liberal lines).

The direction of my analysis is also different from Mehrtens's. He begins with the definition of interpretive categories (modern and countermodern) that he subsequently applies in a series of top-down movements. The impression is that some actors are thus boxed up in ways that do not capture the complexity of their scientific lives. My analysis moved in the opposite direction, focusing on how local communities of mathematicians understood certain problems and addressed them based on available resources and within existing institutional constraints. I built upward from there, toward their mathematical cultures and the solutions that these cultures offered to problems related to modernization. Categories such as "reactionary mathematics" are thus not to be found at the beginning of my historical reconstruction but are rather an outcome of it. They are not clear-cut or rigid; they do not carry out explanatory heavy lifting, which makes it easier to avoid deterministic trajectories and keep the focus on each contingent deployment of mathematical tools and concepts.

Finally, this study is also indebted and complementary to a rich body of literature in the history of science that has studied the emergence of modern analytic machines, techniques and forms of discipline, and their social significance. The Neapolitan analytics had a clear idea of what modernity should look like, and this literature is key to exploring how they understood the nexus of analysis-mechanization-modernity.[21] Building on this literature, I have delved into the vast and under-researched world of those who, around 1800, were *resisting* modernization. Their resistance was not passive. They did not simply cling onto the last remains of defeated traditions, even though they would constantly declare that this was precisely what they were doing. In mathematics, the core of my story, they creatively reactivated and reinterpreted past mathematical knowledge (like Fergola) or pushed innovation in a way that would altogether change the meaning of mathematics itself (like Cauchy). In both cases, the goal was not simply to resist innovation but to reshape mathematics through a process of rigorization that would increase precision and rigor while restricting and disciplining the use of mathematical formalism. They strived, in fact, for a different kind of modernity. And to some extent they succeeded. When we think about our mathematical modernity, from Cauchy onward, we should recognize that some of the assumptions and priorities that shaped it were originally associated with reactionary, anti-modern values.

Sociologically, this recognition reveals the fallacy of seeing a rigid opposition between tradition and modernity, as if the first stood for passivity and immobility and the second for action and innovation. This dichotomy has long been entrenched in social theory, and it still shapes much of contemporary social and political discourse. It is one of the most archetypical Enlightenment myths, and abandoning it is a long and difficult affair.[22] And yet the use of this dichotomy, with all its moral associations, oversimplifies our representation of social processes and clouds our understanding of the social construction of knowledge. The present study is an illustration of how mathematical knowledge, like all other kinds of knowledge, is the outcome of traditions that are creatively reinvented at each single step, even when it looks like they are simply being continued. But there is no such thing as "simply continuing" a knowledge tradition, as if the next step were obvious and predetermined. Not even in mathematics. Continuing a tradition should not be more obvious and natural, in terms of historical understanding, than explicitly attempting to modify it. In this sense deductive formal knowledge is not different from the rest of human knowledge: tradition and change are to be found everywhere in the knowledge-making process.

Historically, these conclusions leave us with an understanding of mathematical modernity as central to the advent of modern political life. As we have seen, mathematical modernity was shaped by distinct, at times opposed, orientations toward natural and social realities. Among them, we paid particular attention to the rise of pure mathematics and the disconnection of mathematical meaning from empirical reality. These phenomena were not logical developments of earlier mathematics; they rather marked a rupture, a shift—a counterrevolution, I argued—and were the results of historically situated choices. These mathematical and often highly technical choices provided the logico-mathematical infrastructure for the age of Restoration and for forms of reason and political life, ranging from reactionary to liberal, that shared a common rejection of revolutionary action and of the empowerment of subordinate social groups. They also rejected a vision of mathematics and science as reflexive practices that could transform social worlds and support new political subjectivities, best exemplified by the revolutionary mathematics of the Jacobins. Political revolution and the questioning of existing social hierarchies had long been condemned as manifestations of moral corruption. In the world of self-restricted modern mathematics, they could also be declared irrational and *technically unviable*. Mathematical restriction entailed a restriction of viable political options.

I conclude by acknowledging that my tools and guiding principles too were forged in the conflict that I describe in this book. It was useful but unfair to point a finger at Gillispie's asymmetrical stance as if I were on epistemological high ground—I am not. The contingent and local notions of reason, knowledge, and mathematical culture that I have deployed to tell this story are obviously rooted in the composite reactionary culture that I have reconstructed, while my understanding of the proximity of mathematics and politics, of mathematical procedures as constitutive of social order, echoes the Jacobins' revolutionary and impure epistemology. Inevitably, we use and adapt the tools we have inherited; and while the meaning of these tools can be transformed by new usages, we can be sure of one thing: that they will never be neutral.

Epilogue

For all its decades-long clamor and drama, the conclusion of the Neapolitan controversy over mathematical methods was rather underwhelming. It ended, or, to be more precise, faded away, in the early 1840s. There was no clear verdict; rather, it slipped into insignificance. The original terms of the controversy had lost their urgency—in fact, they had lost their meaning. Responding to the synthetic challenge of 1839, Padula had proclaimed the failure of Fergola's program and the anachronistic, useless nature of synthetic research. At that point, the analytics had no interest in continuing to give visibility to the synthetics' arguments. Their approach had been institutionalized where it mattered, and they now represented Neapolitan mathematics in the eyes of their Italian and European colleagues.[23]

By the time of the national unification of Italy in 1861, the second and third generation of analytic mathematicians had embraced the liberal-patriotic ideals of the Risorgimento, contributed to the defeat of the Bourbon regime, and even (some of them) entered politics. To them, the Risorgimento had been a process of both political and scientific emancipation, and they saw modern science as central to the life of the new nation. Given the disproportionate political weight of the Italian mathematical community in the new national parliament, it's not surprising that matters like mathematical education would be high on the agenda. It is in this context that the ghost of the by-then-forgotten Neapolitan controversy should make its appearance.[24]

In 1867 a parliamentary commission suggested that Euclid's *Elements* should be adopted as the textbook of geometry in the colleges of the new kingdom. This tactic was seen as the most effective way to replace the congeries of texts used in the pre-unitary states and standardize the teaching of geometry across the nation. Antonio Cremona (1830–1903), who led the commission, authoritatively defended the value of Euclid as "the most logical and rigorous" textbook available and, "in the universal opinion, the most perfect model of geometrical rigor." Leading north-Italian mathematicians believed that the role of mathematical education was not limited to providing a set of useful truths. Rather, it was supposed to support the student's intellectual development "as a gymnastic of thought, aiming to develop the rational faculty, and to support that healthy and correct criterion to distinguish truth from what merely seems true."[25]

The commission blended pedagogical considerations (synthetic geometry is more suitable than "the abstract science of numbers" to instill "the habit to reason with inflexible rigor") with a concern for defending the "purity" of geometrical reasoning (teachers should not "pollute the purity of ancient geometry by transforming geometrical theorems into algebraic formulas"). Euclid is an "inimitable model of logic and clearness" and must be preferred, for pedagogical purposes, to any textbook based on "the mechanism of arithmetical process."[26]

For all its apparent reasonableness, the commission's decision faced the strong and, to most, puzzling opposition of the southern mathematicians. Giuseppe Battaglini (1826–1894), an engineer trained at the School of Application under Padula's supervision, led the critics. An early proponent of non-Euclidean geometries, Battaglini published, in the scientific journal he directed, numerous articles that disparaged this return to Euclid on a national scale.[27] The authors referred to the error of setting geometry apart from arithmetic, the logical difficulties implied by the notion of proportion, and to other well-known obscurities contained in Euclid's text. The north-Italian authors of a new edition of Euclid declared themselves surprised by the "intolerance" of their critics.[28]

In an unexpected twist, the heirs of the Neapolitan analytic school found that their self-proclaimed modernity was being questioned. Their devaluation of the geometrical tradition, radically instrumental conception of mathematics, and exhibited lack of interest in pure mathematics and foundational issues could hardly align with what their northern colleagues thought modern mathematics was all about. They thus found themselves accused of hindering the progress of mathematics by sticking to the methodological tenets that their teachers had defended against the synthetics in the very name of modernization. By centering their teaching on mechanics and applied mathematics, their critics argued, the analytics had kept the best students away from "higher studies"—pure mathematics—thus compromising scientific progress in Naples.[29]

Fergola's mathematics, it turned out, was modern.

Notes

Introduction

1. Wittgenstein, *Remarks on the Foundations of Mathematics*, I, §155.

2. For simplicity, I refer to the "Kingdom of Naples" throughout this book, even if, following the unification of the crowns (1816), its official name turned into "Kingdom of the Two Sicilies."

3. "Mathematical resistance" brings together mathematical knowledge and social action in a way that is emblematic of my overall argument. I use the term "resistance" with an ambivalent rather than obviously emancipatory meaning.

4. Bruno Latour defines modernity in terms of purification, one of the key dynamics of what he calls the "modern constitution" (Latour, *We Have Never Been Modern*). Latour, however, describes the modernist purification process as something that happens spontaneously and inexplicably and as a historical unicum that fully characterizes self-proclaimed modernity. By contrast, I interpret the process of purification as a site of social and political struggle, one not essentially different from other processes of coproduction of scientific and social order.

5. On focusing on "cultures" and their distinctive "epistemic machineries" as a way to disunify the sciences and bring out diversity, see Knorr Cetina, *Epistemic Cultures*, 2–5. For some insightful methodological remarks on studying science as culture, see Pickering, *Science as Practice and Culture*, 1–26. A more recent collection, Chemla and Fox Keller, *Cultures without Culturalism*, offers a useful survey of uses of "culture" as an analytic category in the history of science that avoid the pitfall of "culturalism" (cultural essentialism).

6. The concept of "marginality" enters my narrative in at least three interrelated ways. First, to refer to perceptions and arguments of marginality used by the actors of my story—to fashion the position of an opponent as backward, for example, or to claim proudly their own untimeliness. Second, to refer critically to historiographical traditions that focus on a few mathematical "centers" and that aim to follow the alleged diffusion of ready-made mathematical knowledge across the "peripheries." Third, I refer to marginality as a position of political and economic subordination that conditions action, and especially dynamics of knowledge production and appropriation. On the conceptual dyad of "center" and "periphery" in the study of the circulation of scientific knowledge, see O'Connell, "Metrology"; Grove, *Green Imperialism*; Gavroglu et al., "Science and Technology in the European Periphery"; Schaffer, "Newton on the Beach"; Warwick, "Asia as a Method in Science and Technology Studies"; Patiniotis, "Between the Local and the Global." For a pioneering questioning of the "apparent transparency" of the history of

"European mathematics" and for some insightful historiographical remarks on its origins and spatial boundaries, see Goldstein, Gray, and Ritter, *Europe mathématique*, 8, 15–30, and 207–18.

7. By "mathematical regime" I mean a highly disciplined setting, characterized by institutionalized mechanisms for the production, regulation, distribution, and circulation of mathematical knowledge (see Foucault, "The Political Function of the Intellectual").

8. Histories of mathematics offer an extreme case of what Chakrabarty describes as "historicism" (*Provincializing Europe*, 22–23).

9. Rowe, "New Trends and Old Images."

10. I'm referring to the microhistorical techniques designed to explore specific conditions of experience in order to reveal their "invisible structures." Their aim is to reconstruct how actors experienced constraints and possibilities as well as the presence of resources they could mobilize as part of their strategies. See Ginzburg and Poni, "Il nome e il come"; Levi, "On Microhistory"; and Revel, "L'histoire au ras du sol." For an exemplary microhistorical exercise in the history of science, see Biagioli, *Galileo Courtier*.

11. See Foucault, "Nietzsche, Genealogy, History," and Dreyfus and Rabinow, *Michel Foucault*, 104–25.

12. Warwick, *Masters of Theory*. For a recent study on the globalization of a mathematical theory, see Barany, "Fellow Travelers and Traveling Fellows."

13. Boyer, *History of the Calculus*, 267–98; Grattan-Guinness, "The Emergence of Mathematical Analysis"; Bottazzini, *Il calcolo sublime*, 84–135; Katz, *A History of Mathematics*, 635–58.

14. Boyer, *History of the Calculus*, 224–66; Kline, *Mathematics*, 100–96.

15. Gray, "Anxiety and Abstraction."

16. In fact, biographies of mathematicians have been a genre for centuries, but the relation between the protagonists' lives and their ideas was rarely turned into a significant question.

17. Netz, *The Transformation of Mathematics*, 1; Rowe, "New Trends and Old Images," 3.

18. For exemplary research in this vein, see Fraser, "The Calculus as Algebraic Analysis"; Grabiner, *The Calculus as Algebra*; Bos, *Redefining Geometrical Exactness*; Chemla, "On Mathematical Problems as Historically Determined Artifacts"; Jahnke, "Algebraic Analysis in Germany"; Hanna, Jahnke, and Pulte, *Explanation and Proof in Mathematics*; Mancosu, *Philosophy of Mathematics*; Chemla, *The History of Proof in Ancient Traditions*; Netz, *The Shaping of Deduction*.

19. Rowe, "New Trends and Old Images," 13.

20. See, for example, Guicciardini, *Anachronism in History of Mathematics*. For a recent historiographical survey, see Barany, "Histories of Mathematical Practice."

21. Shapin, "History of Science and Its Sociological Reconstructions," 157–58.

22. My understanding of social life, and hence my use of terms like "world," "object," "society," "structure," "action," and "strategy," is essentially interactionist (see Goffman, *The Presentation of Self* and *Strategic Interaction*).

23. Baxandall, *Patterns of Intention*; Geertz, "Deep Play."

24. The working notion of "reason" deployed in this book is not atemporal, nor do I identify reason with certain core characteristics, like logical rigor. Once one takes seriously the historicity of science and mathematics there is no way back to a ready-made notion of "scientific reason": neither to rely on it nor to blame it. Instead, I invoke a more capacious and inclusive notion, whereby reason can be understood as historical and situated: making sense of the world through discerning patterns of meaning. This ecumenical perspective, inspired by exemplary works in the history of science and science studies, matches my considerations on the historicity of mathematics and the sociohistorical approach to it.

25. Jay, *Reason after Its Eclipse*, 22–23.

26. "Reactionary reason" and "analytic reason" enter my narrative as ideal types positioned at the edges of a spectrum; what is most interesting lies *within* that spectrum. "Reactionary reason" and "reactionary mathematics" are labels that produce an oxymoronic effect, not unlike Jeffrey Herf's "reactionary modernism." Like Herf, I emphasize the conjunction of a tradition of political reaction with what are usually described as distinctive aspects of modernity. Herf studies the Nazi reconciliation of Romantic irrationalism and modern technology. This book shows how counterrevolutionary political reaction, since its emergence, has appropriated and transformed notions of reason and modern science—especially mathematics. See Herf, *Reactionary Modernism*; and Rohkrämer, *Andere Moderne*. On alternative modernities see also Tresch, *Romantic Machine*.

27. See the exemplary MacKenzie, *Statistics in Britain*. This notion of "interest" should not be understood in opposition to the interests in prediction and control; rather, it includes them, as interaction and social organization have priority with respect to instrumental action. For the Edinburgh sociologists, interests are not a ready-made tool for monocausal explanations, but sets of priorities and objectives sustained by the entire culture of a group. See Barnes, *Interests and the Growth of Knowledge*; and the exchange between Barry Barnes, Donald MacKenzie, Steve Woolgar, and Steven Yearley in volumes 11, 13, and 15 of *Social Studies of Science* (1981–1984).

28. Golinski, *Making Natural Knowledge*, 1–46.

29. Richards, *Mathematical Visions*; Latour, *Science in Action*; Latour, "Sur la pratique des théoriciens"; Brian, *La mesure de l'État*; Brian, "Le livre des sciences est-il écrit dans la langue des historiens?"; Pickering and Stephanides, "Constructing Quaternions"; Pickering, *The Mangle of Practice*; MacKenzie, *Mechanizing Proof*; MacKenzie, "Slaying the Kraken"; Warwick, *Masters of Theory*; Porter, *Trust in Numbers*; Daston, *Classical Probability in the Enlightenment*; Erikson, Klein, et al., *How Reason Almost Lost Its Mind*; Rosental, *Weaving Self-Evidence*; Alonso and Starr, *The Politics of Numbers*; Thévenot, "La politique des statistiques"; Desrosières, *The Politics of Large Numbers*; Sætnan, Lomell, and Hammerm, *The Mutual Construction of Statistics and Society*.

30. Rowe, "New Trends and Old Images." See Rabouin, "Styles in Mathematical Practice," for a recent revision of the notion of "mathematical style."

31. Livingston, *The Ethnomethodological Foundations of Mathematics* and "Cultures of Proving"; MacKenzie, *Mechanizing Proof*.

32. See Wittgenstein, *Philosophical Investigations*, and Bloor, *Knowledge and Social Imagery*, 74–140. See also Bloor, "Wittgenstein and Mannheim"; and Bloor, "Polyhedra and the Abominations of the Leviticus."

33. On the Weberian notion of status group I'm deploying here, see Barnes, "Status Group and Collective Action." I use culture and practice as a dyad where the first emphasizes the resources available to the actors, and the second emphasizes their actions, and especially their acts of making/unmaking.

34. Barnes, "Social Life as Bootstrapped Induction"; MacKenzie, *An Engine, Not a Camera*.

35. Davis, *Naples and Napoleon*. See also Franco Venturi's insightful comments about the potential of studying Enlightenment culture in its peripheral manifestations, where its features seem more extreme and its contradictions more radical (Venturi, *Utopia and Reform*, 133).

36. Gillispie, *Science and Polity in France*.

37. Baker, *Inventing the French Revolution*, 153–66, quotation on 164; Baker, "The Establishment of Science in France," 54–55. See also Baker, *Condorcet*.

38. This fruitful line of research has been developed by scholars concerned with the political dimension of engineering and mathematical practices in eighteenth-century France: Picon, *L'invention de l'ingénieur moderne*; Brian, *La mesure de l'État*; Alder, *Engineering the French Revolution*. A more recent book (Tresch, *The Romantic Machine*) complicates our understanding of both Romanticism and Parisian politics while showing the multiple connections between the world of machines and radical politics, especially social utopias.

39. The disinterestedness of science had been constitutive of scientific and political discourses since the seventeenth century but acquired a particular prominence and a renewed appeal around 1800. See especially Shapin, *A Social History of Truth*; and Alder, *Engineering the French Revolution*.

40. There is a vast literature on the historicity of the categories of "pure science" and "applied science" as they were deployed—and contested—in the late nineteenth and, especially, twentieth centuries. See Kline, "Construing Technology as Applied Science"; Clarke, "Pure Science"; McClellan, *The Applied Science Problem*; Grandin, Wormbs, and Widmalm, *The Science-Industry Nexus*; Gooday, "Vague and Artificial"; Leggett and Sleigh, *Scientific Governance*; Shapin, *Scientific Life*.

41. Foucault, *The Archeology of Knowledge*; Foucault, *Power/Knowledge*. On mathematics, note the passage in Foucault, *Dits et écrits*, vol. 1, 1027.

42. MacKenzie, "On Invoking 'Culture.'"

43. This notion of mathematical culture encompasses both the technical and the social dimensions freeing the historiography of mathematics from the persistent dichotomy of internal and external factors. Due to this constant interaction with their environment, mathematical cultures—like all cultures—are fluid, dynamic, and open. It follows that the construction of sameness and universality in mathematics is always situated and contestable—no matter how solid they look (see Chemla and Fox Keller, *Cultures without Culturalism*, 1–26).

44. Barnes, *The Nature of Power*, quotation on 169–70. See also the insightful comparative remarks in Kusch, *Foucault's Strata and Fields*.

45. "Co-production is shorthand for the proposition that the ways in which we know and represent the world (both nature and society) are inseparable from the ways in which we choose to live in it" (Jasanoff, *States of Knowledge*, 3). Research on standards is exemplary of the way scientists and technologists can impose order on modern social life (see Lampland and Star, *Standards*; Timmermans and Epstein, "World of Standards").

46. In this study, I use the term "power" to refer to both power as capacity (power to) and power as domination (power over).

Chapter One

1. Schaffer, "Enlightened Automata," 129. See also Schaffer, "Machine Philosophy"; Ashworth, "Memory, Efficiency, and Symbolic Analysis."

2. D'Alembert, "Discours préliminaire"; Woolf, *Analytic Spirit*; Främgsmyr, Heilbron, and Rider, *Quantifying Spirit*.

3. Diderot and d'Alembert, *Encyclopédie*, vol. 1, 400–03. Hankins, *Science and the Enlightenment*, 21–22.

4. Condillac, *Oeuvres philosophiques*, vol. 2, 402, 469; Diderot and d'Alembert, *Encyclopédie*, vol. 1, 401.

5. Roberts, "Senses in Philosophy and Science," 125–26; Rosenfeld, *Revolution in Language*; Cassirer, *Philosophy of Enlightenment*, 24; Foucault, *Order of Things*.

6. Baker, *Condorcet*, 109–28, quotations at 124–25.

7. Brian, *La mesure de l'Etat*, 45.

8. Paty, "Rapports des mathématiques," 79; d'Alembert, "Discours préliminaire"; Montucla, *Histoire des mathématiques*, vol. 1, 374; Fraser, *Calculus and Analytical Mechanics*.

9. Brian, *La mesure de l'Etat*, 53.

10. Koyré, *Études*, 116; Brian, *La mesure de L'Etat*, 68.

11. Granger, *La mathématique sociale*, 59, 91; Daston, "Condorcet," 118–23, quotation on 120; Daston, "Enlightenment Calculations," 184–93. On Condorcet's awareness of the political and epistemological limits of this calculability, see Bates, *Enlightenment Aberrations*, 85–90. On the historical process that led to the modern separation between intelligence/thinking and calculation, see Jones, *Reckoning with Matter*.

12. Granger, *La mathématique sociale*, 38; Baker, *Condorcet*, 82; Brian, *La mesure de l'Etat*, 47. For a sophisticated rendering of Condorcet's understanding of error and political decision-making, see Bates, Enlightenment Aberrations, 73–97.

13. Granger, *La mathématique sociale*, 61.

14. Jones-Imhotep, "The Unfailing Machine."

15. Riskin, *Science in the Age of Sensibility*, 49–51.

16. Figlio, "Theories of Perception"; Denby, *Sentimental Narrative*, 137; Skinner, *Sensibility and Economics*; Motooka, *The Age of Reasons*; Rosenfeld, "Social Life of the Senses"; Vincent-Buffault, *History of Tears*; Goldstein, *Post-Revolutionary Self*.

17. Riskin, *Science in the Age of Sensibility*, 7; Vila, *Enlightenment and Pathology*, 13–110; Rey, *Naissance et développement du vitalisme*; Vila, "Introduction"; Roberts, "The Senses in Philosophy and Science"; Sheehan and Wahrman, *Invisible Hands*.

18. For a classic example, see Darnton, "Philosophers Trim the Tree of Knowledge."

19. Venturi, *Italy and Enlightenment*; Porter and Teich, *The Enlightenment in National Context*; Whiters, *Placing the Enlightenment*; Pocock, *Barbarism and Religion*; Outram, *The Enlightenment*. For a nuanced and historically sensitive reassertion of the primacy of ideas and the unitary nature of Enlightenment, see Tortarolo, *L'Illuminismo*; Robertson, *Case for the Enlightenment*.

20. Clark, Golinski, and Schaffer, *Sciences in Enlightened Europe*; Golinski, "Science in the Enlightenment, Revisited"; Livingstone, *Putting Science in Its Place*; Whiters, *Placing the Enlightenment*. On the centrality of the quarrel of the ancients and the moderns for the fashioning of the eighteenth-century narrative of Enlightenment, see Edelstein, *Enlightenment*.

21. Venturi, *Settecento Riformatore*; Carpanetto and Ricuperati, *Italy in the Age of Reason*; Rao, "Enlightenment and Reform."

22. Robertson, *Case for the Enlightenment*; Venturi, *Settecento Riformatore*, vol. 1, 523–644; Galasso, *La filosofia in soccorso de' governi*, 15–66; Imbruglia, "Enlightenment in Eighteenth-Century Naples"; Calaresu, "Enlightenment in Naples."

23. For a contextualization of Vico's work in early eighteenth-century Naples, see Stone, *Vico's Cultural History*; Robertson, *Case for the Enlightenment*; Naddeo, *Vico and Naples*.

24. On the centrality of political economy to the Neapolitan—and Scottish—Enlightenment, see Robertson, *Case for the Enlightenment*.

25. Imbruglia, "Enlightenment in Eighteenth-Century Naples," 2.

26. At mid-century, the Neapolitan king was still formally a vassal of Rome, a subordination expressed symbolically through the annual gift of a white palfrey to the pontiff. In their defense of the rights of the state and the crown, Neapolitan reformers relied upon the local tradition of

giurisdizionalismo, a juridical apparatus elaborated by law scholars through centuries of fighting back Roman claims. The issues at stake in this dispute had enormous juridical and economic relevance, including the taxation of ecclesiastical possessions, the investiture of bishops, the boundaries of the ecclesiastical jurisdiction, book censorship, and the control of education. See Lauro, *Il giurisdizionalismo*; Ajello, *Il problema della riforma*; Galasso, *La filosofia*, 171–92; and Chiosi, *Lo spirito del secolo*, 143–96. The anticurial stance had always received support from across Neapolitan society and was hardly a reason for social divide (Chiosi, *Lo spirito del secolo*, 58–70).

27. For a selection of reformist writings, see Venturi, *Illuministi italiani*, vol. 5.

28. Schaffer, "Newtonianism"; Shank, *Newton Wars*; Mandelbrote and Pulte, *Reception of Isaac Newton*.

29. On the relation between Vico and Genovesi, see Zambelli, *Formazione filosofica*, 239–93.

30. Genovesi, *Universae christianae theologiae elementa*; Genovesi, *Dissertatio physico-historica* (which introduces an edition of Pieter van Musschenbroek's Newtonian textbook); Galasso, *La filosofia*, 69–168; Zambelli, *Formazione filosofica*, 427–28, 533. He would nevertheless endure a long trial (1756–63) and his main economic publication would eventually be banned (1817).

31. Ferrone, *Intellectual Roots*; Mazzotti, "Newton in Italy."

32. Genovesi, "Discorso sopra il vero fine," 44–45, 53, 56. See also Genovesi, *Lettere accademiche*, 92–93; Venturi, *Settecento riformatore*, 523–644; Galasso, *La filosofia*, 337–506.

33. Genovesi, *Logica*, 29, 35–36. 75–76, 157, 161.

34. Venturi, "1764"; Genovesi, *Lezioni di commercio*; Pii, *Antonio Genovesi*; Marcialis, "Legge di natura."

35. Filangieri, *La scienza*, vol. 1, 39 ("the tribunal of public opinion"). On the feudal question, see Rao, *L'amaro della feudalità*; Rao, "The Feudal Question"; and Massafra, "Una stagione di studi." On education and the creation of a public sphere, see Villari, "Antonio Genovesi"; Perna, "L'universo comunicativo"; Chiosi, *Spirito del secolo*, 79–85; and Robertson, *Case for the Enlightenment*, 35–37, 382–83.

36. Schipa, *Un ministro napoletano*, 43–44; Brancato, *Il Caracciolo*; Renda, *La grande impresa*.

37. Rao, "Esercito e società."

38. Giarrizzo, *Massoneria e illuminismo*; Melton, *Rise of the Public*, 252–72; Robertson, *Case for the Enlightenment*, 19; Robertson, "Enlightenment and Revolution," 34 (quotation).

39. The Illuminati's resolute opposition to all forms of "despotic" government had attracted the attention of the Bavarian authorities: the original lodge was deemed politically subversive and dispersed in 1784. Some of the foreign lodges, however, including the Neapolitan one, continued to operate for a few more years. See Francovich, *Storia della massoneria*; Giarrizzo, *Massoneria e illuminismo*; Rao, "La massoneria nel Regno di Napoli."

40. Representative of the more philosophical strain are Grimaldi, *Riflessioni sopra l'ineguaglianza tra gli uomini*; Filangieri, *La scienza*; Pagano, *Saggi politici*; and the periodical *Scelta miscellanea*, which published book reviews, excerpts from Genovesi and Vico, and even a commentary on the constitution of the United States of America. See Cortese, *Eruditi e giornali letterari*, 97–103; and Venturi, *Illuministi italiani*, vol. 5, 794–96.

41. Venturi, *Italy and the Enlightenment*, 213; Venturi, *Illuministi italiani*, vol. 5, xv; Imbruglia, "Enlightenment in Eighteenth-Century Naples," 84; Calaresu, "Enlightenment in Naples," 406–08.

42. See, for example, Cuoco, *Saggio*.

43. Quoted in Venturi, *Italy and the Enlightenment*, 219.

44. Fiorentino, *Riflessioni*; Galanti, *Descrizione*; Davis, *Naples and Napoleon*, 42–53; Naddeo, "Cosmopolitan in the Provinces."

45. Salvatori, *Il mito del buongoverno*; Galasso, *L'altra Europa*, 171–225; Schneider, *Italy's 'Southern Question'*; Dickie, *Darkest Hour*.

46. Venturi, *Illuministi italiani*, vol. 5; Di Battista, *Il Mezzogiorno*.

47. Similar considerations could be made for other productive processes, such as silk throwing. See Ciccolella, *La seta*; Ragosta, *Napoli*; Chorley, *Oil, Silk, and Enlightenment*; Salvemini, *L'innovazione precaria*.

48. Grimaldi, *Istruzioni*; Grimaldi, *Piano di riforma*. On Grimaldi, see Basile, "Un illuminista calabrese"; Venturi, *Illuministi italiani*, vol. 5, 411–30; and Luciano, *Domenico Grimaldi*, xi–lvii. On the simultaneous demand for free-trade policies and for government intervention, see Venturi, *Italy and the Enlightenment*, 206.

49. Grimaldi, *Istruzioni*, 60; Grimaldi, *Memoria*, 35; Grimaldi, *Relazione*, 40. On the significant commercial relations between North America and the Italian peninsula in the second half of the eighteenth century, see Codignola, "Relations."

50. Mazzotti, "Enlightened Mills."

51. Chorley, *Oil, Silk, and Enlightenment*; Montaudo, *L'olio nel Regno di Napoli*; Macry, "Ceto mercantile."

52. Presta, *Trattato degli ulivi*, 454–56.

53. Mazzotti, "Enlightened Mills." On the historical bifurcation between ideas and their materialization, see Jones, *Reckoning with Matter*; see also Ingold, "Textility of Making."

54. Palmieri, *Osservazioni*; De Salis Marschlins, *Nel Regno di Napoli*, 40–41; Montaudo, *L'olio nel Regno di Napoli*, 57–131.

55. Palmieri, *Osservazioni*, 99.

56. Venturi, "Napoli capitale."

57. On the notion of "entrepreneur" in the historiography of Naples, see Galasso, *L'altra Europa*, 227–55. These dynamics were typical of the European peripheral highlands. Many ancien régime states functioned through a concentration of the forces of order in the cities, the integration of surrounding plains, and a significant autonomy of mountainous regions. The highlands were thus highly autonomous but not economically isolated; on the contrary, centers and peripheries were interconnected and interdependent. Therefore it would be absurd to think of peripheries as static worlds, and indeed they weren't. As Giovanni Tocci has shown, ancien régime communities and clan networks were dynamic institutions whose stability was sustained by incessant social action. The main forms of this social action were those of kinship and clientage, which in turn reasserted local values and consciousness. Bourbon reformers first and French forces of occupation later were bewildered by the atomized nature of authority in those regions, and by the complexity and informality of communal-clientage networks. They were, above all, baffled by the fierce resistance to technical innovation they encountered, a resistance that they understood in terms of ignorance and isolation—in other words, "backwardness." See Broers, *Napoleonic Empire*, 20–22, 58–60; Tocci, *Comunità negli stati italiani*; Tocci, *Comunità in età moderna*.

58. Lomonaco and Torrini, *Galileo e Napoli*.

59. Around 1700, Naples was the home of a lively school of Cartesian geometry, and in the 1720s and 1730s one finds traces of mathematicians who were sympathetic to—and competent in—the new mathematics of the *Principia*. The Newtonian natural philosophy textbooks of brothers Nicola (1701–1769) and Pietro di Martino (1707–1746) were among the first to be

published in Italy. And yet, overall, their publications and teaching had relatively little impact on Neapolitan academic culture. See N. Di Martino, *Elementa statices*; P. Di Martino, *Philosophiae naturalis*; Amodeo, *Vita matematica*, vol. 1, 1–121; Gatto, "Tradizione e cartesianesimo"; Pepe, "Sulla trattatistica del calcolo infinitesimale"; Mazzotti, "Newton in Italy"; and Addabbo, "*Philosophiae naturalis.*"

60. The marginality of mathematics is confirmed by the fact that the main chair in this discipline at the University of Naples passed from Nicola di Martino to Giuseppe Marzucco (1713–1800), an obscure lecturer who published next to nothing, limited his teaching to elementary geometry and algebra, and used calculus only for squaring geometrical curves. See Marzucco, *Riflessioni*; Amodeo, *Vita matematica*, vol. 1, 98–99; and Torraca et al., *Storia della Università di Napoli*, 457–58.

61. Ferrone, "Riflessioni," 466; Galasso, "Scienze, filosofia e tradizione galileiana"; Rao, "Politica e scienza a Napoli"; Chiosi, "Lo stato e le scienze."

62. Mazzotti, "Newton in Italy"; Maffioli, *Out of Galileo*; Fiocca, Lamberini, and Maffioli, *Arte e scienza.*

63. Saladini, *Compendio d'analisi.*

64. Vito Caravelli (1724–1800), the most prominent mathematician of his generation and author of the first calculus textbook ever published in Naples (1786), never taught at a university. The career of Caravelli, who taught at the Royal Naval Academy and then became the director of the Royal Military Academy, is emblematic of the government's priorities. See Caravelli and Porto, *Trattati di calcolo*; and Amodeo, *Vita matematica*, vol. 1, 105–16.

65. See, for example, the praises of Lagrange and Lavoisier in *Analisi ragionata de' libri nuovi* (1791–1793), a periodical founded by physician Giovanni Leonardo Marugi (1753–1836); see also his *Stato attuale delle scienze* (1792). Rao, "Note sulla stampa" and "La stampa francese."

66. Lagrange, *Oeuvres*, vol. 14, 279–82. On the renewal of the academy, see Chiosi, *Lo spirito del secolo*, 107–42.

67. Salfi, *Saggio*; Pagano, introduction to *Saggi politici* (1783), in Venturi, *Illuministi italiani*, vol. 5, 854–62; Placanica, *Il filosofo e la catastrofe*, 180–98. Starting from the great eruption of 1631, Vesuvius had been unusually active: the last remarkable eruption was as recent as 1779, and it had been accompanied by earthquakes and clouds of ashes that traveled hundreds of miles. See Tortora, *Eruzione vesuviana*; and Tortora, Cassano, and Cocco, eds., *Europa moderna e antico Vesuvio*. On the pictorial and scientific concerns that informed early modern representations of Vesuvius, see also Toscano, "Conoscenza scientifica e pratica artistica."

68. The references to the Neapolitan tradition of Renaissance naturalism were indeed many and explicit—Mario Pagano used the pseudonym Janus Baptista La Porta, an homage to Giovanni Battista della Porta (1535–1615).

69. See also, for example, Longano, *Dell'uomo naturale*; de Giorgi Bertola, *Della filosofia della storia*; and Jerocades's poem *Il Paolo.*

70. On neonaturalism as a radical departure from Enlightenment, see Ajello, "L'estasi della ragione." On neonaturalism as a metamorphosis of enlightened reason and a different "style of thought" from "the science of Galileo," see Ferrone, *I profeti*, 238–300.

71. The continuity with the earlier culture of sensibility explains why a royal committee's report on mesmerism could not easily dispose of it without questioning the key role of sensibility in the production of knowledge and moral order. The solution was found in the identification of a specific kind of sentiment that was the product of a diseased imagination rather than of physical impressions. Darnton, *Mesmerism*; Riskin, *Science in the Age of Sensibility*, 189–225.

72. Genovesi, *Logica*, 214. For a summary of Genovesi's natural-law theory, see his *Della diocesina*, a textbook of moral philosophy that was, in fact, a synthetic guidebook for Neapolitan economic and social development (Guasti, "Antonio Genovesi's *Diocesina*").

73. See Trampus, "La genesi"; and Trampus, *Diritti e costituzione*.

74. Filangieri, *La scienza*, 71, 76; Giarrizzo, *Massoneria e illuminismo*, 275-309; Silvestrini, "Free Trade."

75. Lo Sardo, *Mondo nuovo*, 236-38; Pace, *Franklin and Italy*, 147-66; Ferrone, *La società giusta ed equa*, 24-25. On the debate on constitutionalism in eighteenth-century Italy, see Trampus, *Storia del costituzionalismo*.

76. Filangieri, La *scienza*, vol. 3, 1-309; Robertson, "Enlightenment and Revolution," 35.

77. Tommasi, *Elogio storico*, 168-70; Giarrizzo, *Massoneria e illuminismo*, 281, 284, 288-89.

78. Pagano, *Considerazioni*.

79. Giarrizzo, *Massoneria e illuminismo*, 350-57.

80. Cirillo, *Discorsi accademici*, 29-32.

81. On the eighteenth-century association of Vesuvian eruptions and historical transformations, see Cocco, *Watching Vesuvius*, 228.

Chapter Two

1. On private schools (*studi privati*) in Naples, see Zazo, *L'istruzione pubblica e privata*.

2. The problem, stated by Swiss mathematician Gabriel Cramer (1704-1752), asks that a triangle be inscribed inside a given circle, such that the sides of the triangle, or their extensions, pass through three given points. Jean de Castillon gave the first solution in 1776 and Lagrange provided an analytic solution. See Giordano, "Considerazioni sintetiche"; Chasles, *Aperçu historique*, 328; Carnot, *Géométrie de position*, 383; Ostermann and Wanner, *Geometry*, 175-78.

3. Nicolini, *Luigi de Medici*, 13-14; Croce, *Vite di avventure*, 362-63; Pedio, *Massoni e giacobini*, 15-19. On Carafa, see Nicolini, *La spedizione punitiva*, 99-124, at 103. For an overview of Lauberg's scientific and political activity, see de Lorenzo, *Regno in bilico*, 17-37 and 262-63.

4. Nicolini, *Luigi de Medici*, 8-9; Pedio, *Massoni e giacobini*, 72.

5. Jerocades, *La lira focense* and *Il Paolo*; Nicolini, *Luigi de Medici*, 8, 29; Pedio, *Massoni e giacobini*, 69-73; Ferrone, *I profeti*, 269-77; Lombardi Satriani, *Antonio Jerocades*. On the forms of sociability that prepared the terrain for Jacobin clubs in southern France, see Agulhon, *Pénitents et francs-maçons*; and Vovelle, *Jacobins*, 37-41.

6. Nicolini, *La spedizione punitiva*, 90-93; Pedio, *Massoni e giacobini*, 63-86; Giarrizzo, *Massoneria e illuminismo*, 390-97; Kennedy, *Jacobin Club of Marseille*.

7. Nicolini, *La spedizione punitiva*, 111-15.

8. Pedio, *Massoni e giacobini*, 80; Odazi, *Discorso*; di Battista, "Per la storia della prima cattedra universitaria d'economia"; Perna, "Nota critica," 913-16.

9. Croce, *Vite di avventure*, 364; Nicolini, *La spedizione punitiva*, 9, 93-95.

10. Pedio, *Massoni e giacobini*, 95.

11. Giarrizzo, *Massoneria e illuminismo*, 393; Simioni, *Le origini del Risorgimento*, vol. 2, 61; Nicolini, *La spedizione punitiva*, 108, 115-16, 491-92; Croce, *Vite di avventure*, 366.

12. Pedio, *Massoni e giacobini*, 96; Visconti, *Carteggio*, 5-25.

13. Pedio, *Massoni e giacobini*, 97-98.

14. Vovelle, *Jacobins*, 41-42, 143-44, 176.

15. Pedio, *Massoni e giacobini*, 99, 102.

16. Pedio, *Massoni e giacobini*, 106–07, 108–11; Croce, *La rivoluzione napoletana del 1799*, vol. 1, 59; Croce, "Canti politici."

17. Pedio, *Massoni e giacobini*, 117.

18. After Lauberg's departure the organization split into two factions: one that prioritized proselytizing and regrouping, and one that pressed for popular insurrection. Andrea Vitaliani and his brother Vincenzo were convinced that worsening economic conditions provided the opportunity to win the support of master craftsmen and their workers, key for a successful insurrection in the capital, and focused their conspiratorial activity on the popular neighborhood of Market Square. See Pedio, *Massoni e giacobini*, 108–11, 121–26, 130–31.

19. Among the documents seized at Giordano's house were Mario Pagano's *Political Essays* and a manuscript arguing that "happiness is to be found in equality, because all men are made of one and the same nature" (Pedio, *Massoni e giacobini*, 318).

20. On the procedure *ad modum belli et per horas*, for crimes of *lèse-majesté*, see Pedio, *Massoni e giacobini*, 145–81.

21. Pedio, *Massoni e giacobini*, 185–229.

22. Pedio, *Massoni e giacobini*, 138. Nicolini, *Luigi de Medici*, 104–05.

23. Robison, *Proofs of a Conspiracy*; Morrell, "Robison and Playfair"; Bouloiseau, *Jacobin Republic*, 22; Rao, *Esuli*, 61–65. For a European overview, see Vovelle, *Jacobins*, 59–103. For a transnational approach to the study of Jacobin networks, see Alpaugh, "British Origins."

24. Diaz, "La questione del 'giacobinismo' italiano," 592–602; Spini, Review, 794; Vaccarino, "L'inchiesta," 47. On the Cold War context of Venturi's remarks, see de Francesco, "L'ombra di Buonarroti"; for a broad historiographical survey, see Rao, Introduction. On the historical semantics of "republican," "patriot," and "Jacobin" in France, see Bouloiseau, *Jacobin Republic*, 2–7.

25. Lauberg's detailed plan included storming the royal palace and executing the king, the royal family, and all the members of the government. See Pedio, *Massoni e giacobini*, 103; and Barruel, *Mémoires*, vol. 5, 205–06.

26. Cantimori, "Giacobini italiani," 635; Galasso, *Italia democratica*, 3–15; Rao, *Esuli*, 98–99. See also de Felice, *Triennio giacobino*, 77–94; Pedio, *Massoni e giacobini*, 83–86; and Zaghi, "Il giacobinismo," 740. For some critical considerations, see Venturi, "La circolazione delle idee," 42; Giuntella, *Esperienza rivoluzionaria*, 316; and Guerci, *Istruire*, 13–14. See also Rao, "Sociologia," 231–39. Jacobinism was by no means the only component of the articulated world of Italian democrats and republicans; see Criscuolo, *Albori di democrazia*; and Mastellone, "Dibattito sulla democrazia." For a tentative definition of French Jacobinism in its more mature phase, see Mazauric, *Jacobinisme et Révolution*.

27. Vovelle, *Jacobins*, 41, 177; Ingram, "Jacobinism," 91–92.

28. In 1792 the Neapolitan ambassador in Rome had claimed, not without exaggeration, that "the majority of the Neapolitan Lords [*Signori*] and lawyers [were] all . . . democrats, and they look forward to seeing the French in Naples" (Pedio, *Massoni e giacobini*, 85). That members of the great aristocracy should join a self-proclaimed Jacobin movement is another element of continuity with the age of reforms, which had been championed by the likes of Gaetano Filangieri, cadet of a princely family. Throughout the second half of the century, the Neapolitan aristocracy had experienced a profound crisis of legitimation, its political role diminished and its revenue and juridical privileges threatened by the government's anti-feudal policies. The Masonic lodges and the reform movement offered a space to redesign aristocratic identity and bring new meaning to traditional social stratifications and anti-monarchic sentiments. In this context, around 1792, some young cadet members of the Neapolitan aristocracy turned decisively to

republicanism. See Pedio, *Massoni e giacobini*, 186–87; Pedio, *Giacobini e sanfedisti*, vol. 1, 43–62, 137–49; and Chiosi, *Lo spirito del secolo*, 45–78.

29. Rao, "Sociologia"; Davis, *Naples and Napoleon*, 102–06.

30. Published in Pedio, *Massoni e giacobini*, 231–521.

31. Croce, *La rivoluzione napoletana*, 11.

32. Tataranni, *Catechismo nazionale*, 1–6, 50–78; Tataranni, *Ragionamento*; Guerci, *Istruire*, 71–176.

33. Gramsci, *Quaderni*, 1324–25, 1361–62, 1559–60, 2017, quotation on 1361. See also Candeloro, *Storia dell'Italia moderna*, vol. 1, 180–92, 286–89; Rao, "Sociologia," 232. For critiques of the "passive" revolution, see also Zaghi, "Il giacobinismo," 736; Rao, "Popular Societies"; Robertson, "Enlightenment and Revolution," 22–23. For a comparison with French Jacobinism, see Vovelle, *Jacobins*, 47–51; 94–101.

34. "Even Saint January has become a Jacobin!" announced the *Monitore Napoletano* in May 1799 (see Davis, *Naples and Napoleon*, 101–2; de Lorenzo, *Regno in bilico*, 159; and Viola, "Intellectuels napolitains," 323).

35. Croce, *Vite di avventure*, 370–71. From 1795 on, Lauberg's name is often rendered as "Laubert," especially in French documents.

36. Onnis Rosa, *Filippo Buonarroti*, 13–60; Rao, *Esuli*, 66–75; Conte, "Commissariat."

37. Onnis Rosa, *Filippo Buonarroti*, 19; Bonaparte, *Correspondance*, vol. 3: 486–87 (2288, October 6, 1797); Soriga, *Le società segrete*, 153; Vaccarino, *Patrioti "anarchistes"*; Deleplace, "La notion d'anarchie."

38. Croce, *Vite di avventure*, 378–80, 383; de Felice, *Triennio giacobino*, 95–178.

39. Saitta, *Alle origini del Risorgimento*, vol. 1, vii–xxxv, 263–329 (essay by Matteo Galdi); vol. 2, 167–99 (essay by Leonardo Cesare Loschi, previously attributed to Lauberg himself; see Soriga, *Le società segrete*, 161–64; and Croce, *Vite di avventure*, 373–74); vol. 3, 281–328 (essay by Giuseppe Abbamonti). On the debate on federalism, see Rao, "Unité et fédéralisme"; de Francesco, *Rivoluzione e costituzioni*, 11–28; for France, see Hanson, *Jacobin Republic Under Fire*, and Ozouf, "Federalism." The Jacobin critique of particularism informed their views on economics as well. As in the French case, leading Jacobins initially supported liberal positions inspired to anti-feudal reformism, while more-radical egalitarian views emerged during the concrete experiences of the Jacobin republics. Lauberg, "Sull'alto prezzo delle cose"; Croce, *Vite di avventure*, 392–93; Vovelle, *Jacobins*, 52.

40. "Progressi dello spirito umano nell'Italia," *Il Monitore Italiano* 2 and 5 (January 22 and 28, 1798); reprinted in Soriga, *Le società segrete*, 165–69.

41. "Discorso su la resa di Mantova," *Giornale de' patrioti d'Italia* 15 (February 21, 1797), discussed in Croce, *Vite di avventure*, 374–75 (but the reference listed there for this article is incorrect) and reprinted in Zanoli, ed., *Giornale de' patrioti d'Italia*, vol. 1, 200–02.

42. "Sessione del 9 ventoso [8 March 1798]," *Il Circolo Costituzionale di Milano*; reprinted in Soriga, *Le società segrete*, 177–79. On the social organization of the *circoli costituzionali*, see Schettini, "La fucina dello spirito pubblico." Lauberg's scientific activity has been long dismissed as a mere cover-up for political conspiracy. See, for instance, Croce, *Vite di avventura*, 362 (science was "almost a pretext"); and Pedio, *Massoni e giacobini*, 15 ("chemistry and mathematics were a mere pretext for both maestro and pupils").

43. Conte, "Michele De Tommaso," 38–39, at 39; de Felice, *Triennio giacobino*, 179–204.

44. Tommaso and Orsi, *Catechismo su i dritti dell'uomo*, 4–12 (sensationalism), 18–28 (equality), 56–58 (against excessive concentrations of wealth), 70 (freedom of speech and religion), 124–28

(public education), 134–54 (national sovereignty), 146 (tyrannicide), and 154–62 (resistance against oppression); see also Battaglini, *La Repubblica Napoletana*, 46–56; and Guerci, *Istruire*.

45. Gillispie, "Jacobin Philosophy of Science." On the separation of good science and politics, see also Fayet, *Révolution Francaise* ("the scientific life of the country was independent of its political life," 471).

46. Shaw, *Time and the French Revolution*. For more counterexamples, see Corsi, "Models and Analogies."

47. Hahn, *Anatomy*, 223, 289–90; Guerlac, "Some Aspects of Science." See also Osborne, "Jacobin Science" (a reevaluation of Jacobin *science utile* and its legacy in natural history); and Conner, *Marat*, 42–61.

48. On the *question* of revolution and the conjugation of politics and history as a link between Jacobinism and Marxism, see Ingram, "Jacobinism."

49. Lauberg, *Analisi chimico-fisica*; Lauberg, *Riflessioni sulle operazioni dell'umano intendimento*. I have not been able to inspect copies of these two publications, which are described in Croce, *Vite di avventure*, 355–60 and Gentile, *Storia della filosofia*, vol. 1, 123–27. Lauberg also argued that both matter and mind are active, and that morals should be grounded on physical sensibility through the notion of self-interest (*amor proprio*).

50. Pluquet, *Esame del fatalismo*, vol. 1, 48 and 192; vol. 2, 271. Lauberg's annotated translation of François Pluquet's critique of philosophical fatalism espoused many heretic views on the origin of the universe, the nature of the soul, and the question of free will. In his tongue-in-cheek dedication, Lauberg presented it as a "history of the human mind" and declared that he didn't know whether to pity or admire the creators of such a gallery of contemptible errors. Bruno's "martyrdom" was a leitmotif in the Masonic and reformist tradition of Pagano and Filangieri. On his fortune in eighteenth-century radical circles, see Jacob, *Radical Enlightenment*, 35–40. On François Pluquet (1716–1790), see Coleman, "The Enlightened Orthodoxy of the Abbé Pluquet." Restoration-age profiles of Pluquet should be handled with care; see "Pluquet, François-André Adrien," in *Nouvelle biographie générale*, vol. 40, 502–03.

51. "Laubert, Charles-Jean," in *Nouvelle biographie générale*, vol. 29, 874.

52. Lavoisier, *Trattato elementare*. On this translation and, more generally, on chemistry in Naples during the revolutionary period, see Abbri, "Chimici e artiglieri," and Guerra, *Lavoisier e Parthenope*.

53. Lauberg, *Memoria*, unpaged dedication, 4.

54. Giordano and Lauberg, *Principi analitici*, 1–2.

55. Giordano and Lauberg, *Principi analitici*, 1–2. The idea that reason should be allowed to develop naturally at both the individual and the social levels was central to Lauberg's scientific pedagogy. Instead, Lauberg lamented, teachers "oppress their students with never-ending series of experiments," in the misguided pursuit of a noncommittal empiricism. See Lauberg's letter in *Giornale de' patrioti d'Italia*, 27 and 28 (March 21 and 23, 1797), discussed in Croce, *Vite di avventure*, 377–78 and reprinted in Zanoli, *Giornale de' patrioti d'Italia*, vol. 1, 298–99, 302–03.

56. Ozouf, "Regeneration," 781; Ozouf, *Homme regenerée*.

57. Pedio, *Massoni e giacobini*, 101, 504, 482, 439.

58. Mathiez, *Théophilantropie*; Vovelle, *Revolution against the Church*; Guerci, "Incredulità e rigenerazione"; Schettini, "Teofilantropia."

59. "Sessione del 5 nevoso [25 dicembre 1796]," *Il Circolo costituzionale di Milano*. Reprinted in Croce, *Vite di avventure*, 386–88.

60. [Voltaire], *Istoria dello stabilimento del cristianesimo*, 91 (Massa on Saint January's miracle); Soriga, *Le società segrete*, 239; Croce, *Vite di avventure*, 297. On the miracle, see de Ceglia, *Segreto di San Gennaro*.

61. Croce, *Vite di avventure*, 389–90, 394–99. Lauberg collaborated with a Neapolitan Jacobin, Raffaele Netti (1775–1863), who had settled in Milan as a bookseller and publisher. Netti published primarily anti-religious books in compact, affordable editions. Between 1798 and 1799, he published Italian-language editions of texts like Holbach's *Good Sense*, Maréchal's *Ritual and Laws of a Society of Men without God*, a version of the *Treatise of the Three Impostors*, and *Hell Destroyed*. Netti also published Lauberg's annotated translation of Helvétius's *On the Mind* to illustrate the "horrors caused by [the alliance of] miter and throne." (See [Holbach], *L'inferno distrutto*, unpaged preface; Helvétius, *Dello spirito*, vol. 1, 151; *Traité des trois imposteurs*). Netti also published the periodical of the "Italian patriots," and his collaboration with Lauberg's circle spurred other important publications, such as Girolamo Bocalosi's *Istituzioni democratiche*, "the most thoughtful and complete treatise of democratic education" (Hazard, *La Révolution française*, 100–02), and Lauberg's translation of French scientific essays (*Lezioni ad uso delle scuole normali di Francia*). See Criscuolo, "Girolamo Bocalosi fra libertinismo e giacobinismo."

62. The term "Jacobin priests" has been used loosely to refer to clergymen who held a variety of positions, from sympathizers of France to moderate unitary patriots to radical egalitarians like Orsi and Tommaso (Stella, *Il prete piemontese*, 10). On the connections between Jansenism and Jacobinism in Naples, see Ambrasi, "Il clero a Napoli."

63. About 14 percent of those exiled to France at the fall of the Republic were ecclesiastics (Rao, "Sociologia," 223). In Turin, they made up a similar percentage of the local Jacobins, the second largest group after lawyers and notaries (Favaro, "Consistenza del clero giacobino," 190). See also Pepe, *Il clero giacobino*.

64. Cestari was known for his 1788 book on the right of the king to nominate bishops (Gennaro Cestari, *Lo spirito della giurisdizione*; Ambrasi, "Giuseppe e Gennaro Cestari," 231–57). The book denied the pontiff's supremacy over the episcopal community, reducing pontifical authority to a matter of habit. His brother Giuseppe had scorned a book that defended the politically charged devotion of the Sacred Heart of Jesus. This devotion had become emblematic of the alliance of throne and altar, and all that the Jansenists fought against. It did not simply express a longing for ancien régime absolutism; it pointed toward the restoration of a mythical *res publica christiana* and thus represented the Catholic Church's new hierocratic orientation (Menozzi, *Sacro Cuore*, 19–106; Jonas, *France and the Cult of the Sacred Heart*, 1–146).

65. Cestari, *Tentativo sulla rigenerazione*; Darnton, "Philosophers Trim the Tree of Knowledge."

66. Cestari, *Tentativo sulla rigenerazione*, 8–10, 15, 19, 23–24, 104, 110.

67. Cestari, *Tentativo sulla rigenerazione*, 141–43. Note the references to Genovesi's logic (136). Note also how the faculty of memory is reduced entirely to sensibility and perception, and to the capacity of associating ideas (152). Natural history is not a "history of nature" (the definition given by the *Encyclopédie*), but rather the basis of physics. Arts, crafts, and manufacture are not parts of natural history, but rather of experimental physics, which originates from reason, rather than from memory.

68. Cestari, *Tentativo sulla rigenerazione*, 179, 185–87, 195. Again, the target is d'Alembert, who had claimed that the objects of the sciences were space, time, spirit, and matter. Cestari's remarks on the misleading dualism of matter and spirit are useful in making sense of these Jacobin thinkers' reluctance to embrace full-fledged materialism and their attempt to abandon the

dichotomy of matter and spirit. On the connection between hylozoism and political subversion, see Shapin, "Social Uses of Science" (for the Jacobin case, 121–22). For a broader reconstruction of the emergence of the language of self-organization in the seventeenth and eighteenth centuries, see Sheehan and Wahrman, *Invisible Hands* (on the flexibility of the politics of self-organization, see especially 271–95).

69. Cestari, *Discorsi due*, 12–21, 65.

70. Cestari, *Discorsi due*, 39–40.

71. Cestari, *Discorsi due*, 42, 77–83; Ambrasi, "Giuseppe e Gennaro Cestari," 254–57. Religion is "the sentiment of subjection to the Author of our being"; it expresses the "relation between man and God," and consists in "feeling the strength of this link" (Cestari, *Tentativo sulla rigenerazione*, 202–05, 250–53). It follows that "those who, for whatever reason, lack this religious sentiment cannot but regard the forms of religious life as human inventions, opinions, tales, chimeras." Grouping the theological systems and forms of cults of the different nations together with the sciences is therefore a categorical error. Instead, cults should be classified under history, like other opinions and practices. A "separating wall" must be erected between human reason and religious sentiment. This wall would protect both science, which will not be misled by extraneous questions, and religion, which is independent from empirical reality. The constitutional project for the 1799 Neapolitan Republic reflected this clear-cut separation (Cestari, *Tentativo sulla rigenerazione*, 267–73).

72. Cestari, *Tentativo sulla rigenerazione*, 244–49; Cestari, *Tentativo secondo*, 36–37, 50–55, 77, 84–85; Tataranni, *Filosofo politico*, vol. 5, 358–60.

73. Lomonaco, *Rapporto al cittadino Carnot*, 288.

74. Napoli Signorelli, "Discorso istorico," xxv–xxvi, xxxii; Capialbi, *Nuovi motivi*, 58; Ferber, *Briefe aus Wälschland*, 120–21. On the Neapolitan seminar, see Ambrasi, "Seminario e clero," in *Campania Sacra*.

75. Fiorentino, *Saggio sulle quantità infinitesime*, 16, 51.

76. Fiorentino, *Riflessioni*, ix, xi, 2, 17, 19, 56–57, 122, 124, 163, 181. See also d'Andria, "Nicola Fiorentino."

77. De Filippis, *De' terremoti della Calabria Ultra*.

78. Mazzei, "Un calabrese del '700"; "De Filippis, Vincenzo," in *Dizionario biografico degli italiani*.

79. Aracri, *Elementi del diritto naturale*; Aracri, *Dell'amor proprio*.

80. "Aracri, Gregorio," in *Dizionario biografico degli italiani*.

81. Battaglini, *La Repubblica Napoletana*, 73–167; Rao and Villani, *Napoli 1799–1815*, 9–123.

82. Davis, *Naples and Napoleon*, 75–81, at 81.

83. On the lavishness of Neapolitan church silver and its cultural, religious, and political meaning, see Hills, "Silver's Eye."

84. Battaglini, *Atti, leggi, proclami*, vol. 2, 1230.

85. Galasso, *La filosofia in soccorso de' governi*, 633–60; Battaglini, *Repubblica Napoletana*, 199–264.

86. Battaglini, *Mario Pagano*, 68, 76–78, 78–85, 134–35, 174–76. On de Fonseca Pimentel (1752–1799), see Croce, *La rivoluzione napoletana*, 25–102; Rao, "Eleonora de Fonseca Pimentel."

87. Davis, *Naples and Napoleon*, 94–106.

88. Croce, *Riconquista*; Cingari, *Giacobini e Sanfedisti*; Davis, "1799"; Davis, *Naples and Napoleon*, 116–21.

89. Burke, "Virgin," 20; Blackbourn, *Marpingen*, 363; Broers, *Politics and Religion*, 59; Meyer, "Cristiada."

90. Robertson, "Enlightenment and Revolution," 19.

Chapter Three

1. Francesco De Sanctis remembered this phase as the "curious graft" of sensationalism onto scholasticism (*Giovinezza*, 10).

2. Broers, *Politics and Religion*, 25-26; Broers, "Cultural Imperialism"; Davis, *Naples and Napoleon*, 115-16, 138. On "orientalism within one country" in the *Risorgimento*, see Schneider, *Italy's 'Southern Question,'* 1-26; dal Lago, "Italian National Unification."

3. Rao and Villani, *Napoli 1799-1815*, 179-284; Davis, *Naples and Napoleon*, 129-330 ; Spagnoletti, *Storia del Regno delle Due Sicilie*; Schroeder, *Transformation of European Politics*.

4. Galdi, *Saggio*, 3.

5. Davis, *Naples and Napoleon*, 161-62. On the connections between the technical aspects of the new cadastre and its sociopolitical meaning, see de Lorenzo, "Catasti napoleonici."

6. De Martino, *Nascita delle intendenze*, 30-31.

7. Candeloro, *Storia dell'Italia moderna*, 324-39.

8. Davis, *Naples and Napoleon*, 209-31.

9. De Martino, *Nascita delle intendenze*, 220.

10. De Martino, *Nascita delle intendenze*, 360-79. On the long-term nature of these conflicts, see Davis, "Naples during the French decennio"; Davis, *Merchants, Monopolists and Contractors* (investment strategies).

11. Giampaolo, *Lezioni e catechismo di agricoltura*, 10-11; Oldrini, *La cultura filosofica*, 27-50.

12. Scirocco, "Corpi rappresentativi"; Rao and Villani, *Napoli 1799-1815*, 179-282; de Martino, *Nascita delle intendenze*; de Lorenzo, *Proprietà fondiaria e fisco*. On the continuities between Jacobin sociability and later Masonic and conspiratorial networks, see Giarrizzo, "Alla ricerca del giacobinismo italiano" and "Massoneria e Risorgimento." On the myth of the 1799 Revolution and its later reinterpretations, see de Lorenzo, *Regno in bilico*, 357-71.

13. Godechot, *Commissaires*, vol. 1, 245; Godechot, "Saliceti."

14. De Martino, *Nascita delle intendenze*, 58-79; Davis, *Naples and Napoleon*, 165-73.

15. Emblematic of this phase are anecdotes in which prestigious foreign professors, analytic and Francophile, visited Fergola only to be humiliated by his students in some geometrical challenge. See, for example, Telesio, *Elogio*, 106-09.

16. Galdi, *Pensieri sull'istruzione pubblica*, 112-33. For Galdi's political repositioning, see his *Quadro politico* (1809). See also Rao, "Espace méditerranéen"; Guerci, "Democrazia e costituzione democratica"; Imbruglia, "Illuminismo e politica."

17. Galdi, *Saggio*, 3-4.

18. Galdi, *Pensieri sull'istruzione pubblica*, 3-4, 38, 52, 68.

19. Galdi, *Pensieri sull'istruzione pubblica*, 47-48, 142-66, 174-87, 212-21; Davis, *Naples and Napoleon*, 251.

20. Galdi, *Pensieri sull'istruzione pubblica*, 198-99, 231-68.

21. Galdi, *Pensieri sull'istruzione pubblica*, 112-33.

22. For some insightful comments on this point, see de Lorenzo, *Regno in bilico*, 273-76.

23. The continuities between the Jacobin experience of the triennium and the Napoleonic phase are emphasized, from different perspectives, in Criscuolo, *Pietro Custodi*, 391; and through de Francesco, *Storie dell'Italia rivoluzionaria e napoleonica*. For a *longue durée* perspective, see Rao, *Esuli*, 517-85.

24. Cagnazzi, *Memoria*, 5, 22; *Elementi dell'arte statistica*; and *Elementi di economia politica*.

25. Cuoco, *Saggio storico*, 5.

26. Cuoco, *Saggio storico*; "Cuoco, Vincenzo," in *Dizionario biografico degli italiani*; de Francesco, *Vincenzo Cuoco*. On Cuoco's complex historiographical trajectory, see Biscardi and de Francesco, *Vincenzo Cuoco*.

27. Cuoco, *Scritti vari*, vol. 1, 300.

28. Cuoco, *Platone*, vol. 1, 87 (quotation). Cuoco presented many of these ideas in his *Platone* (1806), which appeared shortly before J. G. Fichte's *Reden an die deutsche Nation* (1808).

29. Cuoco, *Scritti vari*, vol. 2, 32–33, 65–68, 156–57; Cuoco, *Scritti di statistica*; Martirano, "Cuoco e la scienza."

30. Cabanis, "Rapport," 396. The term *idéologie* was coined by Destutt de Tracy in 1796, meaning the science of ideas. On the origin and meaning-shifts of this term, see Mannheim, *Ideology and Utopia*, 59–75; Kennedy, "Ideology"; Eagleton, *Ideology*, 63–70. On French *idéologie*, see Moravia, *Pensiero degli idéologues*; Kennedy, *Philosophe in the Age of Revolution*; Head, *Ideology and Social Science*; Staum, *Cabanis*; Azouvi, *Institution de la raison*; and Chappey, "Idéologues." On the reception of French ideology in Naples, see Rascaglia, "Filosofia e scienza."

31. Chappey, "Idéologues," 75.

32. Cabanis, *Rapporti del fisico e del morale*.

33. Delfico, *Pensieri*; Delfico, "Discorso." See also Liberatore, "Necrologia"; and Gentile, *Storia della filosofia*, vol. 1, 25–120.

34. Carletti, *Melchiorre Delfico*, 174.

35. Delfico, *Ricerche*.

36. Delfico, *Memorie*, 239.

37. There is much in Delfico that tastes like mid-eighteenth century. Science is essentially calculus; the reality of sensations transmutes into the evidence of corresponding ideas; and there should be no other guides for human action than "analysis, experience, and observation" (*Pensieri*, 39, 144–59). His sensationalism, however, was up to date, built on the notion of "imitative sensibility," a model of mechanical imitation of external inputs ("Ricerche"). The entire human civilization is the result of imitative sensibility, which he described as a basic phenomenon of animal physiology, originated by the organic interplay between sensory organs and the central nervous system ("internal sensibility"). Language, storytelling, the fine arts, and religious rituals all developed naturally from the human physical constitution, and more specifically from the mechanism of imitative sensibility (on the basis of a similar "principle of imitation," Mario Pagano in his "Discorso" had explained the origin of language, dance, and music). The very sentiment of compassion is an outcome of this mechanism, and therefore a "disposition of the [human] machine"; morals are just a branch of physiology; and there exist necessary connections between the ideas of pleasure, moral good, and virtue ("Ricerche," 343–45, 367).

38. On his pedagogy, see also Delfico, "Memoria"; and Delfico, "Seconda memoria."

39. Alder, *Engineering the French Revolution*, 257–317. But see also Tresch, *Romantic Machine*.

40. Amodeo, *Vita matematica napoletana*, vol. 2, 46.

41. Amodeo, *Vita matematica napoletana*, vol. 1, 167; Zazo, *L'istruzione pubblica e privata*.

42. Zazo, "L'ultimo periodo borbonico."

43. Russo, *La scuola d'ingegneria di Napoli*, 35–75, 445–93; Buccaro and de Mattia, *Scienziati artisti*, 217–35; Foscari, "Ingegneri e territorio."

44. The Military Academy changed its name and structure numerous times between 1769 and 1860 (Amodeo, *Vita matematica napolitana*, vol. 1, 168–74). For the sake of consistency I refer to it as the Military Academy throughout the book, except when I emphasize the significance of a particular denomination, as in this case.

45. *Saggio di un corso di matematiche*, in twelve volumes, covered arithmetic, algebra, plane and solid geometry, two- and three-coordinate analysis, plane trigonometry, differential and integral calculus, mechanics, descriptive geometry, and mathematical geography. See Amodeo, *Vita matematica napoletana*, vol. 1, 168–75.

46. Russo, *La scuola d'ingegneria di Napoli*, 48, 53.

47. Liberatore, "Sulla Scuola di Applicazione," 330 (description of the curriculum).

48. Luigi Malesci (1774–1853), professor of civil architecture, presented his course in the form of an analytic table, where the subject is progressively divided into its elementary constituents. "The materials presented in the table," he commented, "can be taught in two years, as it's not necessary to insist on details but rather to show the most general and essential principles." Malesci aimed to transmit to his students what engineers called "the spirit of the analysis," thus connecting the practice of engineering to the alleged core of the scientific method. Russo, *La scuola d'ingegneria di Napoli*, 56–57.

49. Rivera, *Considerazioni sul progetto di prosciugare il lago Fucino*, 29, 38.

50. Matarazzo, "La stampa periodica," 289–92; Ozouf, "Public Spirit"; Leso, *Lingua e rivoluzione*, 119–20; Zazo, *Il giornalismo a Napoli*, 41–42.

51. *Biblioteca analitica di scienze, letteratura e belle arti* (from 1812: *Biblioteca analitica d'istruzione e di utilità pubblica*, from 1816: *Nuova biblioteca analitica di scienze, lettere ed arti*); Trombetta, *L'editoria a Napoli*, 116–17, 204–08; Matarazzo, "La stampa periodica"; Dotti, *Il Progresso*, 13–60.

52. [De Ritis], "Breve memoria filosofica."

53. [De Ritis], "Programma," 39, 65–67, 70.

54. [De Ritis], "Programma," 137, 192.

55. [De Ritis], "Programma," 194–96, 203.

56. [De Ritis], "Programma," 18, 22–24, 42. The author uses Cabanis as a resource for conceptualizing the relation between morals and physics, and follows Destutt de Tracy's division of philosophy into ideology, grammar, and logic, with ideology being the study of our sensible perceptions and the foundation of every other science, as every intellectual operation is a function of perception.

57. Borrelli, *Introduzione*; Borrelli, *Principi della genealogia del pensiero*; Bozzelli, *Essai*; Galluppi, *Saggio filosofico*.

58. Oldrini, *La cultura filosofica*, 73–82.

59. [De Ritis], "Programma"; Ottonello, "Vincenzo De Ritis."

60. In the words of philosopher and deputy Pasquale Borrelli, a moderate liberal who spoke to the newly minted Neapolitan Parliament brandishing a copy of the constitution: "I bring you the Code of your liberty, the object of your forbears' desires and your fellow citizens' wishes, the main foundation of their hopes" (Oldrini, *La cultura filosofica*, 57).

61. Oldrini, *La cultura filosofica*, 76–77.

62. Bozzelli was interested in "public sentiments" and theorized a monarchic-constitutional government that could guarantee social order by understanding and controlling the "forces" (understood in sensationalist-sentimental terms) that permeate society. Critical of Jacobin abstractness and democracy, Bozzelli exemplifies the convergence of physiological-ideological theories and the early study of crowd behavior that, he argued, could become as scientific as that of physics (Bozzelli, *Esquisse politique*; *Essai*). In this phase, Victor Cousin's conscientialism was also perceived as an important resource; see Galluppi, *Cousin*, 1831–1832.

63. Davis, *Naples and Napoleon*, 304 (orderly revolution).

64. For their interpretation of Vico, see Jannelli, *Saggio*.

65. Spaventa, *Dal 1848 al 1861*, 29; texts by Stanislao Gatti and Stefano Cusani in Oldrini, *Primo Hegelismo*, 138–51, 167–98; Berti, *Democratici*, 210–11; Gallo, "Gli hegeliani di Napoli."

66. Sabatini, *Ottavio Colecchi*, 28–31.

67. Colecchi, "Riflessioni sopra alcuni opuscoli," cited in Amodeo, *Vita matematica napoletana*, vol. 1, 147–48.

68. The three-year course enrolled about one hundred students per year. For a description of the curriculum, see Amodeo, *Vita matematica napoletana*, vol. 2, 121.

69. Amodeo, *Vita matematica napoletana*, vol. 2, 137; Tucci, "Soluzione di alcuni problemi relativi alle curve coniche." On Tucci as a member of the synthetic school, see also degli Uberti, *Esercitazioni geometriche*, 10.

70. Gergonne, "Géométrie de la règle"; Poncelet, "Philosophie mathématique"; Amodeo, *Vita matematica napoletana*, vol. 2, 138–42; Lorenat, "Figures Real, Imagined, and Missing"; Tucci, *Il problema del cerchio e de' tre punti*.

71. Miławicki, *Dominicans*, 210–12.

72. Sabatini, *Ottavio Colecchi*, 39–40; Capograssi, "Nuovi documenti," 78 (Russian period), 84 (report against Colecchi).

73. Zazo, "La nomina del Galluppi," 110–12; Capograssi, "Nuovi documenti," 80–83.

74. Sabatini, *Ottavio Colecchi*, 47–57 (sonnets); Romano, "Un antagonista del Galluppi," 159 (epitaph); Gentile, *Storia della filosofia italiana*, vol. 2, 137–249; Tessitore, *Da Cuoco a De Sanctis*, 139–75.

75. Galluppi, *Lezioni di logica e metafisica*; Galluppi, *Sull'analisi e sulla sintesi*; Gentile, *Storia della filosofia italiana*, vol. 2, 27–109; Oldrini, *La cultura filosofica*, 191–256; Ottonello, *Cultura filosofica*, vol. 5 (on Colecchi and Galluppi); Oldrini, *Gli hegeliani di Napoli*.

76. "Delle teorie kantiane," 5. The article was a polemical response to a positive review of the partial edition of Colecchi's collected works (1843) that had appeared in the *Progress*.

77. "Delle teorie kantiane," 7, 22, 27–28. On Colecchi's use of Vichian historicism, see Cacciatore, "Vico e Kant."

78. *Science and Faith*, the voice of Neapolitan Neo-Scholasticism, aimed to demonstrate that all knowledge supports Catholic religion. Its founder, the ecclesiastic Gaetano Sanseverino (1811–1865), had contributed to transforming Neo-Scholastic philosophy into the Neapolitan Church's official response to the secularization of science and society. The political meaning of Neo-Scholasticism—which reshaped Catholic culture worldwide—was clear. To Sanseverino, "the relation of the sovereign to his subjects is similar to the relation of God to the world, and of the soul to the body." Orlando, *Tomismo*, 42–107, at 54. On this periodical, see also Palmisciano, "Filosofia e intransigenza."

79. Mathematical induction is a proof technique generally used to prove statements involving whole numbers. It proves a particular rule or pattern, usually infinite, and uses two steps: a base step and an induction step. The idea behind it is that to show that one can get to any rung on a ladder, it suffices to show that one can get on the first rung, and then show that one can climb from any rung to the next.

80. Colecchi, "Sull'induzione matematica"; *Opuscoli matematici*, 54 (Fergola replaces induction with geometric demonstrative techniques). The literature on Kant's philosophy of mathematics is vast, but see Parsons, "Infinity," 108 (mathematical induction); and Carson, "Kant on the Method" and "Metaphysics" (intuition, sensibility and intellect).

81. Amodeo, *Vita matematica napoletana*, vol. 2, 265–317, at 312.

82. Gentile, *Storia della filosofia italiana*, vol. 2, 299–302; Sabatini, *Ottavio Colecchi*, 61; Romano, *Un antagonista del Galluppi*, 159; Piccone, *Italian Marxism*, 11–32; de Lorenzo, *Regno in bilico*, 357.

Chapter Four

1. Emilio Sereni describes the agrarian landscape as a continuously changing "shape" (*forma*), which is the outcome of the interaction among modes of production, juridical systems, and social conflicts (Sereni, *Storia del paesaggio*). Baudelaire's suggestive reference to the fast-changing "shape of a city" is elaborated in Gracq, *Shape of a City*. For a cultural history of Neapolitan landscape in the early modern period, see Cocco, *Watching Vesuvius*, and Fehrenbach and van Gastel, *Nature and the Arts*.

2. Broers, *Napoleonic Empire*, 12.

3. Sutherland, *Chouans*, 218 (moral unity). On the plurality of resistances, see Lebrun and Dupuy, *Résistances à la Révolution*.

4. Botta, *Storia d'Italia*, vol. 3, 158. Vaccarino comments on this passage with a reference to Cartesianism (Vaccarino, *Giacobini piemontesi*, vol. 2, 867). As we know, however, it wasn't Cartesianism but algebraic analysis that captured the Italian Jacobins' imagination.

5. Marino, *Formazione*, ch. 5 (ideology of progress, neutrality of science).

6. Massafra, *Il Mezzogiorno preunitario*; Spagnoletti, *Storia del Regno delle Due Sicilie*, 173–206, 221–70.

7. I use the term "Corps" to refer to the various incarnations of this institution.

8. Russo, *La scuola d'ingegneria di Napoli*, 77–110, 491–93.

9. De Mattia and de Negri, "Il Corpo di Ponti e Strade," 450, 462.

10. Di Biasio, *Carlo Afan de Rivera*; di Biasio, *Politica e amministrazione*, 9–99; *Memoria e progetto intorno all'organizzazione del Corpo reale d'ingegneri di acque e strade*, 1825, ASNA, Ministero delle finanze, 4962 (reform plan); *Modello dell'uniforme degli ingegneri di Acque e strade, accordata dal re su proposta del direttore Afan de Rivera*, March 22, 1826, ASNA, Ministero delle finanze, 4996/1117 (new uniforms).

11. Rivera, *Considerazioni sui mezzi*, vol. 2: 461–80; Picon and Yvon, *L'ingénieur artiste*.

12. ASNA, Archivio Borbone I, 860, f.412 (Corps as machine); Maiuri, *Delle opere pubbliche*, 12.

13. Rivera, *Considerazioni sui mezzi*, vol. 2, 461; Picon, *L'Invention de l'ingénieur moderne*; Giustiniani, *Dizionario geografico-ragionato*; Valerio, *Società, uomini, e istituzioni*; Brancaccio, *Geografia, cartografia e storia*; Demarco, *La statistica del Regno di Napoli*.

14. Rivera, *Del bonificamento del lago Salpi*; di Biasio, *Politica e amministrazione*, 101–270; di Biasio, *Carlo Afan de Rivera*, 153–83.

15. Rivera's hostility to private monopolies explains his otherwise surprising opposition to railways in the 1830s. Rivera was concerned by the role of private investors in the development of the railway system, and by the fact that the system would not fall under the Corps' control. On this point he clashed with politicians who favored contracting public works, especially railways, to private companies. See ASNA, Direzione generale di ponti e strade, Appendice, 1163 bis (railways); Ostuni, *Iniziativa privata e ferrovie*; di Biasio, *Carlo Afan de Rivera*, 115–16; di Biasio, *Politica e amministrazione*, 49–70 (against monopolies).

16. Davis, *Naples and Napoleon*; dal Lago, *Agrarian Elites*.

17. Di Biasio, *Politica e amministrazione*, 223–24 (Rivera quotation), 84–85, 90–91.

18. Pigonati, *La parte di strada degli Abruzzi*; Pigonati, *Le strade antiche e moderne*; ASNA, Direzione generale di ponti e strade, II s., 1178 (critique of road system); Ostuni, *Le comunicazioni stradali*; ASNA, Archivio Borbone I, 857 (critique of mathematical training).

19. Quotation in de Biasio, *Politica e amministrazione*, 340.

20. Rivera, *Circolari concernenti il servizio degl'ingegnieri.*

21. De Mattia and de Negri, "Il Corpo di Ponti e Strade," 468.

22. Davis, *Merchants, Monopolists and Contractors.*

23. L[iberatore], "Sulla Scuola di Applicazione," 328, 330, 332–33.

24. Neri, *Giuseppe Ceva Grimaldi*; "Ceva Grimaldi Pisanelli, Giuseppe," in *Dizionario biografico degli italiani*; Ceva Grimaldi, *Del lavoro degli artigiani* (medieval guilds); Ceva Grimaldi, *Riflessioni sulla polizia.*

25. Giannetti, "L'ingegnere moderno," 937.

26. Rivera, *Considerazioni sui mezzi.*

27. Ceva Grimaldi, *Considerazioni sulle opere pubbliche.*

28. Ceva Grimaldi, *Considerazioni sulle opere pubbliche*, 173.

29. Kolnai, "Notes sur l'utopie réactionnaire."

30. Ceva Grimaldi, *Considerazioni sulle opere pubbliche*, 151–79.

31. Giannetti, "L'ingegnere moderno," 942.

32. *Memoria sulla storia delle strade del Regno e su i vizi dell'amministrazione di esse al Parlamento nazionale*, [1820], ASNA, Ministero della polizia generale, Seconda numerazione, 120.

33. Ceva Grimaldi, *Sulla riforma della Direzione generale dei ponti e strade, delle acque e foreste.*

34. *Osservazioni del direttore generale di Ponti e strade sul progetto di riordinamenteo del servizio di Acque e strade complilato dal procurator generale della Gran corte dei conti*, 1831, ASNA, Archivio Borbone, 857; Giustino Fortunato al Comm. Caprioli, March 27, 1837, in ASNA, Archivio Borbone I, 860.

35. Rivera, *Libro de' costumi e dell'abilità e condotta nel servizio degli ingegneri di Acque e Strade*, in ASNA, Archivio Borbone I, 859; di Biasio, *Carlo Afan de Rivera*, 110–11.

36. Ceva Grimaldi, *Considerazioni sulle opere pubbliche*, 169.

37. Rivera, *Tavole di riduzione dei pesi e delle misure*; di Biasio, *Politica e amministrazione*, 70–77. For an instructive comparison with the French case, see Alder, "A Revolution to Measure."

38. Rivera, *Della restituzione del nostro sistema*, 13.

39. Ceva Grimaldi, *Considerazioni sulla riforma*, xlvii.

40. Ceva Grimaldi, *Considerazioni sulla riforma*, xlviii 1.

41. Ceva Grimaldi, *Considerazioni sulla riforma*, li–lvi.

42. Ceva Grimaldi, *Considerazioni sulla riforma*, liv–lv.

43. Ceva Grimaldi, *Considerazioni sulla riforma*, iv.

44. Ceva Grimaldi, *Considerazioni sulla riforma*, ix, xvi, xvii. For a biographical sketch of Visconti, see Valerio, "Earth Sciences."

45. Ceva Grimaldi, *Considerazioni sulla riforma*, xxv.

46. Rivera, *Tavole di riduzione dei pesi e delle misure*; di Biasio, *Politica e amministrazione*, 70–77.

47. Di Biasio, *Politica e amministrazione*, 77.

48. Rivera, *Della restituzione del nostro sistema*, 114–15; Lugli, *Unità di misura*, 121–37.

49. Galanti, *Descrizione*, vol. 1, xiii; Naddeo, "Cosmopolitan in the Provinces."

50. Sofia, *Scienza per l'amministrazione*. See also Martuscelli, *Popolazione del Mezzogiorno*; de Lorenzo, *Organizzazione dello stato*, 129–85.

51. Patriarca, *Numbers and Nationhood*; Davis, *Conflict and Control.*

52. Porter, *Rise of Statistical Thinking*; Romagnosi, "Questioni," vol. 14, 291.

53. On Cagnazzi, see Salvemini, *Economia politica.* See also Lombardo, "Primo trattato italiano di statistica."

54. Romani, "Un popolo da disciplinare."

55. Patriarca, *Numbers and Nationhood*, 30 (quotations).

56. Patriarca, *Numbers and Nationhood*, 31–32; Gioia, *Filosofia*, vol. 1, 12 (*mostra all'occhio*); Quaini, "Appunti" (*colpo d'occhio*).

57. Bourguet, *Dechiffrer la France*; Gioia, *Filosofia*, vol. 1, 3 (descriptive logic); vol. 1, 6–12 (structure).

58. Gioia, *Filosofia*, vol. 1, 6–11; Patriarca, *Numbers and Nationhood*, 73–75. For a comparison between different national approaches to statistics at the opening of the nineteenth century, see Desrosières, *Politics of Large Numbers*, 16–44.

59. Patriarca, *Numbers and Nationhood*, 83–84. The vicissitudes of the Central Directorate of Statistics, established in Palermo in 1832, with an ambitious but problematic program of data collection and quantification, illustrates well this collaboration and its limits, as well as the tensions within Sicilian society as it faced modernization (88–95).

60. Cantù, Review, 167.

61. Cagnazzi, *Elementi*, 1–36; Patriarca, *Numbers and Nationhood*, 37–38. On the emergence—and contested nature—of this notion of neutrality of numbers and calculations in modern political life, see Deringer, *Calculated Values*; and Porter, *Trust in Numbers.*

62. Cagnazzi, *Elementi*, 31; Cantù, Review, 167; Gioia, *Filosofia*, vol. 1, 5–6; Patriarca, *Numbers and Nationhood*, 36–38, 163–65.

63. Valerio, *Società*, 73–117.

64. Napoli Signorelli, "Discorso istorico," xxiv–v.

65. Valerio, *Società*, 121–30.

66. Valerio, *Costruttori di immagini.*

67. Valerio, *Società*, 205–94; Valerio, "Occhio mutevole."

68. My interpretation is inspired by exemplary research on the ideological underpinnings of eighteenth-century English landscape painting: Barrell, *Dark Side*; Solkin, *Richard Wilson*; Bermingham, *Landscape and Ideology*; Hemingway, *Landscape Imagery*; Payne, *Toil and Plenty*; Klonk, *Science.*

69. Pliny, *Natural History*, III, 40–41, 59–62. See also Montone, "Campania felix," and Fehrenbach and van Gastel, *Nature and the Arts*, viii. On the Renaissance version of the topos and its connection to the myth of the Golden Age, see Vecce, "Maestra natura" and Barreto, "Virgile."

70. Spinosa, *Vedute napoletane*; Fino, *Il vedutismo*; Alisio, *Napoli*; de Seta, *Golfo di Napoli*; Thoenes, "Unendliche Leben."

71. Ortolani, *Gigante*; Causa, "Scuola di Posillipo"; Causa, *Napoli*; Martorelli, *Gigante.*

72. Villari, *Pittura moderna*, 47.

73. Causa, *Napoli*, 12.

74. Causa, "Scuola di Posillipo," 793.

75. Labrot and Delfino, *Collections*; de Maio, *Pittura e Controriforma*; Spinosa, *Pittura napoletana*; Pavone, *Pittori napoletani.*

76. Napier, *Notes*, 135. War, confiscations, and "the fatal introduction of the democratic law of inheritance," continued Napier, "have rapidly destroyed the fortunes, the manners, and the tastes, which nourished so much elegance and grandeur" (157–58).

77. Napier, *Notes*, 163. On other contemporary trends like neoclassicism and religious romanticism, see Morelli and Dalbono, *Scuola napoletana*; and Greco, *Pittura napoletana*.

78. Napier, *Notes*, 64, 69.

79. Causa, *Pitloo*; Pitloo, *Boschetto di Francavilla al Chiatamone*, Naples, Museo di Capodimonte, collezione Banco di Napoli; Ortolani, *Gigante*, 141.

80. Napier, *Notes*, 69. Also relevant in this early phase was the figure of Wilhelm Hüber, who promoted the technique of watercolor and the use of the camera lucida.

81. Napier, *Notes*, 92.

82. Napier, *Notes*, 70.

83. Pitloo's production has been divided into different periods according to his chromatic representation of atmospheric data (Causa, "Scuola di Posillipo," 796).

84. Pitloo, *Tramonto sul Castello di Baja*, Sorrento, Museo Correale di Terranova, inv. no. 2844.

85. Napier, *Notes*, 75.

86. Napier, *Notes*, 95–96, 106–09, 164.

87. Napier, *Notes*, 76.

88. Napier, *Notes*, 77–79.

89. Spinosa, *Collezione Angelo Astarita*, 10; Martorelli, "Aspects of Collecting," 59–64.

90. For an example of this procedure, see *Marina di Napoli* (Naples, Museo di San Martino, inv. no. 6315), where Gigante transformed a view taken with the camera lucida into a romantic landscape.

91. Naples, Museo di San Martino, inv. no. 19012; and Museo di Capodimonte, inv. no. 1705 and 8528.

92. Naples, Museo di Capodimonte, inv. no. 5214 and 5219.

93. Naples, Museo di San Martino, inv. no. 18981.

94. For reconstruction of Gigante's use of the camera lucida, see Fiorentini, "Nuovi punti di vista," 551; on Gigante at the Topographical Office, see Valerio, "Occhio mutevole," 239; more broadly on the different forms of expertise that coexisted at the Topographical Office, see Valerio, *Costruttori di immagini*.

95. Gigante, *Via dei Sepolcri a Pompei*, Naples, Museo di San Martino, inv. no. 23105; *Paesaggio di Cava*, Naples, Museo di Capodimonte, inv. no. 5373; *Bacoli*, Naples, Museo di San Martino, inv. no. 21279; *Il Duomo di Amalfi*, Naples, Museo di Capodimonte, inv. no. 2945.

96. Giampaolo, *Lezioni e catechismo di agricoltura*, vol. 1, 10–11.

97. Sereni, *Storia del paesaggio*, 295–304.

98. Davis, *Naples and Napoleon*, 319; Sereni, *Storia del paesaggio*, 301.

99. Napier, *Notes*, 149–65.

100. Causa, *Napoli e la Campania Felix*, 14.

Intermezzo

1. On the problems of the 1839 challenge, see Gatto, "Discussione sul metodo."

2. Mancosu, *Philosophy of Mathematics*; Mahoney, *Mathematical Career*; Gingras, "What Did Mathematics Do to Physics?"

3. Loria, *Fergola*, 5–6.

4. See, for instance, Flauti, *Elogio di Fergola*, 16, on the "corruption" of geometry.

5. Panza and Otte, *Analysis and Synthesis*, ix–xi. In the early modern period the two terms were routinely understood to be in opposition to each other, although this was not necessarily the case in ancient Greek culture (367–68).

6. This is what Panza and Otte call the "directional interpretation" of the terms "synthesis" and "analysis."

7. Knorr, *Ancient Tradition*; Fowler, *Mathematics of Plato's Academy*.

8. Pappus is key to the synthetics because of his detailed descriptions of geometrical procedures, including the twofold procedure of analysis and synthesis seen above (Knorr, *Ancient Tradition*, 110).

9. Knorr, *Ancient Tradition*, 339–81.

10. Mahoney, *Mathematical Career*, 26–142.

11. Behboud, "Greek Geometrical Analysis"; Mahoney, "Another Look." For an engaging interpretation of the different meanings of mathematical construction in the history of mathematics, see Lachterman, *Ethics of Geometry*, where mathematical modernity is in fact defined through this meaning shift. Lachterman also explores the mathematical dimension of the *querelle des anciens et des modernes* as it took place in late seventeenth- and early eighteenth-century Naples, and especially Giambattista Vico's understanding of synthetic geometry and his critique of Cartesian analysis (Lachterman, "Vico, Doria, e la geometria sintetica" and "Mathematics and Nominalism").

12. Fergola, *Della invenzione geometrica*, 198–99 (with some minor changes to the original notation). This text had circulated in manuscript form since 1809 as *The Heuristic Art*.

13. Fergola, *Della invenzione geometrica*, 193–99.

14. Flauti, *Elogio di Fergola*, 11.

15. Fergola, *Della invenzione geometrica*, 239.

16. Joseph-Louis Lagrange's work was taken as exemplary of the "very modern method," especially his famous 1773 essay on triangular pyramids ("Solutions analytiques de quelques problèmes sur les pyramides triangulaires," in *Oeuvres*, vol. 3, 661–92). Here Lagrange tackles an extremely general problem: to find a series of properties—such as the surface, the circumscribed spheres, the center of gravity—of "all the triangular pyramids whose six sides are known." The solutions he provides are "purely analytic and can be understood without figures." The inspection of figures is irrelevant, as the geometer is not interested in particulars but in the general relations between different classes of objects and between different parts of some typical object. The preliminary geometric construction is eliminated, together with the final composition of the problem. As Lagrange remarked proudly, geometric problem solving is thus reduced to a matter of calculation (vol. 3, 661–92).

17. Fergola, *Della invenzione geometrica*, 194–96.

18. On the practice of constructing equations, see Bos, "Arguments."

19. Fergola, *Della invenzione geometrica*, 245–48.

20. Padula, *Raccolta di problemi*.

21. Padula, *Raccolta di problemi*, 13, 145–46.

22. On the contested notion of "genius" in Romantic natural philosophy and on its social significance see Schaffer, "Genius."

23. Lagrange himself was very clear on this point: in his essay on triangular pyramids he was not interested in investigating the geometrical properties of such figures. Rather, he wanted to show how the analytic method could be fruitfully used in solving even those problems that were considered most suitable for a purely geometrical treatment. Lagrange, "Solutions analytiques," in Oeuvres, vol. 3, 692.

24. On abstraction in the scholastic tradition, see Kluge, "Abstraction"; Mauer, "Thomists."

25. On the relevance and multiple meanings of vision in mathematics, see Mancosu and Pedersen, *Visualization, Explanation and Reasoning Styles in Mathematics*.

26. Cagnazzi, *Memoria dell'uso della sintesi*, 21–30.

27. Scotti Galletta, *Osservazioni*, 4, 26. Synthetics are devoted to "scholastic abstractions" (133).

28. I keep referring to the "analytics" for simplicity, but we have seen the important episte-mological differences between Jacobin mathematicians and the engineer-mathematicians of the Restoration.

29. Amodeo, *Vita matematica napoletana*, vol. 2, 188.

30. Loria, *Fergola*, 21, 131.

31. These lines matched Loria's misogynist views in a way that Mario Praz would have found exemplary (Praz, *The Romantic Agony*).

32. Volterra, "Le matematiche in Italia," 57, 61.

33. Galluzzi, "Geometria algebrica," 1016.

34. Ferraro and Palladino, *Il calcolo sublime*.

35. Palladino, *Metodi matematici*.

36. Casini, *L'antica sapienza italica*; de Francesco, *The Antiquity of the Italian Nation*.

Chapter Five

1. Ventura, *Elogi funebri*, 51–90; [Telesio], *Elogio*; Marchi, *Elogio funebre di Nicola Fergola*; Flauti, *Elogio storico di Nicola Fergola*.

2. Marchi, *Elogio funebre di Nicola Fergola*, 10–11, 17; Ventura, *Elogi funebri*, 72–75; de Liguori, *Le glorie di Maria*; Rosa, *Settecento religioso*, 38, 126–27, 221–23; Giannantonio, *Alfonso M. De Liguori*. For a political contextualization of de Liguori's religiosity, see Palmieri, *Taumaturghi*.

3. [Telesio], *Elogio*, 16–17.

4. Amodeo, *Vita matematica napoletana*, vol. 1, 104–16.

5. *Inventario della libreria*, 3 vols., 1822, ms.XVIII.19–20; *Catalogue raisonné*, 6 vols., 1826, manuscript, ms.XVIII.13–18, BNN (Berio's library); [Telesio], *Elogio*, 22 (philosophical studies).

6. "Solutiones novorum quorundarum problematum geometricorum" ([Telesio], *Elogio*, 31–32).

7. Genovesi, *Elementa*.

8. "[E]st enim natura physica quasi basis et fundamentum moralis" (Genovesi, *Elementa*, 4).

9. Fergola, "Risoluzione di alcuni problemi ottici"; Fergola, "La vera misura delle volte a spira."

10. Genovesi, *Elementa*, 4, 45.

11. On the "right" and "left" components of Genovesi's school, see Zambelli, *Formazione filosofica*, 435.

12. Fergola, "Nuovo metodo da risolvere alcuni problemi di sito e di posizione," 120.

13. Fergola, "Nuove ricerche sulle risoluzioni dei problemi di sito."

14. Giordano, "Considerazioni sintetiche sopra un celebre problema piano"; Chasles, *Aperçu historique*, 328; Carnot, *Géométrie de position*, 383. Carnot refers to him as "Oltajano," a mis-spelling of his birthplace, Ottaviano, near Naples. A revised version of this paper opened an important 1811 collection of synthetic works (*Opuscoli matematici*, 1–11).

15. Quoted in Amodeo, *Vita matematica napoletana*, vol. 2, 12.

16. Amodeo, *Vita matematica napoletana*, vol. 2, 13.

17. [Telesio], *Elogio*, 59–61.

18. Croce, "Il Marchese Caracciolo."

19. Schipa, "Il secolo decimottavo," 463.

20. [Telesio], *Elogio*, 199–200.

21. [Telesio], *Appendicetta*, 28; Amodeo, *Vita matematica napoletana*, vol. 2, 60.

22. [Fergola], *Elementi di geometria sublime*.

23. [Telesio], *Elogio*, 36.

24. F[ergola], *Trattato analitico delle sezioni coniche*, xiv.

25. Fergola, *Trattato analitico de' luoghi geometrici*.

26. Published posthumously as Fergola, *Della invenzione geometrica*.

27. Naples, Biblioteca Nazionale (BNN): III.C.31–36; XIII.B.52; XVI.A.27; XVIII, 13–20. For a description of Fergola's manuscripts at BNN, see Ferraro and Palladino, "Sui manoscritti di Nicola Fergola."

28. BNN, Ms.III.C.32, 99r.

29. [Telesio], *Elogio*, 201, 80, 15.

30. [Telesio], *Elogio*, 27 ("highest esteem"), 106–08.

31. [Telesio], *Elogio*, 82, 83; Amodeo, *Vita matematica*, vol. 2, 141.

32. Zazo, "L'ultimo periodo borbonico," 469, 470–72.

33. [Telesio], *Elogio*, 84, 86.

34. Flauti, *Servizi scientifici resi*, 3.

35. Piolanti, *L'Accademia di Religione Cattolica*.

36. [Telesio], *Elogio*, 177.

37. *Opuscoli matematici*, v.

38. Giordano, "Risoluzioni di alcuni difficilissimi problemi geometrici," in *Opuscoli matematici*, 1–11, at 1.

39. "Giudizio degli editori sulle varie soluzioni del problema del cerchio e de' tre punti prodotte da geometri illustri," in *Opuscoli matematici*, 13–23, at 13, 17, 23.

40. Scorza, Giuseppe, "Nuove speculazioni sull'istesso argomento," in *Opuscoli matematici*, 25–35, at 33–34; Loria, *Fergola*, 50–55.

41. "Estratto dall'*Arte Euristica* di un nostro geometra," in *Opuscoli matematici*, 127–94.

42. "Estratto dall'*Arte Euristica* di un nostro geometra," in *Opuscoli matematici*, 182 ("monstrous analytical solutions"), 190–93 (Euclidean spirit).

43. *Opuscoli matematici*, vii.

44. "Estratto da un manoscritto di analisi sublime di un nostro geometra," in *Opuscoli matematici*, 37–95, at 37.

45. "Estratto da un manoscritto di analisi sublime di un nostro geometra," in *Opuscoli matematici*, 69–71.

46. Forte, "Risoluzione del celebre problema della cilindroide di Wallis con altre ricerche affini" [1797], in *Opuscoli matematici*, 97–110. Stefano Forte (1770–1818), a lecturer of philosophy, was highly regarded by Fergola, who asked him to write the "prenozioni geometriche" to his *Prelezioni* (vol. 1, vii-xxiv; vol. 2, v-xiv). Amodeo, *Vita matematica, napoletana*, vol. 2, 32. This problem, studied among others by John Wallis and d'Alembert, had been derived from Archimedes's treatment of conoids and spheroids. See Flauti and Vincenzo, "Continuazione dello stesso argomento della cilindroide," in *Opuscoli matematici*, 111–22; and "Esame delle varie soluzioni del problema della cilindroide wallisiana," in *Opuscoli matematici*, 123–26. Solutions to this problem by the synthetics were published up to 1839. See Sangro, "Nuova soluzione del noto problema sul cilindroide wallisiano"; Fergola, "Continuazione della precedente memoria sul cilindroide wallisiano"; Flauti, "La vera misura del cilindroide wallisiano"; Amodeo, *Vita matematica, napoletana*, vol. 2, 32–34; and Loria, *Fergola*, 55–61.

47. Fergola, *I problemi delle tazioni*. See also [Telesio], *Elogio*, 78–79.

48. Fergola, "Dal teorema tolemaico"; Fergola, "Su la rettificazione dell'ellisse e gli integrali che ne dipendono."

49. *Opuscoli matematici*, 128–29.

50. *Opuscoli matematici*, 145–46. Amodeo, *Vita matematica napoletana*, vol. 1, 146. For another similar episode, see [Telesio], *Elogio*, 78–79.

51. Montucla, *Histoire des mathématiques*. The first part appeared in two volumes in 1758 (repr. 1798); the first volume covered Greek, Roman, and Eastern traditions; the second, geometry, mechanics, and optics up to the seventeenth century. A second edition, expanded by Jérôme Lalande, was published posthumously in four volumes (1799–1801), and covered the entire eighteenth century. On the persisting influence of this landmark book see, for example, Chasles, *Aperçu historique*; and Loria, *Fergola*, 16. On d'Alembert's and Montucla's views of the history of mathematics and the crisis of this enlightened tradition in France, see Richards, "Historical Mathematics."

52. De Luca, *Memoria*, 11–12. De Luca was primarily a topographer and geographer. Like other mathematicians who had studied with Fergola and came of age in the French decade, de Luca argued for a compromise between synthetic and analytic methods, which he described as two different but equally legitimate languages. Synthesis should be valued for its intellectual import; analysis, for its applicative power. He thus joined the cult of ancient geometry with an eminently operative view of mathematics, which he connected directly to machine construction, industrial development, and the production of national wealth (*Nuovo sistema*, xii-xiv). He was a moderate liberal who played active roles in the revolutions of 1820 and 1848, but he would not be recognized and gratified as a true patriot by the new Italian regime.

53. Cuoco, *Platone in Italia*; Casini, *L'antica sapienza italica*; de Francesco, *Antiquity of the Italian Nation*.

54. Corcia, *Storia delle Due Sicilie*.

55. De Francesco, *Antiquity of the Italian Nation*, 85–113.

56. Flauti, *Elementi di geometria di Euclide*, lii, xlvii, xvii ("arrogant desire to innovate"). The sixth edition also contains materials from Archimedes, *On the Sphere and the Cylinder*, which Flauti had edited in 1804 (Flauti, *Elementi di geometria solida*), arguing that a modern course of mathematics—including infinitesimal analysis—could still be taught using solely the texts of Archimedes (Flauti, *Elementi di geometria di Euclide*, 359–60). On Flauti's edition of Euclid, see Ferraro, "Manuali di geometria," 120–23.

57. Loria, *Fergola*, 97.

58. Scorza, *Euclide vendicato*, 50, 91–95. Scorza later presented a third proof based on the method of limits (Scorza, "Nuova e semplice dimostrazione"). The fifth postulate, or parallel postulate, states: If a line segment intersects two straight lines forming two interior angles on the same side that are less than two right angles, then the two lines, if extended indefinitely, meet on that side on which the angles sum to less than two right angles. The postulate is not self-evident, and many attempts were made over the centuries to prove it using Euclid's first four postulates.

59. Scorza, *Euclide vendicato*, 1–4, 35, 45.

60. Flauti, *Elogio di Giuseppe Scorza*.

61. Scorza, *Divinazione della geometria analitica degli antichi*. For another exercise of divination, see Maresca, *Memoria*.

62. Flauti, *Elogio di Giuseppe Scorza*, 10–11.

63. Flauti, "Soluzioni geometriche"; Bruno, "Ricerche geometriche"; Loria, *Fergola*, 72. The following year, the French Jean Hachette (1769–1834) provided new analytic solutions for the problems solved by Bruno. To Flauti, Hachette's solutions confirmed the validity of an approach to mathematical research as an endless refinement of the art of problem solving (see his appendix to Hachette, "Solution algébrique").

64. Bruno, *Soluzioni geometriche*, iii–vi. Vincenzo degli Uberti (1791–1877), a professor of fortifications at the Military School, later claimed priority over Bruno's solutions (degli Uberti, *Esercitazioni geometriche*). He offers an interesting reconstruction of student life at Fergola's school around 1809–11, particularly of the influence of synthetic teaching on young army officers. With Giannattasio and Sangro as teachers, many officers engaged in synthetic problem solving as an agonistic activity (degli Uberti, *Esercitazioni geometriche*, 7–8).

65. Flauti, "Elogio dell'abate Felice Giannattasio," 343–44.

66. Flauti, "Elogio dell'abate Felice Giannattasio," 352; Amodeo, *Vita matematica napoletana*, vol. 2, 99–102.

67. Flauti was a member of numerous foreign academies, including those of Copenhagen (1817), Modena (1821), Berlin (1829), and Bologna (1845). Flauti, *Prospetto*.

68. Bézout, *Elementi del calcolo differenziale* (I have not been able to consult this book); Flauti, *Analisi algebrica elementare*.

69. Flauti, *Corso di geometria elementare e sublime*. On Flauti's textbooks of elementary geometry, see Ferraro, "Manuali di geometria."

70. Flauti, *Elementi di geometria descrittiva*, for the School of Engineering and Artillery, which had replaced the Military Academy in 1801.

71. Flauti, *Elementi di geometria descrittiva*, vi vii, 34.

72. Flauti, *Geometria di sito*, reprinted in 1821 and 1842; Fiocca, "La geometria descrittiva in Italia," 219.

73. Flauti, *Geometria di sito*, 173 75; Allman, *Greek Geometry from Thales to Euclid*, 119–20; Loria, *Fergola*, 105–6.

74. Trudi, *Teoria de' determinanti*, vii; Amodeo, *Vita matematica napoletana*, vol. 2, 193–213; Ferraro, "Nicola Trudi."

75. Flauti, *Tentativo di un progetto*, 4, 20.

76. Flauti, *Tentativo di un progetto*, 6–8.

77. Flauti, *Tentativo di un progetto*, 12.

78. Flauti, *Tentativo di un progetto*, 15–22.

79. Flauti, *Tentativo di un progetto*, 26.

80. Flauti, *Del metodo*, 30.

81. Flauti, *Del metodo*, 8.

82. Flauti, *Del metodo*, 10, 14.

83. Flauti, *Del metodo*, 12.

84. Flauti, *Del metodo*, 19–21, 24–25.

85. Flauti, *Del metodo*, 26–27.

86. Flauti, *Del metodo*, 29–31, 38.

87. Flauti, *Del metodo*, 43–44, 40–41, 46–67 (detailed description and bibliography of Flauti's course).

88. Some of the most important usages of the term "scientific counterrevolution," and of other related terms, are to be found in the scholarship on Jesuit science. They signal sophisticated

historiographical attempts to reconstruct the significant contribution of the Society of Jesus to modern science. See Gorman, *Scientific Counter-Revolution*, and Romano, *La contre-réforme mathématique*. Amir Alexander presents the mathematical practice of seventeenth-century Jesuit scholars as guided by a different vision of modernity (Alexander, *Infinitesimal*).

Chapter Six

1. Mannheim, "Conservative Thought"; Bloor, "Wittgenstein and Mannheim." On the notion of scientific persona and how it can be used to explore the "features of scholarly self-conceptions, professional identities, and the labour and institutions in which these are embedded," see Daston and Sibum, "Scientific Personae and Their Histories"; and Niskanen and Barany, *Gender, Embodiment, and the History of the Scholarly Persona* (quotation on 5).

2. Oldrini, *La cultura filosofica*, 50.

3. Cassitto, *Ne' funerali di Agostino Gervasio*, 6–7, 9–10; [Guarini], *Catalogo*, 63–67; Schipa, "Il secolo decimottavo," 466; Oldrini, *La cultura filosofica*, 23–24.

4. Storchenau, *Institutiones logicae*; Storchenau, *Institutiones metaphysicae*; Baumeister, *Institutiones metaphysicae*; Mako, *Compendio di logica*; Mako, *Compendiaria metaphysicae institutio*.

5. Capocasale, *Alloquutio*, 7–9; Capocasale, *Cursus philosophicus*, vol. 1, v–xv; vol. 2, 297–98; Panvini, *Capocasale*; Lovejoy, *The Great Chain of Being*; Tonelli, "The Law of Continuity."

6. See, for example, Semmola, *Istituzioni di filosofia*; Ciampi, *Elementi di filosofia*; Giampaolo, *Lezioni di metafisica*; Troisi, *Istituzioni metafisiche*.

7. See, for example, Mazzarella, *Corso d'ideologia elementare*.

8. Jacyna, "Immanence or Transcendence"; Desmond, *Politics of Evolution*.

9. Troisi, *Istituzioni metafisiche*, vol. 1, 8–9 (proper limits); Semmola, *Istituzioni di filosofia*, vol. 2, 244 (proof of the immateriality of the soul).

10. Colangelo, *L'irreligiosa libertà di pensare*, 165; Capocasale, *Codice eterno*, vol. 1, 58–60; Capocasale, *Saggio di politica*.

11. Gatti, "Lettera critica," 94–95.

12. Giampaolo, *Dialoghi sulla religione*, vol. 1, 2.

13. Capocasale, *Codice eterno*, vol. 2, 92–93. The first Italian translation of Karl Ludwig von Haller's *Restoration of Political Science* appeared in Naples in 1826–1828 (Haller, *Ristaurazione*).

14. Troisi, *Istituzioni metafisiche*, vol. 1, 5.

15. Among Colangelo's mentors were Nicola Spedalieri and cardinals Stefano Borgia and Sigismondo Gerdil (Parise, "Lettere inedite"); Cagnazzi, *La mia vita*, 203 (hangman or assassin); Colangelo, *Riflessioni storico-politiche*, 9, 21–22.

16. Parise, "Lettere inedite," 58–59. Colangelo's main work was *Storia dei filosofi e matematici napoletani*; but see also his *Apologia della religione cristiana* and *Saggio di alcune considerazioni sull'opera di G.B. Vico*.

17. Colangelo, *L'irreligiosa libertà*, 1–9.

18. Colangelo, *L'irreligiosa libertà*, 19; Holbach, *Systeme de la nature*, vol. 2, 107; Hobbes, *Leviathan*, 50.

19. Colangelo, *L'irreligiosa libertà*, 43. See, for example, his discussion of materialism, 74–84, 322–23. The connection to politics is immediate: religious truths are necessary instruments of government and have the function of "protecting society"; conveniently enough, they can also be grounded on factual evidence. The authority of religion is indeed based on matters of fact, and on a few intuitive principles. The facts are the evangelists' historical reports: why should they be less

credible than, say, Julius Caesar's *De Bello Gallico*? Miracles too are supported by historical evidence, and their possibility is guaranteed by the nature of God's rule. To Colangelo, miracles are temporary suspensions of the physical order, possible because God rules the universe like "an absolute master" rather than "a mechanical agent." To deny the historical evidence of religion and miracles would therefore mean to bring a radical and destructive skepticism into history—and into the sciences. Colangelo, *L'irreligiosa libertà*, 266, 288–89. The example he uses is in Genovesi, *Logica*, 98.

20. Colangelo, *L'irreligiosa libertà*, 313–19. Colangelo enthuses about the theological implications of Roger Cotes's preface to the 1713 edition of the *Principia* (314). Colangelo had been directed to study the Geneva edition of the *Principia* by the Jesuit mathematician Virgilio Cavina, who also recommended the works of Colin Maclaurin (Parise, "Lettere inedite," 54–55). Similarly Valsecchi, *La religion vincitrice*, vol. 1, 109, argues that Newton was a theologian *because* he was a mathematician.

21. Colangelo, *L'irreligiosa libertà*, 324. Colangelo often describes the physical universe in the Renaissance language of the fine arts to emphasize the role of God as creator-artist (*ornato, varietà, disegno, leggiadria*). Natural philosophers, he argued, have always seen nature as a pyramid, or a stairway that takes the contemplator to God. Similarly, the fathers of the Church moved from the contemplation of nature to its first cause. Science and religion share many principles, like the assumption that an effect is always commensurable with its cause, which—among other things—sustains the human understanding of the system of punishments and rewards in the afterlife.

22. Colangelo, *L'irreligiosa libertà*, 35, 333 (estasi scientifiche), 385.

23. Colangelo, *Vico*, 29, 46, 66, 138, 147.

24. Colangelo, *L'irreligiosa libertà*, 38, 381–98. On Colangelo's use of Baconian themes, see Parise, "Francis Bacon."

25. Colangelo, *L'irreligiosa libertà*, 27, 41–149.

26. Colangelo, *L'irreligiosa libertà*, 393.

27. Aquinas, *Summa Theologiae*, I–II, q. 96, a.1; Aristotle, *Nicomachean Ethics*, I, 3; Pasnau, *After Certainty*, 21–45.

28. Colangelo, *L'irreligiosa libertà*, 356.

29. While in mathematics one deals with "clear and distinct" ideas, in fields like law or history all one can do is try to "penetrate the darkness of uncertain phenomena, make conjectures, and risk predictions" (Colangelo, *L'irreligiosa libertà*, 357).

30. Apologetic empiricism can be interestingly compared with earlier phenomenal conceptions of scientific knowledge and especially with the notion of "mitigated skepticism," an outcome of the *crise phyrronienne* explored by Richard Popkin. Apologetic empiricism emphasizes the connection between religiously grounded skepticism and reactionary politics, consistently with Ian Hampster-Monk's study of Edmund Burke's religious sources. See Popkin, *History of Scepticism*, 129–50; Popkin, "Scepticism and Anti-Scepticism"; Tonelli, "The Weakness of Reason"; Bongie, "Hume and Scepticism"; Reill, "Analogy, Comparison, and Active Living Forces"; and Hampster-Monk, "Burke and the Religious Sources of Sceptical Conservatism."

31. Colangelo, *Galileo*, 26, 29–30, 115.

32. Colangelo, *Galileo*, 5, 6–7; [Fergola], *Prelezioni*, vol. 2, 319.

33. Colangelo, *Galileo*, 4, 9–10, 31, 77.

34. Colangelo, *Galileo*, 9, 78–81, 93. Colangelo devoted an entire chapter to Euclid's *Elements*, which—he contended— effectively illustrates "the naked, beautiful, simple, and ordered truth" (Colangelo, *Galileo*, 23). Through Euclidean geometry, Colangelo argued, Galileo learned the correct way to argue and draw conclusions, but he also had an innate proclivity for mathematics. Colangelo did not believe that all minds are equal: "the equality of intellectual faculties,"

he commented wryly, is just a dream. In fact, trying to teach the art of geometry to a mind not naturally predisposed to receiving it is tantamount to committing "violence against nature." Colangelo, *Galileo*, 1–6; Gerdil, *Précis d'un cours d'instruction*, 126–27.

35. Gerdil had praised Galileo's "philosophical spirit" in the 1750s (Gerdil, *Opere edite e inedite*, vol. 2, 46–49, 477). See also Colangelo, *Galileo*, 96; and Parise, "Francis Bacon," 375–78.

36. Colangelo, *Galileo*, xi–xii, 113.

37. Colangelo, *L'irreligiosa libertà*, 398.

38. [Telesio], *Elogio*, 156–59, 188; Cattaneo, *Gli occhi di Maria sulla rivoluzione*; Liguori, *Le glorie di Maria*, 137–45 (on Mary's gaze); Broers, *Napoleonic Empire*, 67 (patron of guerrillas).

39. Broers, *Politics and Religion* (war against God).

40. Jemolo, *Chiesa e stato*, 20; Broers, *Politics and Religion*, 180–82.

41. Fergola, *Teorica de' miracoli*, 108; Ventura, *Elogi funebri*, 72–74, 82–83; Orlov, *Mémoires*, 5:90.

42. Here, as in many other similar arguments, Fergola is clearly trying to strike a balance between the recognition that mathematical certainty does not pertain to empirical and historical knowledge and the defense of modern science against full-blown skeptical attacks from within the ranks of reactionary Catholics. See Fergola, *Teorica de' miracoli*, 91; Ventura, *Elogi funebri*, 75–76; de Ceglia, *Segreto di San Gennaro*, 300–03.

43. Politicization, notes Michael Broers, can come from the right as well as the left; it depends on the source of repression (Broers, *Politics and Religion*, 186–87). Lucas, "Résistances populaires"; Dupuy, *De la Révolution à la chouannerie*; Sutherland, *Chouans*, 309–12.

44. Caffiero, "Rivoluzione e millennio."

45. Princeton University Library, *Nicola Fergola Papers, 1769–1824* (NFP), undated letters (ailments); letter dated August 2, 1815 (Fergola appointed a member of the Giunta); undated letter (heterodoxy; see also [Telesio], *Elogio*, 147–48); undated letter (conspirators).

46. *NFP*, letter dated January 8, 1819 (mystic hell); Ventura, *Elogi funebri*, 84–88; Sarnelli, *L'anima desolata*, 173–81.

47. [Telesio], *Elogio*, 188–97; Ventura, *Elogi funebri*, vi–vii, 51–54, 89.

48. Ventura, *Elogi funebri*, 55–59, 83. Modesty (*humilitas*) is another virtue necessary for the acquisition of knowledge. Pride is indeed "the epidemic disease of the modern young." Mathematicians are particularly prone to the sin of pride because their science alone can be considered a human creation, and can therefore inspire the cult of one's reason. Those who fall to hubris end up ignoring both the limits of human reason and the mysteries of nature and religion; they aspire to "subjugate everything to the rigor of calculations." Not Fergola, though, who rather intensified his Catholic devotion precisely when religion came under attack, and who—against the dominant secularizing trend—aimed to "divinize science." Simple and docile in his piety, Fergola was also a paragon of scientific humility, to the point that he resisted publishing his works, and when he did, it was mostly anonymously. He also regularly declined honors and titles, including an invitation to tutor the heir to the throne. Nothing flattered him, we are told, least of all the "profaned crosses" that the French offered him. See Ventura, *Elogi funebri*, 60–64, 78–79; *NFP*, "Rinuncia degl'impieghi repubblicani" and letters of Campredon dated June 1811.

49. Ventura, *Elogi funebri*, 65, 77.

50. Ventura, *Elogi funebri*, 67–69.

51. Ventura, *Elogi funebri*, 70–71.

52. See, for example, Chateaubriand, *Génie du christianisme*, vol. 3, 29–44, on the contrast between an "intellectual" and a "material" geometry and on modern mathematicians as

"geometrical machines" (40–41). See also Ventura, *Elogi funebri*, 97, 113; NFP, undated letter ("offron pure *la tranquillità politica* degli stati").

53. Fergola, *Prelezioni*, vol. 1, 298; vol. 2, 205; Amodeo, *Vita matematica napoletana*, vol. 2, 348.

54. Fergola, *Prelezioni*, vol. 1, 24–25, 26–27, 195. Like Maupertuis, Fergola believed that the principle of least action was "the most powerful argument against atheism." See Terrall, *Maupertuis*, 272–74, 284–86.

55. Fergola, *Prelezioni*, 337.

56. [Telesio], *Elogio* 87–88; Colangelo, *L'irreligiosa libertà*, 399.

57. Ventura, *Elogi funebri*, 76; Fergola, *Teorica de' miracoli*, 8–9.

58. Fergola, *Teorica de' miracoli*, 48–56.

59. Fergola, *Teorica de' miracoli*, 14–15, 63.

60. Ventura, *Elogi funebri*, 1–50; Shapin, "Of Gods and Kings."

61. Gurr, *Principle of Sufficient Reason*.

62. Storchenau, *Institutiones metaphysicae*, vol. 3, 641–68 (immortality of the soul, apologetic strategies), vol. 3, 584–613 (body and soul relationship), vol. 3, 613–37 (sensationalist epistemology). On influxionism see also Genovesi, *Elementa metaphysicae*, vol. 2, 119–51. On the political and religious significance of geometry in the Jesuit tradition in an earlier period, see Alexander, *Infinitesimal*.

63. The *locus classicus* is Aquinas, *Super Boetium*, q. 5, a. 1–4.

64. Fergola, *Invenzione geometrica*, 1; *Opuscoli matematici*, 34.

65. On the rejection of a distinction between spiritual soul and consciousness, see also Genovesi, *Elementa metaphysicae*, vol. 2, 46–118.

66. Jéhan, *Dictionnaire*, 1047–48.

67. Dellucci, "Il trionfo della fede"; Bellucci, "La biblioteca dei Girolamini di Napoli"; Bellucci, "Giambattista Vico," 185–90; Mandarini, *I codici manoscritti*, viii–x.

68. Morelli di Gregorio, "Cav. Giuseppe Saverio Poli"; Gatti, *Elogio*; Giampaolo, *Elogio del commendatore Giuseppe Saverio Poli*; Olivier Poli, *Cenno biografico*.

69. Poli, *Testacea utriusque siciliae*. The third volume was published posthumously, in 1826–1827, by his student and collaborator Stefano delle Chiaje. The wax models are today part of the malacological collections of the Muséum nationale d'Histoire naturelle in Paris. See Gatti, *Elogio*, 26–32; Tëmkin, "The Art and Science"; Toscano, "Il Museo Poliano" and "Giuseppe Saverio Poli"; and Torino, "The Dismantling."

70. Poli, *Elementi*; Oldrini, *La cultura filosofica*, 28, 91.

71. Poli, *Elementi*, vol. 5, 435 (electrical fluids); Cotugno, *Opere*, 309–10 (letter on electrical discharge); Poli, *Elementi*, vol. 5, 316 (Galvani's experiments). See also Poli, *La formazione del tuono* (on Franklin); Schettino, "L'insegnamento," 374–76; Schettino, "Franklinists," 170–72; de Frenza, "The 'Poles' of Healing."

72. Stevens, *Development of Biological Systematics*; Scotti, *Iosephi Xaverii Poli Elogium*, engraving by Francesco Pagliuolo.

73. Poli, *Elementi*, vol. 1, xiv–xv (tenebre dell'incertezza); Poli, *Ragionamento*; Poli, *Breve ragionamento*, xix.

74. Poli, *Elementi*, vol., 1, 1–90, 167; vol. 5:437–38; Poli, *Ragionamento*, xxxii; Poli, *Breve ragionamento*, xvi–xviii.

75. Poli, *Elementi*, vol. 1, 68.

76. Poli, *Breve ragionamento*, vi; Poli, *Elementi*, vol. 1, 2–8; vol. 5, 437.

77. Poli, *Elementi*, vol. 1, xvi; Gatti, *Elogio*, 24–26; Giampaolo, *Elogio del commendatore Giuseppe Saverio Poli*, 17–18.

78. Morelli di Gregorio, "Cav. Giuseppe Saverio Poli," 4–5 (estasi scientifiche), 18–21; Giam-paolo, *Elogio del commendatore Giuseppe Saverio Poli*, 9–13, 34–45; Gatti, *Elogio*, 14–16, 19, 34, 37–41. During his studies at Padua, Poli had been close to the intransigent Dominican theologian Antonio Valsecchi (1708–1791).

79. Morelli di Gregorio, "Cav. Giuseppe Saverio Poli," 17–18; Landi, *Istituzioni*, vol. 1, 151–73 (Consiglio di stato ordinario); Zazo, "L'ultimo periodo borbonico," 471; Zazo, *L'istruzione pubblica e privata*, 76.

80. Poli, *Viaggio celeste*, xxvii (calculations), 5; Giampaolo, *Elogio del commendatore Giuseppe Saverio Poli*, 29–33. See also Poli, "Inno al sole," reprinted in Borrelli, "Il *Viaggio celeste*," 55–56.

81. Cotugno, *De acquaeductibus*. Cotugno's observational skills are also evident in his study of the sciatic nerve (Cotugno, *De ischiade nervosa*). See Magliari, *Elogio istorico di Domenico Cotugno*; Scotti, *Elogio*; Borrelli, "Carteggio di Domenico Cotugno"; and Romagnoli, "La scoperta degli acquedotti." Hermann von Helmholtz associated the phenomenon of hearing with the dynamics of labyrinthine fluid in 1863 (Helmholtz, *On the Sensations of Tone*).

82. Cotugno, *Dello spirito della medicina*, 14–34; Scotti, *Elogio*, 15; Borrelli, *Istituzioni scientifiche*, 113–23.

83. On the social dimension of similar anti-systematic, neo-Hippocratic conceptions of the medical practice see, for instance, Wolfe, "Sydenham and Locke"; Cunningham, "Thomas Sydenham"; and Lawrence, "Incommunicable Knowledge."

84. Cotugno, *De animorum*; Borrelli, *Istituzioni scientifiche*, 123–31.

85. On the early modern connection between natural theology and brain physiology, see Bynum, "The Anatomical Method"; Borrelli, *Istituzioni scientifiche*, 129–30 (meditation).

86. Iacovelli, "Brownismo"; Cosmacini, *Il medico giacobino*; Borrelli, *Istituzioni scientifiche*, 194–95; Frasca, "Dalla Scozia all'Italia," 45–48.

87. Cotugno had met and admired Alfonso de Liguori, whose sentimental religiosity he embraced. See Ventura, *Elogi funebri*, 190–91, 220, 223–24; Scotti, *Elogio*, 20, 61–62, 97.

88. Borrelli, "Introduzione," 11.

89. Ferrone, *I profeti*, 112–13, 253. For an updated account of Catholic Enlightenment as a global movement, see Lehner, *Catholic Enlightenment*.

Chapter Seven

1. In his philosophical reconstruction, David Lachterman interpreted the emergence of a creative and autonomous mind in mathematics as the core of modernity itself (*Ethics of Geometry*).

2. Geertz, "Common Sense as a Cultural System"; Rosenfeld, *Common Sense: A Political History*.

3. Salvatorelli, *Chiesa e stato*, 5.

4. Armenteros, *French Idea of History*, 82–114.

5. Sternell, *Anti-Enlightenment Tradition*, 98, 182–83, 441. See also Berlin, *Against the Current*; Hirschman, *Rhetoric of Reaction*; and Lukács's very undialectical *Destruction of Reason*.

6. McMahon, *Enemies of the Enlightenment*; Garrard, *Counter-Enlightenments*; Schmidt, "Inventing the Enlightenment."

7. Constant, *Réactions politiques*; Starobinski, *Action and Reaction*, 299–371.

8. Starobinski, *Action and Reaction*, 344–347.

9. Lilla, *The Shipwrecked Mind*, xiii–xvi.

10. Tomasi di Lampedusa, *Leopard*, 28.

11. Compagnon, *Les antimodernes*, 7, 448.

12. Robin, *Reactionary Mind*, 32–33. In this sense, "reaction" is a sharper conceptual tool than "conservatism," albeit one that has received less theoretical attention.

13. Menozzi, "Tra riforma e restaurazione"; Menozzi, "Intorno alle origini del mito della cristianità."

14. Bona, *Le "Amicizie."*

15. Ferrand, *Les conspirateurs démasqués*; Barruel, *Mémoires.*

16. Spedalieri, *De' diritti dell'uomo*; Tamburini, *Lettere teologico-politiche*; Salvatorelli, *Il pensiero politico italiano*, 110.

17. Verri, quoted in Candeloro, *Storia dell'Italia moderna*, vol. 1, 168–69; Leoni, *Storia della controrivoluzione*, 5–22; Guerci, *Spettacolo*, 189–249; del Corno, "Reazione."

18. Fontana, *La controrivoluzione*, 13 (spiritual spring); Hocedez, *Histoire de la théologie*, vol. 1, 72–91 ; Ravera, *Tradizionalismo francese.*

19. Maistre, *Considerations sur la France*, in *Oeuvres*, vol. 1, 1–184; Bonald, *Théorie du pouvoir*, in *Oeuvres*, vols. 13–15; Lamennais, *Réflexions sur l'état de l'église*, in *Oeuvres*, vol. 6; Godechot, *The Counter-Revolution*, 3–135; Fontana, *La controrivoluzione*, 13–64.

20. See George Orwell's short but insightful note on the "neo-pessimists," in *Essays*, 500–01.

21. Maistre, *Lettres d'un royaliste savoisien* [1793], in *Oeuvres*, vol. 7, 39.

22. Maistre, *Considerations sur la France*, in *Oeuvres*, vol. 1, 111; Maistre, *Lettres d'un royaliste savoisien* [1793], in *Oeuvres*, vol. 7, 166.

23. Maistre, *Souveraineté*, in *Oeuvres*, vol. 1, 318, 344, 358.

24. Maistre, *Lettres d'un royaliste savoisien* [1793], in *Oeuvres*, vol. 7, 154–55.

25. Maistre, *Principe générateur*, in *Oeuvres*, vol. 1, 258.

26. Maistre, *Pape*, in *Oeuvres*, vol. 2. 2–3 (no sovereignty without infallibility).

27. Bonald, *Théorie du pouvoir*, in *Oeuvres*, vol. 13, 1–3 (natural constitution).

28. Bonald, *Théorie du pouvoir*, in *Oeuvres*, vol. 14, 45.

29. Bonald, *Essai analytique sur les lois naturelles de l'ordre social* [1800], in *Oeuvres*, vol. 1, 53; Bonald, *Recherches philosophiques sur les premiers objets des connaissances morales* [1818], in *Oeuvres*, vol. 8, 68 ; Klinck, *The French Counterrevolutionary Theorist*. Friedrich Hayek attributed to Bonald the title of his essay and book *The Counter-Revolution of Science* (Hayek, *Counter-Revolution*, 123). In fact, it's dubious that Bonald ever used the expression "une contrerévolution de la science" (Sombart, *Sozialismus*, 54), which is instead emblematic of early-twentieth-century perceptions.

30. Fontana, *La controrivoluzione*, 45.

31. Lamennais, *Essai sur l'indifférence en matière de religion* [1817–24], in *Oeuvres*, vol. 1, xxix–xxx; "Indifférence religieuse" in *Dictionnaire de Théologie Catholique.*

32. Lamennais, *Essai sur l'indifférence en matière de religion* [1817–24], in *Oeuvres*, vol. 2, 3–34.

33. Lamennais, *De la religion considérée dans ses rapports avec l'ordre politique et civil* [1825–26], in *Oeuvres*, vol. 7.

34. Fontana, *La controrivoluzione*, 65–124; Verucci, "Chiesa e società"; Leoni, *Storia della controrivoluzione*, 159–87, 239–76.

35. Fontana, *La controrivoluzione*, 106.

36. *Breve saggio*, 56–58; Bellucci, "La biblioteca dei Girolamini di Napoli."

37. *Breve saggio*, 6; *Argomenti di discorsi*; de Maio, *Società e vita religiosa*, 295–97; Chiosi, *Lo spirito del secolo*, 34–44.

38. Rossi, *Dottrina di Gesù Cristo*; de Maio, *Società e vita religiosa*, 298.

39. Galdi, *Instruzioni*; Chiosi, *Lo spirito del secolo*, 253 (anchor of public safety).

40. Crocenti, *Meditazioni*, vol. 1, 43.

41. Chiosi, *Lo spirito del secolo*, 253-58.

42. Maturi, *Canosa*; del Corno, *Italia reazionaria*, 25-45.

43. Capece Minutolo, *Epistola*, 183-84.

44. Vitale, *Canosa*, 14 (quotation); Maturi, *Canosa*, 2.

45. Barruel, *Mémoires*; Capece Minutolo, *Epistola*, 101, 181. For the conspiratorial interpretation of the Revolution, see also Lefranc, *Voile levé*.

46. Capece Minutolo, *Trinità*, 36, 79, 73.

47. Capece Minutolo, *L'utilità della monarchia*, 8, 50-51; Maturi, *Canosa*, 38-40.

48. Vitale, Canosa, 26; Maturi, *Canosa*, 13.

49. Maturi, *Canosa*, 15-24; Croce, "Il principe di Canosa."

50. Maturi, *Canosa*, 26; Capece Minutolo, *Memoria dilucidativa*, 16.

51. Capece Minutolo, *La passione e morte*; Capece Minutolo, *La natività*; Vitale, *Canosa*, 33 (satire); Maturi, *Canosa*, 38-39 (Genovesi).

52. Maturi, *Canosa*, 117-84.

53. Capece Minutolo, *Piffari*; Capece Minutolo, *Epistola*, 204.

54. Leopardi, *Catechismo filosofico* (natural inequality, submission to authority); Leopardi, *Dialoghetti*; Themelly and Lo Curto, *Scrittori cattolici*, 14; del Corno, *Italia reazionaria*, 47-60. Compare with Basil Willey's "Cosmic Toryism" in Willey, *The Eighteenth Century Background*, 43-56.

55. Leopardi, *Dialoghetti*, 75-77.

56. Capece Minutolo, *Epistola*, 136.

57. Piergili, *Leopardi*, 45.

58. Mannheim, "Conservative Thought," 95, 115.

59. Robin, *Reactionary Mind*, 3-37.

60. Foucher, *Philosophie catholique*, 252; Masnovo, *Neotomismo*; Orlando, *Tomismo*; Fontana, *La controrivoluzione*, 86-105, 142-44; Maturi, *Canosa*, 185-344; Verucci, "Cattolicesimo intransigente," 251-85; Guccione, *Ventura*; Scarpato, "Fatiche."

61. Ventura, *Opere*, vol. 12, 428 (quotation), 489-96 (social contract), 530-36 (freedom of the press).

62. Ventura, *Opere*, vol. 12, 1-11, 491; Fontana, *La controrivoluzione*, 142-45.

63. Fontana, *La controrivoluzione*, 125-209; Chenaux, "All'insegna del neotomismo."

64. Fontana, *La controrivoluzione*, 172 (quotation), 198-200; Haller, *Ristaurazione*.

65. Ventura, *Methodo philosophandi*.

66. Ventura, *Methodo philosophandi*, xlii (substantial unity), lxxxvi (quotation); Chenaux, "All'insegna del neotomismo," 24.

67. Taparelli d'Azeglio, "Due filosofie," 484 (quotation); Ventura, *Osservazioni*, 3-23.

Chapter Eight

1. Mehrtens, Bos, and Schneider, *Social History of Nineteenth Century Mathematics*, 3.

2. Kline, *Mathematics* (rigorization as self-explanatory); Wittgenstein, *Remarks on the Foundations of Mathematics*, 3, §56.

3. Grattan-Guinness, *Convolutions*, vol. 1, 107; Mehrtens, Bos, and Schneider, *Social History of Nineteenth Century Mathematics*, esp. the essays by Hans Jahnke and Michael Otte, Luke

Hodgkin, Ivo Schneider, Gert Schubring, and Philip Enros; Belhoste, Dahan-Dalmedico, and Picon, *La formation polytechnicienne*; Belhoste, *La formation d'une technocratie*; Richards, *Mathematical Visions*; Becker, "Radicals, Whigs and Conservatives"; Craik, "Calculus and Analysis"; Ackenberg-Hastings, "Analysis and Synthesis"; Phillips, "Robert Woodhouse"; Pyenson, *Neohumanism*; Jahnke, *Mathematik und Bildung*.

4. Gray, *Plato's Ghost*; Daston and Sibum, "Introduction: Scientific Personae and Their Histories"; Daston and Galison, *Objectivity*.

5. Kusch, "Objectivity and Historiography," 131.

6. Ruffini, *Teoria generale delle equazioni*; Ruffini, *Della immaterialità*.

7. Ruffini, *Riflessioni*, unpaged preface (quotation).

8. Taddei, *Consulti medici*; Piola, *Lettere di Evasio ad Uranio*.

9. Piola, preface to *Opuscoli matematici e fisici di diversi autori*, iii–iv (quotations); Piola, "Saggio sulla metafisica dell'analisi pura"; Redondi, "Cultura e scienza," 711–18.

10. Ruffini, *Opere matematiche*, vol. 3, 88–89 (letter dated September 20, 1821).

11. Grabiner, *Origins*, 1–15. On the negotiations, rhetorical and philosophical, that made it possible for Cauchy to articulate his rigorization program, and for the diverse possible readings of his work by contemporaries, see Barany, "God, King, and Geometry."

12. Cauchy, *Cours d'analyse* [1821], in *Oeuvres*, ser. 2, vol. 3, ii–iii; Barany, "God, King, and Geometry."

13. Grabiner, *Origins*, 6, 30–31.

14. For an example of this transformation, and of its historiographical readings, see the case of the Intermediate Value Theorem (Cauchy, *Cours d'analyse* [1821], in *Oeuvres*, ser. 2, vol. 3, iii; Grabiner, *Origins*, 47–76; Barany, "God, King, and Geometry," 6–8).

15. Berkeley, *Analyst*; Maclaurin, *Treatise*, vol. 1, i–vi; Cantor, "Anti-Newton"; d'Alembert, "Essai sur les éléments de philosophie," 288–99; Gerdil, "De l'infini absolu"; Lagrange, "Note sur la métaphysique du calcul infinitésimal," in *Oeuvres*, vol. 7, 597–99.

16. Valson, *Cauchy*, 108–21, 169–213; Belhoste, *Cauchy*, 142 (quotations); Barany, "God, King, and Geometry," 14–17.

17. Ruffini, *Opere matematiche*, vol. 3, 88–89; Cauchy, "Sept leçons de physique général faites à Turin en 1833," in *Oeuvres*, ser. 2, vol. 15, 413; Belhoste, *Cauchy*, 216.

18. Cauchy, *Oeuvres*, ser. 2, vol. 15, 415; Cauchy, "Sur les limites des connaissances humaines," in *Oeuvres*, ser. 2, vol. 15, 5–7; Belhoste, *Cauchy*, 218–21.

19. Gray, *Plato's Ghost*; Henderson, *Fourth Dimension*; Henderson, "Editor's Introduction."

20. Mehrtens grapples with the question of explaining historically the features of modern mathematics mostly through discourse analysis. The countermoderns' use of the term *Anschauung*, for example, can be interpreted as taking position within "a dynamical system of cultural, political and scientific orientations." This term was a powerful rhetorical symbol that immediately summoned a specific set of associations outside the field of mathematics, and therein lay its significance. Mehrtens traces the roots of *anschauliche* mathematics to the work of Felix Klein (1849–1925) and, in one of the strongest claims of the book, argues for an essential continuity between this strain of Wilhelmian mathematics and the anti-modern and anti-Semitic campaign of the proponents of Aryan mathematics under the Nazi regime. See Mehrtens, *Moderne-Sprache-Mathematik*, 79 (quotation); Rowe, Review of Gray's *Plato's Ghost*; Epple, "Styles of Argumentation." Also centered on an original redefinition of mathematical modernity is Lachterman, *Ethics of Geometry*.

21. Schaffer, "Machine Philosophy"; Schaffer, "Enlightened Automata"; Ashworth, "Memory, Efficiency, and Symbolic Analysis"; Alder, *Engineering the French Revolution*; Tresch, *Romantic*

Machine. See Jones, *Reckoning with Matter*, for a genealogy of the shifting relations between thinking, calculation, mathematics, and mechanical invention.

22. Versions of this myth can be found in Habermas, *Theory of Communicative Action*, vol.1; and Giddens, *The Consequences of Modernity*.

23. This new role became clear, for example, during the visit of prominent German mathematicians Jakob Steiner (1796–1863) and Carl Gustav Jacobi (1804–1851) in 1844, and at the seventh Congress of the Italian Scientists in Naples in 1845. See Loria, *Fergola*, 132–35; and *Diario del Settimo Congresso*.

24. On the participation of scientists and mathematicians in the Risorgimento and in Italian political life, see Bottazzini, *Va' pensiero*; Bottazzini and Nastasi, *Patria*.

25. Bottazzini and Nastasi, *Patria*, 270–74; Bottazzini, *Va' pensiero*, 146; Besana and Galluzzi, "Geometria e latino," 1291 (last quotation), 1292. A fundamental study on the turn to Euclid in nineteenth-century pedagogy is Richards, *Mathematical Visions*, which reconstructs the broader meaning of this choice in Victorian England.

26. Besana and Galluzzi, "Geometria e latino," 1291 (quotations).

27. *Giornale di matematiche ad uso degli studenti delle università italiane* 1 (1863)–31 (1893).

28. Brioschi and Cremona, "Al signor direttore."

29. Amodeo, *Vita Matematica*, vol. 2, 122.

Bibliography

Archival Materials

Archivio di Stato di Napoli (ASNA): Naples, Italy
    Archivio Borbone
    Direzione generale di ponti e strade
    Ministero della polizia generale
    Ministero delle finanze
Biblioteca Nazionale di Napoli (BNN): Naples, Italy
    III, XIII, XVI, XVIII
Princeton University Library (PUL): Princeton, NJ
    Nicola Fergola Papers, 1769–1824 (NFP)

Printed Sources

Abbri, Ferdinando. "Chimici e artiglieri: Lavoisier e la cultura napoletana." In *Atti del VI Convegno nazionale di Storia e fondamenti della chimica*, 245–57. Rome: Accademia delle Scienze, 1995.

Ackenberg-Hastings, Amy. "Analysis and Synthesis in John Playfayr's *Elements of Geometry.*" *British Journal for the History of Science* 35 (2002):43–72.

Addabbo, Claudia. "I *Philosophiae naturalis institutionum libri tres* di Pietro Di Martino e il newtonianesimo napoletano." In *La circolazione dei saperi scientifici tra Napoli e l'Europa nel XVIII secolo*, edited by Roberto Mazzola, 77–109. Naples: Diogene Edizioni, 2013.

Agulhon, Maurice. *Pénitents et francs-maçons dans l'ancienne Provence: Essai sur la sociabilité méridionale.* Paris: Fayard, 1968.

Ajello, Raffaele. *Il problema della riforma giudiziaria e legislative nel Regno di Napoli durante la prima metà del secolo XVIII.* 2 vols. Naples: Jovene, 1961.

Ajello, Raffaele. "L'estasi della ragione: dall'illuminismo all'idealismo. Introduzione alla *Scienza* di Filangieri." In Raffaele Ajello, *Formalismo medievale e moderno*, 39–184. Naples: Jovene, 1990.

Alder, Ken. "A Revolution to Measure: The Political Economy of the Metric System in France." In *The Values of Precision*, edited by M. Norton Wise, 39–71. Princeton, NJ: Princeton University Press, 1995.

Alder, Ken. *Engineering the French Revolution: Arms and Enlightenment in France.* Chicago: University of Chicago Press, 1997.

Alembert, Jean Lerond d'. "Discours préliminaire." In *Encyclopédie, ou dictionnaire raisonné des sciences, des arts et des métiers, par une société de gens de lettres,* vol. 1, edited by Denis Diderot and Jean Lerond d'Alembert, i –xlv. [Geneva]: 1751.

Alembert, Jean Lerond d'. "Essai sur les éléments de philosophie, ou sur les principes des connaissances humaines, avec les éclarcissemens." In d'Alembert, *Œuvres complètes,* vol. 1, 114–348. Paris: Belin, 1821–1822.

Alexander, Amir. *Infinitesimal: How a Dangerous Mathematical Theory Shaped the Modern World.* New York: Farrar, Straus and Giroux, 2014.

Alisio, Giancarlo. *Napoli com'era nelle gouaches del Sette e Ottocento.* Rome: Newton Compton, 1990.

Allman, George. *Greek Geometry from Thales to Euclid.* Dublin: Hodges, Figgis, and Co., 1889.

Alonso, William, and Paul Starr, eds. *The Politics of Numbers.* New York: Sage, 1987.

Alpaugh, Micah. "The British Origins of the French Jacobins: Radical Sociability and the Development of Political Club Networks, 1787–1793." *European History Quarterly* 44 (2014): 593–619.

Ambrasi, Domenico. "Giuseppe e Gennaro Cestari dal gallicanesimo al giacobinismo rivoluzionario." In Domenico Ambrasi, *Riformatori e ribelli a Napoli nella seconda metà del Settecento,* 172–289. Naples: Regina, 1979.

Ambrasi, Domenico. "Il clero a Napoli nel '99 tra rivoluzione e restaurazione." *Campania sacra* 22 (1991): 52–81.

Ambrasi, Domenico. "Seminario e clero a Napoli dalla nascita dell'istituzione alla fine del Settecento." *Campania sacra* 15–17 (1984–1986): 7–95.

Amodeo, Federico. *Vita matematica napoletana: Studio storico, biografico, bibliografico.* 2 vols. Naples: Giannini, 1905–1924.

Aquinas, Thomas. *Summa theologiae.* In *Opera omnia,* vol. 4–12. Rome: Ex typographia polyglotta S. C. de Propaganda Fidei, 1888–1906.

Aquinas, Thomas. *Super Boetium De Trinitate: Expositio libri Boetii De ebdomadibus.* In *Opera omnia,* vol. 50. Paris: Editions du Cerf, 1992.

Aracri, Gregorio. *Dell'amor proprio.* Naples: Orsino, 1789.

Aracri, Gregorio. *Elementi del diritto naturale.* Naples: 1787.

Argomenti di discorsi destinati in questo terzo anno nell'accademia di materie Ecclesiastiche eretta dentro la Congregazione de' Padri dell'Oratorio di Napoli nell'anno MDCCXLI, sotto la protezione dell'Eminentissimo Signor Cardinale Spinelli arcivescovo, presidente della medesima. Naples: Stamperia Muziana, 1744.

Aristotle. *Nicomachean Ethics.* Edited by H. Rackham. Cambridge, MA: Harvard University Press, 1934.

Armenteros, Carolina. *The French Idea of History: Joseph De Maistre and His Heirs, 1794–1854.* Ithaca, NY: Cornell University Press, 2011.

Ashworth, William. "Memory, Efficiency, and Symbolic Analysis: Charles Babbage, John Herschel, and the Industrial Mind." *Isis* 87 (1996): 629–53.

Atti della Reale Accademia delle Scienze e Belle-Lettere di Napoli dalla fondazione sino all'anno MDCCLXXXVII. Naples: Campo, 1788. Single volume. Henceforth *ARAS* 1788.

Atti della Reale Accademia delle Science, sezione della Società Reale Borbonica. 6 vols. Naples: Stamperia reale, 1819–1851. Henceforth *ARAS* 1–6, 1819–1851.

Azouvi, François, ed. *L'institution de la raison.* Paris: Vrin, 1992.

Baker, Keith. *Condorcet: From Natural Philosophy to Social Mathematics.* Chicago: University of Chicago Press, 1975.

Baker, Keith. *Inventing the French Revolution.* Cambridge: Cambridge University Press, 1990.

Baker, Keith. "The Establishment of Science in France." *Science* 214 (1981): 54–55.

Barany, Michael. "Fellow Travelers and Traveling Fellows: The Intercontinental Shaping of Modern Mathematics in Mid-Twentieth Century Latin America." *Historical Studies in the Natural Sciences* 46 (2016): 669–709.

Barany, Michael. "God, King, and Geometry: Revisiting the Introduction to Cauchy's *Cours d'analyse*." *Historia Mathematica* 38 (2011): 368–88.

Barany, Michael. "Histories of Mathematical Practice: Reconstruction, Genealogy and the Unruly Pasts of Ruly Knowledge." *ZDM—Mathematics Education* 52 (2020): 1075–86.

Barnes, Barry. *Interests and the Growth of Knowledge.* London: Routledge, 1977.

Barnes, Barry. "Social Life as Bootstrapped Induction." *Sociology* 17 (1983): 524–45.

Barnes, Barry. "Status Group and Collective Action." *Sociology* 26 (1992): 259–70.

Barnes, Barry. *The Nature of Power.* Cambridge: Polity Press, 1988.

Barrell, John. *The Dark Side of the Landscape: The Rural Poor in English Painting, 1730–1840.* Cambridge: Cambridge University Press, 1980.

Barreto, Joana. "Virgile, source artistique et scientifique 'locale' dans la Naples aragonaise." In *Nature and the Arts in Early Modern Naples*, edited by Frank Fehrenbach and Joris van Gastel, 7–23. Berlin: De Gruyter, 2020.

Barruel, Augustin. *Mémoires pour servir à l'histoire du jacobinisme.* 4 vols. London: Le Boussonier, 1797–1798.

Dasile, Antonio. "Un illuminista calabrese: Domenico Grimaldi da Seminara," *Archivio storico per la Calabria e la Lucania* 13 (1943–1944): 16–31.

Bates, David. *Enlightenment Aberrations: Error and Revolution in France.* Ithaca, NY: Cornell University Press, 2002.

Battaglini, Mario. *Atti, leggi, proclami ed altre carte della Repubblica napoletana, 1798–1929.* 3 vols. Salerno: Società editrice meridionale, 1983.

Battaglini, Mario. *La Repubblica napoletana: Origini, nascita, struttura.* Rome: Bonacci, 1992.

Battaglini, Mario. *Mario Pagano e il progetto di costituzione della Repubblica napoletana.* Rome: Archivio Guido Izzi, 1994.

Baumeister, Friedrich. *Institutiones metaphysicae, ontologiam, cosmologiam, psychologiam, theologiam denique naturalem complexae, methodo Wolfii adornatae* [1754]. Venice: Zatta, 1797.

Baxandall, Michael. *Patterns of Intention: On the Historical Explanation of Pictures.* New Haven, CT: Yale University Press, 1985.

Becker, Harvey. "Radicals, Whigs and Conservatives: The Middle and Lower Classes in the Analytical Revolution at Cambridge in the Age of Aristocracy." *British Journal for the History of Science* 28 (1995): 405–26.

Behboud, Ali. "Greek Geometrical Analysis." *Centaurus* 37 (1994): 52–86.

Belhoste, Bruno. *Augustin-Louis Cauchy: A Biography.* New York: Springer 1991.

Belhoste, Bruno. *La formation d'une technocratie: L'École polytechnique et ses élèves de la Révolution au Second Empire.* Paris: Belin, 2003.

Belhoste, Bruno, Amy Dahan-Dalmedico, and Antoine Picon, eds. *La formation polytechnicienne, 1794–1994.* Paris: Dunod, 1994.

Bellucci, Antonio. "Giambattista Vico e la biblioteca dei Girolamini." In *Giambattista Vico nel terzo centenario della nascita*, 181–205. Naples: Edizioni Scientifiche Italiane, 1971.

Bellucci, Antonio. "Il trionfo della fede sulle scienze e sulle virtù attraverso le allegorie decorative del soffitto della biblioteca dei girolamini." *Roma*, July 2, 1927: 3.

Bellucci, Antonio. "La biblioteca dei girolamini di Napoli." *Accademie e biblioteche d'Italia* 4
(1930): 38–64.

Berkeley, George. *The Analyst, or a Discourse Addressed to an Infidel Mathematician*. London:
Tonson, 1734.

Berlin, Isaiah. *Against the Current: Essays in the History of Ideas*. Princeton, NJ: Princeton University Press, 2013.

Bermingham, Anne. *Landscape and Ideology: The English Rustic Tradition, 1740–1860*. Berkeley:
University of California Press, 1986.

Berti, Giuseppe. *I democratici e l'iniziativa meridionale nel Risorgimento*. Milan: Feltrinelli, 1962.

Besana, Luigi, and Massimo Galluzzi. "Geometria e latino: Due discussioni per due leggi." In
*Storia d'Italia. Annali 3: Scienza e tecnica nella cultura e nella società dal Rinascimento ad
oggi*, edited by Gianni Micheli, 1285–1306. Turin: Einaudi, 1980.

Bézout, Étienne. *Elementi del calcolo differenziale, tradotti ed illustrati da Vinc. Flauti*. Naples: 1801.

Biagioli, Mario. *Galileo, Courtier: The Practice of Science in the Culture of Absolutism*. Chicago:
University of Chicago Press, 1994.

Biscardi, Luigi, and Antonino de Francesco. *Vincenzo Cuoco nella cultura di due secoli*. Bari:
Laterza, 2002.

Blackbourn, David. *Marpingen: Apparitions of the Virgin Mary in Bismarckian Germany*. Oxford: Clarendon, 1993.

Bloor, David. *Knowledge and Social Imagery*. London: Routledge, 1976.

Bloor, David. "Polyhedra and the Abominations of the Leviticus." *British Journal for the History
of Science* 11 (1978): 245–72.

Bloor, David. "Wittgenstein and Mannheim on the Sociology of Mathematics." *Studies in History
and Philosophy of Science Part A* 4 (1973): 173–91.

Bocalosi, Girolamo. *Istituzioni democratiche per la rigenerazione del popolo italiano*. 2 vols. Milan: Raffaele Netti, 1798.

Bona, Candido. *Le "Amicizie": Società segrete e rinascita religiosa, 1770–1830*. Turin: Deputazione
Subalpina di Storia Patria, 1962.

Bonald, Louis de. *Oeuvres*. 14 vols. Paris: Le Clère, 1817–1843.

Bonaparte, Napoleon. *Correspondance de Napoléon Ier, publiée par ordre de l'empereur Napoléon
III*. 32 vols. Paris: Imprimerie Impériale, 1858–1869.

Bongie, Laurence. "Hume and Scepticism in Late Eighteenth-Century France." In *The Skeptical
Tradition around 1800*, edited by Johan van der Zande and Richard Popkin, 15–30. Dordrecht: Kluwer, 1998.

Borrelli, Antonio. "Carteggio di Domenico Cotugno." *Nuncius* 1 (1986): 93–101.

Borrelli, Antonio. "Il *Viaggio celeste* by Giuseppe Saverio Poli." In *Atti del XXXVI Convegno
annuale della Società Italiana degli Storici della Fisica e dell'Astronomia*, edited by Salvatore
Esposito, 53–60. Pavia: Pavia University Press, 2017.

Borrelli, Antonio. "Introduzione." In Domenico Cotugno, *Dello spirito della medicina*, 7–17.
Naples: Procaccini, 1988.

Borrelli, Antonio. *Istituzioni scientifiche, medicina e società: Biografia di Domenico Cotugno
(1736–1822)*. Florence: Olschki, 2000.

Borrelli, Pasquale. [Pirro Lallebasque, pseud.]. *Introduzione alla filosofia naturale del pensiero*
[1824]. Lugano: Ruggia, 1830.

Borrelli, Pasquale. [Pirro Lallebasque, pseud.]. *Principi della genealogia del pensiero*. 3 vols. Lugano: Vanelli, 1825–1829.

Bos, Henk. "Arguments on Motivation in the Rise and Decline of a Mathematical Theory: The Construction of Equations, 1637-ca.1750." *Archive for the History of Exact Sciences* 30 (1984): 331–80.

Bos, Henk. *Redefining Geometrical Exactness: Descartes' Transformation of the Early Modern Concept of Construction*. New York: Springer, 2001.

Botta, Carlo. *Storia d'Italia dal 1789 al 1814*. 6 vols. Capolago: Tipografia Elvetica, 1833–1838.

Bottazzini, Umberto. *Il calcolo sublime: Storia dell'analisi da Euler a Weierstrass*. Turin: Boringhieri, 1981.

Bottazzini, Umberto. *Va' pensiero: Immagini della matematica nell'Italia dell'Ottocento*. Bologna: Mulino, 1994.

Bottazzini, Umberto, and Pietro Nastasi. *La patria ci vuole eroi: Matematici e vita politica nell'Italia del Risorgimento*. Bologna: Zanichelli, 2013.

Bouloiseau, Marc. *The Jacobin Republic, 1792–1794*. Cambridge: Cambridge University Press, 1983.

Bourguet, Marie-Noëlle. *Déchiffrer la France: La statistique départementale à l'époque napoléonienne*. Paris: Éditions des archives contemporaines, 1988.

Boyer, Carl. *The History of the Calculus and Its Conceptual Development*. New York: Dover, 1949.

[Bozzelli, Francesco]. *Esquisse politique sur l'action des forces sociales dans les différentes espèces de gouvernement*. Bruxelles: Lacrosse, 1827.

Bozzelli, Francesco. *Essai sur les rapports primitifs qui liens ensemble la philosophie et la morale*. Paris: Grimbert, 1825.

Brancaccio, Giovanni. *Geografia, cartografia e storia nel Mezzogiorno*. Naples: Guida, 1981.

Brancato, Francesco. *Il Caracciolo e il suo tentativo di riforme in Sicilia*. Naples: Palumbo, 1946.

Breve saggio dell'Accademia di Materie Ecclesiastiche eretta dentro la Congregazione de'Padri dell'Oratotio di Napoli nell'anno MDCCXLI, sotto la protezione dell'Eminentissimo Signor Cardinale Spinelli arcivescovo, presidente della medesima. Naples: Stamperia Muziana, 1741.

Brian, Éric. *La mesure de l'État: Administrateurs et géomètres au XVIIIe siècle*. Paris: Albin Michel, 1995.

Brian, Éric. "Le livre des sciences est-il écrit dans la langue des historiens?" In *Les formes de l'expérience: Une autre histoire sociale*, edited by Bernard Lepetit, 85–98. Paris: Albin Michel, 1994.

Brioschi, Francesco, and Luigi Cremona, "Al signor direttore." *Giornale di matematiche ad uso degli studenti delle università italiane* 7 (1869): 51–54.

Broers, Michael. "Cultural Imperialism in a European Context? Political Culture and Cultural Politics in Napoleonic Italy." *Past and Present* 170 (2001): 152–80.

Broers, Michael. *Politics and Religion in Napoleonic Italy, 1801–1814*. London: Routledge, 2007.

Broers, Michael. *The Napoleonic Empire in Italy, 1796–1814: Cultural Imperialism in a European Context?* Basingstoke: Palgrave Macmillan, 2005.

Bruno, Francesco. "Ricerche geometriche sopra un difficile problema di sito." In *ARAS* 2, 1825: 29–40.

Bruno, Francesco. *Soluzioni geometriche di alcuni difficili problemi solidi condotte a fine col metodo degli antichi geometri, per servir di guida alla gioventù matematica che desidera in esso esercitarsi*. Naples: 1824.

Buccaro, Alfredo, and Fausto de Mattia, eds. *Scienziati artisti: Formazione e ruolo degli ingegneri nelle fonti dell'Archivio di Stato e della Facoltà di Ingegneria di Napoli*. Naples: Electa Napoli, 2003.

Burke, Peter. "The Virgin of the Carmine and the Revolt of Masaniello." *Past and Present* 99 (1983): 3–21.

Bynum, William. "The Anatomical Method, Natural Theology, and the Functions of the Brain." *Isis* 64 (1973): 445–68.

Cabanis, Pierre Jean George. "Rapport fait au conseil de Cinq-Cents sur l'organisation des écoles de médicine" [1798]. In Cabanis, *Oeuvres complètes*, vol. 1, 361–97. Paris: Firmin Didot, 1823.

Cabanis, Pierre Jean George. *Rapporti del fisico e del morale dell'uomo*. Naples: Sangiacomo, 1807.

Cacciatore, Giuseppe. "Vico e Kant nella filosofia di Ottavio Colecchi." *Bollettino del Centro di Studi Vichiani* 12 (1982): 63–99.

Caffiero, Marina. "Rivoluzione e millennio: Tematiche millenaristiche in Italia nel periodo rivoluzionario." *Critica storica* 24 (1987): 584–602.

Cagnazzi, Luca de Samuele. *Elementi dell'arte statistica*. 2 vols. Naples: Flautina, 1808–1809.

Cagnazzi, Luca de Samuele. *Elementi di economia politica*. Naples: Sangiacomo, 1813.

Cagnazzi, Luca de Samuele. *La mia vita (1764–1852)*. Milan: Hoepli, 1944.

Cagnazzi, Luca de Samuele. *Memoria dell'uso della sintesi e dell'analisi nell'istruzione delle scienze matematiche*. [Naples]: [1813].

Calaresu, Melissa. "The Enlightenment in Naples." In *A Companion to Early Modern Naples*, edited by Tommaso Astarita, 405–26. Leiden: Brill, 2013.

Candeloro, Giorgio. *Storia dell'Italia moderna*. Vol. 1: *Le origini del Risorgimento*. Milan: Feltrinelli, 1989.

Cantimori, Delio. "Giacobini italiani [1956]." In Delio Cantimori, *Studi di storia*, 629–38. Turin: Einaudi, 1959.

Cantor, Geoffrey. "Anti-Newton." In *Let Newton Be!*, edited by John Fauvel et al., 203–22. Oxford: Oxford University Press, 1988.

Cantù, Cesare. Review of Cagnazzi, *Saggio sulla popolazione del Regno di Puglia ne' passati tempi e nel presente*. In *Annali universali di statistica* 67 (1841): 161–68.

Capece Minutolo, Antonio. *Epistola, ovvero riflessioni critiche sulla moderna storia del Reame di Napoli del generale Pietro Colletta* [1834]. Reprinted in Silvio Vitale, *Il principe di Canosa e l'epistola contro Pietro Colletta*, 74–249. Naples: Berisio, 1969.

Capece Minutolo, Antonio. *I piffari di montagna, ossia cenno estemporaneo di un cittadino imparziale sulla congiura del principe di Canosa e sopra i carbonari: Epistola critica diretta all'estensore del foglio letterario di Londra*. Dublin [Lucca]: 1820.

Capece Minutolo, Antonio. *La natività del Divino Nostro Redentore, dimostrata con le autorità degli etnici filosofi*. [Naples]: [1802].

Capece Minutolo, Antonio. *La passione e morte del Divino Nostro Redentore, predetta dai profeti, confermata dai prodigi e dalle testimonianze degli eterodossi scrittori*. Naples: Orsino, 1802.

Capece Minutolo, Antonio. *La trinità, orazione dogmatico-filologica*. Naples: Zambraja, 1795

Capece Minutolo, Antonio. *L'utilità della monarchia nello stato civile, orazione diretta contro i novatori del secolo*. Naples: Zambraja, 1796.

Capece Minutolo, Antonio. *Memoria dilucidativa di vari articoli da aversi in considerazione nella abolizione da farsi dei feudi e della feudalità*. [Naples]: [1799].

Capialbi, Vito. *Nuovi motivi comprovanti la dualità della Mesma e della Medama, Edizione seconda: Si uniscono le notizie dell'ab. Nicola Maria Pacifico*. Naples: Porcelli, 1849.

Capocasale, Giuseppe. *Alloquutio in privato suo auditorio de logicae ac metaphysicae utilitate habita*. Naples: 1804.

Capocasale, Giuseppe. *Cursus philosophicus sive universae philosophiae institutiones* [1789]. 3 vols. Naples: Zambraja, 1824–1825.

Capocasale, Giuseppe. *Il codice eterno, ridotto in sistema secondo i veri principi della ragione e del buon senso* [1793]. 2 vols. Naples: Zambraja, 1822.

Capocasale, Giuseppe. *Saggio di politica privata per uso de' giovanetti*. Naples: Zambraja, 1791.

Capograssi, Antonio. "Nuovi documenti sull'accusa di ateismo ad Ottavio Colecchi." *Samnium* 13 (1940): 73–89.

Caravelli, Vito, and Vincenzo Porto. *Trattati di calcolo differenziale e del calcolo integrale, per uso del regale collegio militare*. Naples: Raimondi, 1786.

Carletti, Gabriele. *Melchiorre Delfico: Riforme politiche e riflessione teorica di un moderato meridionale*. Pisa: ETS, 1996.

Carnot, Lazare. *Géométrie de position*. Paris: Duprat, 1803.

Carpanetto, Dino, and Giuseppe Ricuperati. *Italy in the Age of Reason, 1685–1789*. London: Longman, 1987.

Carson, Emily. "Kant on the Method of Mathematics." *Journal of the History of Philosophy* 37 (1999): 629–52.

Carson, Emily. "Metaphysics, Mathematics and the Distinction Between the Sensible and the Intelligible in Kant's Inaugural Dissertation." *Journal of the History of Philosophy* 37 (2004): 652–79.

Casini, Paolo. *L'antica sapienza italica: Cronistoria di un mito*. Bologna: Il Mulino, 1998.

Cassirer, Ernst. *The Philosophy of Enlightenment*. Princeton, NJ: Princeton University Press, 1951.

Cassitto, Luigi Vincenzo. *Ne' funerali di monsignor fr. Agostino Gervasio, agostiniano, arcivescovo di Capua, e cappellano maggiore del regno*. [Naples]: [1806].

Cattaneo, Massimo. *Gli occhi di Maria sulla rivoluzione: "Miracoli" a Roma e nello Stato della Chiesa (1796–1797)*. Rome: Istituto Nazionale di Studi Romani, 1995.

Cauchy, Augustin-Louis. *Oeuvres complètes*. 27 vols. Paris: Gauthier-Villars, 1882–1938.

Causa, Raffaello. "La Scuola di Posillipo." In *Storia di Napoli*, vol. 9: *Dalla Restaurazione al crollo del Reame*, 783–832. Naples: Società Editrice Storia di Napoli, 1972.

Causa, Raffaello. *Napoli e la Campania Felix: Acquarelli di Giacinto Gigante*. Naples: Società editrice napoletana, 1983.

Causa, Raffaello. *Pitloo*. Naples: Mele, 1956.

Cestari, Gennaro. *Discorsi due, relativi alla scienza dell'uomo* [1807–1808]. Naples: Ursino, 1810.

Cestari, Gennaro. *Lo spirito della giurisdizione ecclesiastica sulle consagrazioni de' vescovi*. Naples: Orsino, 1788.

Cestari, Gennaro. *Tentativo secondo sulla rigenerazione delle scienze*. Milan: Stamperia del Genio Tipografico, 1804.

Cestari, Gennaro. *Tentativo sulla rigenerazione delle scienze*. Milan: Pirotta e Maspero, 1803.

Ceva Grimaldi, Giuseppe. *Considerazioni sulla riforma de pesi e delle misure ne reali dominii di qua dal faro*. [Naples]: [1838?].

Ceva Grimaldi, Giuseppe. *Considerazioni sulle opere pubbliche della Sicilia di qua del Faro dai Normanni sino ai nostri tempi* [1839]. In Giuseppe Ceva Grimaldi, *Opere*, vol. 1, 1–244. Naples: Stamperia reale, 1847.

Ceva Grimaldi, Giuseppe. *Del lavoro degli artigiani*. Naples: Puzziello, 1845.

Ceva Grimaldi, Giuseppe. *Riflessioni sulla polizia*. Aquila: Rietelliana, 1817.

Ceva Grimaldi, Giuseppe. *Sulla riforma della Direzione generale dei ponti e strade, delle acque e foreste*. Naples: 1838.

Chakrabarty, Dipesh. *Provincializing Europe: Postcolonial Thought and Historical Difference*. Princeton, NJ: Princeton University Press, 2000.

Chappey, Jean-Luc. "Les Idéologues et l'Empire: Étude des transformations entre savoirs et pouvoir (1799–1815)." In *Da Brumaio ai cento giorni: Cultura di governo e dissenso politico nell'Europa di Bonaparte*, edited by Antonino de Francesco, 211–27. Milan: Guerini, 2007.

Chasles, Michel. *Aperçu historique sur l'origine et le développement des méthodes en géométrie, particulièrement de celles qui se rapportent à la géométrie modern, suivi d'un mémoire de géométrie sur deux principes généraux de la science, la dualité et l'homographie*. Bruxelles: Hayez, 1837.

Chateaubriand, François-René de. *Génie du christianisme, ou beautés de la religion chrétienne*. 4 vols. Paris: Migneret, 1802.

Chemla, Karine. "On Mathematical Problems as Historically Determined Artifacts: Reflections Inspired by Sources from Ancient China." *Historia Mathematica* 36 (2009): 213–46.

Chemla, Karine, ed. *The History of Proof in Ancient Traditions*. Cambridge: Cambridge University Press, 2012.

Chemla, Karine, and Evelyn Fox Keller, eds. *Cultures without Culturalism: The Making of Scientific Knowledge*. Durham, NC: Duke University Press, 2017.

Chenaux, Philippe. "All'insegna del neotomismo e dell'intransigentismo nella Roma di Leone XIII." *Ioannes XXIII* 5 (2017): 15–24.

Chiosi, Elvira. *Lo spirito del secolo: Politica e religione a Napoli nell'età dell'Illuminismo*. Naples: Giannini, 1992.

Chiosi, Elvira. "Lo stato e le scienze: L'esperienza napoletana nella seconda metà del Settecento." In *La politca della scienza: Toscana e stati italiani nel tardo Settecento*, edited by G. Barsanti, V. Becagli, and R. Pasta, 531–49. Florence: Olschki, 1996.

Chorley, Patrick. *Oil, Silk, and Enlightenment: Economic Problems in XVIIIth Century Naples*. Naples: Istituto Italiano di Studi Storici, 1965.

Ciampi, Angelo. *Elementi di filosofia* [1812–1813]. Naples: Garruccio, 1820.

Ciccolella, Daniela. *La seta nel Regno di Napoli nel XVIII secolo*. Naples: Edizioni Scientifiche Italiane, 2003.

Cingari, Gaetano. *Giacobini e Sanfedisti in Calabria nel 1799* [1957]. Reggio Calabria: Casa del Libro, 1978.

Cirillo, Domenico. *Discorsi accademici*. [Naples]: 1789.

Clark, William, Jan Golinski, and Simon Schaffer, eds. *The Sciences in Enlightened Europe*. Chicago: University of Chicago Press, 1999.

Clarke, Sabine. "Pure Science with a Practical Aim: The Meanings of Fundamental Research in Britain circa 1916–1950." *Isis* 101 (2010): 285–311.

Cocco, Sean. *Watching Vesuvius: A History of Science and Culture in Early Modern Italy*. Chicago: University of Chicago Press, 2012.

Codignola, Luca. "Relations between North America and the Italian Peninsula, 1763–1799: Tuscany, Genoa and Naples." In *Rough Waters: American Involvement in the Mediterranean in the Eighteenth and Nineteenth Centuries*, edited by Silvia Marzagalli, James Sofka, and John McCusker, 25–42. St John's: International Maritime Economic History Association, 2010.

Colangelo, Francesco. *Apologia della religione cristiana compilata dalle risposte degli antichi padri della chiesa alle accuse fatte al cristianesimo*. Naples: Orsino, 1818.

Colangelo, Francesco. *Galileo come guida per la gioventù studiosa*. Naples: Orsino, 1815.

Colangelo, Francesco. *L'irreligiosa libertà di pensare, nemica del progresso delle scienze*. Naples: Orsino, 1804.

Colangelo, Francesco. *Riflessioni storico-politiche su la rivoluzione accaduta in Napoli.* Naples: Orsino, 1799.

Colangelo, Francesco. *Saggio di alcune considerazioni sull'opera di Gio. Battista Vico intitolata Scienza nuova.* Naples: Trani, 1822.

Colangelo, Francesco. *Storia dei filosofi e dei matematici napolitani e delle loro dottrine da' pitagorici sino al secolo 17. dell'èra volgare.* 3 vols. Naples: Trani, 1833–34.

Colecchi, Ottavio. "Riflessioni sopra alcuni opuscoli che trattano delle funzioni fratte e del loro risolvimento in funzioni parziali." *Biblioteca analitica di lettere, scienze e belle arti* 2 (1810): 249–69, 329–76.

Colecchi, Ottavio. *Sopra alcuni quistioni le più importanti della filosofia.* 3 vols. [only two were published]. Naples: Manuzio, 1843. Reprinted as Colecchi, *Quistioni filosofiche*, Naples: Procaccini, 1980.

Colecchi, Ottavio. "Sull'induzione matematica." *Il progresso delle scienze delle lettere e delle arti* 17 (1837): 55–73.

Coleman, Patrick. "The Enlightened Orthodoxy of the Abbé Pluquet." In *Histories of Heresy in Early Modern Europe: For, Against, and Beyond Persecution and Toleration*, edited by John Christian Laursen, 223–38. New York: Palgrave, 2002.

Compagnon, Antoine. *Les antimodernes: De Joseph de Maistre à Roland Barthes.* Paris: Gallimard, 2016.

Condillac, Etienne Bonnot de. *Oeuvres philosophiques.* 3 vols. Edited by Georges Le Roy. Paris: Presses Universitaires de France, 1947–1951.

Conner, Clifford. *Jean Paul Marat: Scientist and Revolutionary.* Amherst, MA: Humanity Books, 1998.

Constant, Benjamin. *Des réactions politiques.* [n.p.]: 1796.

Conte, Paolo. "Le Commissariat d'Oneille: Au-delà de Buonarroti (1794–1796)." *Annales historiques de la Révolution Française* 2 (2017): 75–105.

Conte, Paolo. "Michele De Tommaso: Tra costituzione montagnarda e sistema napoleonico (1792–1804)." *Il Risorgimento* 63 (2016): 21–54.

Corcia, Nicola. *Storia delle Due Sicilie dall'antichità più remota al 1789.* 4 vols. Naples: Virgilio, 1843–1852.

Corry, Leo. "Calculating the Limits of Poetic License: Fictional Narrative and the History of Mathematics." *Configurations* 15 (2007): 195–226.

Corsi, Pietro. "Models and Analogies for the Reform of Natural History: Features of the French Debate, 1790–1800." In *Lazzaro Spallanzani e la biologia del Settecento: Teorie, esperimenti, istituzioni scientifiche*, edited by Walter Bernardi and Antonello La Vergata, 381–96. Florence: Olschki, 1982.

Cortese, Nino. *Eruditi e giornali letterari nella Napoli colta del Settecento.* Naples: Ricciardi, 1922.

Cosmacini, Giorgio. *Il medico giacobino: La vita e i tempi di Giovanni Rasori.* Bari: Laterza, 2002.

Cotugno, Domenico. *De acquaeductibus auris humanae anatomica dissertatio.* Naples: Simoniana, 1761.

Cotugno, Domenico. *De animorum ad optimam disciplinam praeparatione* [. . .]. Naples: Simoni, 1778.

Cotugno, Domenico. *De ischiade nervosa commentarius.* Naples: Simoni, 1764.

Cotugno, Domenico. *Dello spirito della medicina: Raggionamento accademico.* Naples: Morelli, 1783.

Cotugno, Domenico. *Opere.* Edited by Antonio Iurilli. Manduria: Lacaita, 1986.

Craik, Alex. "Calculus and Analysis in Early 19th-Century Britain: The Work of William Wallace." *Historia Mathematica* 26 (1999): 239–67.

Criscuolo, Vittorio. *Albori di democrazia nell'Italia in rivoluzione, 1792–1802.* Milan: Franco Angeli, 2006.

Criscuolo, Vittorio. "Girolamo Bocalosi fra libertinismo e giacobinismo." *Critica storica* 27 (1990): 557–642.

Criscuolo, Vittorio. *Il giacobino Pietro Custodi: Con un'appendice di documenti inediti.* Rome: Istituto storico italiano per l'età moderna e contemporanea, 1987.

Croce, Benedetto. *Canti politici del popolo napoletano.* Naples: Priore, 1892.

Croce, Benedetto. "Gaspare Selvaggi." *Quaderni della critica* 3 (1947): 80–87.

Croce, Benedetto. "Il Marchese Caracciolo." In Benedetto Croce, *Uomini e cose della vecchia Italia*, vol. 2. Bari: Laterza, 1956.

Croce, Benedetto. "Il principe di Canosa." In Benedetto Croce, *Uomini e cose della vecchia Italia*, vol. 2, 225–53. Bari: Laterza, 1956.

Croce, Benedetto. *La Riconquista del Regno di Napoli nel 1799: Lettere del cardinale Ruffo, del re, della regina e del ministro Acton.* Bari: Laterza, 1943.

Croce, Benedetto. *La rivoluzione napoletana del 1799: Biografie, racconti, ricerche.* 2 vols. Naples: Bibliopolis, 1998.

Croce, Benedetto. *Vite di avventure, di fede, e di passione.* Bari: Laterza, 1936

Crocenti, Domenico. *Meditazioni filosofico-politiche sopra l'anarchico sistema giacobino della libertà ed eguaglianza.* 3 vols. Messina: Fratelli del Nobolo, 1794.

Cunningham, Andrew. "Thomas Sydenham: Epidemics, Experiment and the Good Old Cause." In *The Medical Revolution of the Seventeenth Century*, edited by Roger French and Andrew Wear, 164–90. Cambridge: Cambridge University Press, 1989.

Cuoco, Vincenzo. *Platone in Italia: Traduzione dal greco* [1804–1806]. Parma: Carmignani, 1820.

Cuoco, Vincenzo. *Saggio storico sulla rivoluzione napoletana del 1799, seguito dal Rapporto al cittadino Carnot di Francesco Lomonaco.* Edited by Fausto Nicolini. Bari: Laterza, 1913.

Cuoco, Vincenzo. *Scritti di statistica e di pubblica amministrazione.* Edited by Antonino de Francesco and Luigi Biscardi. Bari: Laterza, 2009.

Cuoco, Vincenzo. *Scritti Vari.* 2 vols. Edited by Nino Cortese and Fausto Nicolini. Bari: Laterza, 1924.

Dal Lago, Enrico. *Agrarian Elites: American Slaveholders and Southern Italian Landowners, 1815–1861.* Baton Rouge: Louisiana State University Press, 2005.

Dal Lago, Enrico. "Italian National Unification and the *Mezzogiorno*: Colonialism in One Country?" In *The Shadow of Colonialism on Europe's Modern Past*, edited by Róisín Healy and Enrico Dal Lago, 57–72. London: Palgrave, 2014.

D'Andria, Antonio. "Per un profilo biografico-culturale di Nicola Fiorentino." *Bollettino storico della Basilicata* 29 (2013): 209–27.

Darnton, Robert. *Mesmerism and the End of Enlightenment in France.* Cambridge, MA: Harvard University Press, 1986.

Darnton, Robert. "Philosophers Trim the Tree of Knowledge: The Epistemological Strategy of the *Encyclopédie*." In Robert Darnton, *The Great Cat Massacre, and Other Episodes in French Cultural History*, 191–213. New York: Basic Books, 1985.

Daston, Lorraine. *Classical Probability in the Enlightenment.* Princeton, NJ: Princeton University Press, 1988.

Daston, Lorraine. "Condorcet and the Meaning of Enlightenment." *Proceedings of the British Academy* 151 (2007): 113–34.

Daston, Lorraine. "Enlightenment Calculations." *Critical Inquiry* 21 (1994): 182–202.

Daston Lorraine, and Otto Sibum. "Introduction: Scientific Personae and Their Histories." *Science in Context* 16 (2003): 1–8.

Daston, Lorraine, and Peter Galison. *Objectivity*. Cambridge, MA: MIT Press, 2007.

Davis, John. "1799: The *Santafede* and the Crisis of the *Ancien Régime* in Southern Italy." In *Society and Politics in the Age of the Risorgimento: Essays in Honour of Denis Mac Smith*, edited by John Davis and Paul Ginsborg, 1–25. Cambridge: Cambridge University Press, 1991.

Davis, John. *Conflict and Control: Law and Order in Nineteenth-Century Italy*. Basingstoke: Macmillan, 1988.

Davis, John. *Merchants, Monopolists and Contractors: A Study of Economic Activity and Society in Bourbon Naples, 1815–1860*. New York: Arno Press, 1981.

Davis, John. *Naples and Napoleon: Southern Italy and the European Revolutions*. Oxford: Oxford University Press, 2006.

Davis, John. "Naples during the French *decennio*: A Problem Unresolved." *Publications de l'École française de Rome* 96 (1987): 327–54.

Davis, John. "The Mezzogiorno and Modernization: Changing Contours of Public and Private during the 'French Decennio.'" In *Fra storia e storiografia: Scritti in onore di Pasquale Villani*, edited by Paolo Macry and Antonio Massafra, 691–708. Bologna: Il Mulino, 1994.

De Ceglia, Francesco Paolo. *Il segreto di San Gennaro: Storia naturale di un miracolo napoleano*. Turin: Einaudi, 2016.

De Felice, Renzo. *Il triennio giacobino in Italia (1796–1799)*. Rome: Bonacci, 1990.

De Francesco, Antonino. "L'ombra di Buonarroti." *Storica* 15 (1999): 7–67.

De Francesco, Antonino. *Rivoluzione e costituzioni: Saggi sul democratismo politico nell'Italia napoleonica, 1796–1821*. Naples: Edizioni Scientifiche Italiane, 1996.

De Francesco, Antonino. *Storie dell'Italia rivoluzionaria e napoleonica (1796–1814)*. Milan: Bruno Mondadori, 2016.

De Francesco, Antonino. *The Antiquity of the Italian Nation: The Cultural Origins of a Political Myth in Modern Italy, 796–1943*. Oxford: Oxford University Press, 2013.

De Francesco, Antonino. *Vincenzo Cuoco: Una vita politica*. Bari: Laterza, 1997.

De Frenza, Lucia. "The Poles of Healing: Mineral Magnetism vs. Animal Magnetism." In *Atti del XXXVI Convegno annuale della società italiana degli storici della fisica e dell'astronomia*, edited by Salvatore Esposito, 71–79. Pavia: Pavia University Press, 2017.

Del Corno, Nicola. *Italia reazionaria: Uomini e idee dell'antirisorgimento*. Milan: Bruno Mondadori, 2017.

Del Corno, Nicola. "Reazione." In *Atlante culturale del Risorgimento: Lessico del linguaggio politico dal Settecento all'Unità*, edited by A. M. Banti, A. Chiavistelli, L. Mannori, and M. Meriggi, 163–75. Bari: Laterza.

Deleplace, Marc. "La notion d'anarchie pendant la Révolution française (1789–1801)." *Annales historiques de la Révolution française* 287 (1992): 17–45.

Delfico, Melchiorre. "Discorso della importanza di far precedere le cognizioni fisiologiche allo studio della filosofia intellettuale" [1823]. *Rivista abruzzese di scienze e lettere* 2 (1887): 1–11, 63–75.

Delfico, Melchiorre. "Memoria su la perfettibilità organica considerata come il principio fisico dell'educazione; con alcune vedute sulla medesima" [1814]. *ARAS* 1, 1819: 377–415.

Delfico, Melchiorre. *Memorie storiche della Repubblica di San Marino*. Milan: Sonzogno, 1804.

Delfico, Melchiorre. *Pensieri sulla storia e su l'incertezza ed inutilità della medesima* [1806]. Naples: Nobile, 1814.

Delfico, Melchiorre. "Ricerche su la sensibilità imitativa considerata come il principio fisico della sociabilità della specie e del civilizzamento de' popoli, e delle nazioni" [1813]. In *ARAS* 1, 1819: 343–76.

Delfico, Melchiorre. *Ricerche sul vero carattere della giurisprudenza romana e de' suoi cultori*. Naples: Porcelli, 1791.

Delfico, Melchiorre. "Seconda memoria su la perfettibilità organica considerata come il principio fisico dell'educazione" [1816]. *ARAS* 1, 1819: 417–45.

"Delle teorie kantiane difese da Ottavio Colecchi nella sua opera che ha per titolo: *Sopra alcune quistioni le più importanti della filosofia.*" *La scienza e la fede* 8 (1844): 5–32.

De Lorenzo, Renata. "I catasti napoleonici nel Mezzogiorno d'Italia tra strumento fiscale e rappresentazione cartografica." In *Atti della 7° Conferenza nazionale ASITA: L'informazione territoriale e la dimensione tempo*, 955–60. Varese: Artestampa, 2003.

De Lorenzo, Renata, ed. *L'organizzazione dello stato al tramonto dell'Antico Regime*. Naples: Morano, 1990.

De Lorenzo, Renata. *Proprietà fondiaria e fisco nel Mezzogiorno: La riforma della tassazione nel Decennio francese (1806–1815)*. Salerno: Centro Studi per il Cilento e il Vallo di Diano, 1984.

De Lorenzo, Renata. *Un regno in bilico: Uomini, eventi, e luoghi del Mezzogiorno preunitario*. Rome: Carocci, 2001.

De Maio, Romeo. *Pittura e Controriforma a Napoli*. Bari: Laterza, 1983.

De Maio, Romeo. *Società e vita religiosa a Napoli nell'età moderna (1656–1799)*. Naples: Edizioni Scientifiche Italiane, 1971.

Demarco, Domenico, ed. *La statistica del Regno di Napoli nel 1811*. 4 vols. Rome: Accademia Nazionale dei Lincei, 1988.

De Martino, Armando. *La nascita delle intendenze: Problemi della amministrazione periferica nel Regno di Napoli (1806–1815)*. Naples: Jovene, 1984.

De Mattia, Fausto, and Felicita de Negri. "Il Corpo di Ponti e Strade dal decennio francese alla riforma del 1826." In *Il mezzogiorno preunitario: Economia, società e istituzioni*, edited by Angelo Massafra, 449–68. Bari: Dedalo, 1988.

Denby, David. *Sentimental Narrative and the Social Order in France, 1760–1820*. Cambridge: Cambridge University Press, 1994.

Derringer, William. *Calculated Values: Finance, Politics, and the Quantitative Age*. Cambridge, MA: Harvard University Press, 2018.

De Sanctis, Francesco. *La giovinezza: Frammento autobiografico*. Florence: Le Monnier, 1957.

De Seta, Cesare. *Il golfo di Napoli da una serie di gouaches del primo Ottocento*. Milan: Il Polifilo, 1995.

Desmond, Adrian. *The Politics of Evolution: Morphology, Medicine, and Reform in Radical London*. Chicago: University of Chicago Press, 1989.

Desrosières, Alain. *The Politics of Large Numbers: A History of Statistical Reasoning*. Cambridge, MA: Harvard University Press, 1998.

Diario del settimo congresso degli scienziati italiani in Napoli. Naples: Nobile, [1845].

Diaz, Furio. "La questione del 'giacobinismo' italiano." *Critica storica* 3 (1964): 577–602.

Di Battista, Francesco, ed. *Il Mezzogiorno alla fine del Settecento*. Bari: Laterza, 1992.

Di Battista, Francesco. "Per la storia della prima cattedra universitaria d'economia: Napoli 1754–1866." In *Le cattedre di economia politica in Italia. La diffusione di una disciplina 'sospetta'*

(1750–1900), edited by Massimo Augello, Marco Bianchini, Gabriella Gioli, and Piero Roggi, 31–46. Milan: Franco Angeli, 1988.

Di Biasio, Aldo. *Carlo Afan de Rivera e il Corpo di Ponti e Strade*. Latina: Amministrazione Provinciale, 1993.

Di Biasio, Aldo. *Politica e amministrazione del territorio nel Mezzogiorno d'Italia tra Settecento e Ottocento*. Naples: Edizioni Scientifiche Italiane, 2004.

Dickie, John. *Darkest Hour: The Nation and the Stereotypes of Mezzogiorno, 1860–1900*. New York: St. Martin's, 1999.

Dictionnaire de Théologie Catholique, contenant l'exposé des doctrines de la théologie catholique, leurs preuves et leur histoire. 30 vols. Edited by Alfred Vacant, Eugène Mangenot and Émile Amann. Paris: Letouzey et Ané, 1902–1950.

Dizionario biografico degli italiani. 100 vols. Rome: Istituto della Enciclopedia italiana, 1960–2020.

Dotti, Ugo, ed. *Il Progresso delle scienze, delle lettere e delle arti (1832–1834)*. Rome: Edizioni dell'Ateneo, 1970.

Dreyfus, Hubert, and Paul Rabinow. *Michel Foucault: Beyond Structuralism and Hermeneutics*. Chicago: University of Chicago Press, 1983.

Dupuy, Roger. *De la Révolution à la Chouannerie*. Paris: Flammarion, 1988.

Eagleton, Terry. *Ideology*. London: Routledge, 2014.

Edelstein, Dan, *The Enlightenment: A Genealogy*. Chicago: University of Chicago Press, 2010.

Encyclopédie, ou dictionnaire raisonné des sciences, des arts et des métiers, par une société de gens de lettres. Edited by Denis Diderot and Jean Lerond d'Alembert. 17 vols. [Geneva]: 1751–1765.

Epple, Moritz. "Styles of Argumentation in Late 19th-Century Geometry and the Structure of Mathematical Modernity." In *Analysis and Synthesis in Mathematics: History and Philoso phy*, edited by Michael Otte and Marco Panza, 177–98. Kluwer: Dordrecht, 1997.

Erikson, Paul, Judy L. Klein, Lorraine Daston, Rebecca Lemov, Thomas Sturm, and Michael D. Gordin. *How Reason Almost Lost Its Mind: The Strange Career of Cold War Rationality*. Chicago: University of Chicago Press, 2013.

Favaro, Oreste. "Consistenza del clero giacobino nella diocesi di Torino, ed opera riformatrice degli arcivescovi Rorà e Costa negli anni precedenti alla rivoluzione francese." *Bollettino storico-bibliografico subalpino* 89 (1991): 189–263.

Fayet, Joseph. *La Révolution Francaise et la Science, 1789–1795*. Paris: Rivière, 1960.

Fehrenbach, Frank, and Joris van Gastel, eds. *Nature and the Arts in Early Modern Naples*. Berlin: De Gruyter, 2020.

Ferber, Johann Jacob. *Briefe aus Wälschland über natürliche Merkwürdigkeiten*. Prague: Gerle, 1773.

Fergola, Nicola. "Continuazione della precedente memoria sul cilindroide wallisiano ove un tal problema vien risoluto analiticamente con metodo diretto e generale." In *ARAS* 1, 1819: 97–104.

Fergola, Nicola. "Dal teorema tolemaico ritraggonsi immediatamente i teoremi delle sezioni angolari di Vieta e Wallis e le principali verità proposte nella trigonometria analitica dei moderni." In *ARAS* 1, 1819: 205–47.

Fergola, Nicola. *Della invenzione geometrica: Opera postuma di Nicola Fergola ordinata compiuta e corredata d'importanti note dal prof. V. Flauti*. Naples: Stamperia privata dell'autore, 1842.

[Fergola, Nicola]. *Elementi di geometria sublime, parte prima: Le istituzioni su i conici illustrate dal rev. sacerdote D. Felice Giannattasio*. Naples: Raimondi, 1791. Reworked and republished under different titles in the early nineteenth century.

Fergola, Nicola. "I problemi delle tazioni risoluti con nuovi artifizi di geometria" [1809]. In *ARAS* 1, 1819: 1–19.

Fergola, Nicola. "La verà misura delle volte a spira." In *ARAS* 1788, 65–84.

Fergola, Nicola. "Nuove ricerche sulle risoluzioni dei problemi di sito." In *ARAS* 1788, 157–67.

Fergola, Nicola. "Nuovo metodo da risolvere alcuni problemi di sito e di posizione." In *ARAS* 1788, 119–38.

[Fergola, Nicola]. *Prelezioni sui principi matematici della filosofia naturale del cavaliere Isacco Newton.* 2 vols. Naples: Porcelli, 1792–1793.

Fergola, Nicola. "Risoluzione di alcuni problemi ottici." In *ARAS* 1788, 1–14.

Fergola, Nicola. "Su la rettificazione dell'ellisse e gli integrali che ne dipendono, estratta dai manoscritti del defunto Nicola Fergola e presentata all'Accademia da Vincenzo Flauti." In *ARAS* 4, 1839: 13–23.

Fergola, Nicola. *Teorica de' miracoli esposta con metodo dimostrativo seguita da un discorso apologetico sul miracolo di S. Gennaro e da una raccolta di pensieri su la filosofia e la religione.* Naples: Stamperia per le opere del prof. Flauti, 1839.

F[ergola], N[icola]. *Trattato analitico delle sezioni coniche.* Naples: Chianese, 1814.

Fergola, Nicola. *Trattato analitico de' luoghi geometrici.* Naples: Accademia di Marina, 1818.

Ferrand, Antoine-François-Claude de. *Les conspirateurs démasqués, par l'auteur de Nullité et Despotisme.* [Turin]: 1790.

Ferraro, Giovanni. "*Excellens in art non debet mori:* Nicola Trudi da napolitano a italiano." Working paper, STAT—Dipartimento di Scienze e Tecnologie per l'Ambiente e il Territorio, 2012. https://halshs.archives-ouvertes.fr/halshs-00682088.

Ferraro, Giovanni. "Manuali di geometria elementare nella Napoli preunitaria (1806–1860)." *History of Education & Children's Literature* 3 (2008): 103–39.

Ferraro, Giovanni, and Franco Palladino. *Il calcolo sublime di Eulero e Lagrange, esposto col metodo sintetico, nel progetto di Nicola Fergola.* Naples: La Città del Sole, 1995.

Ferraro, Giovanni, and Franco Palladino. "Sui manoscritti di Nicola Fergola, 1753–1824." *Bollettino di storia delle scienze matematiche* 13 (1993): 147–97.

Ferrone, Vincenzo. *I profeti dell'illuminismo: Le metamorfosi della ragione nel tardo Settecento italiano.* Bari: Laterza, 1989.

Ferrone, Vincenzo. "Riflessioni sulla cultura illuministica napoletana e l'eredità di Galilei." In *Galileo e Napoli,* edited by Fabrizio Lomonaco and Maurizio Torrini, 429–48. Naples: Guida, 1987.

Ferrone, Vincenzo. *The Intellectual Roots of the Italian Enlightenment: Newtonian Science, Religion, and Politics in the Early Eighteenth Century.* Atlantic Highlands, NJ: Humanity Books, 1995.

Ferrone, Vincenzo. *La società giusta ed equa: Repubblicanesimo e diritti dell'uomo in Gaetano Filangieri.* Bari: Laterza, 2008.

Fichte, Johann Gottlieb. *Reden an die deutsche Nation.* Berlin: Realschulbuchhandlung, 1808.

Fino, Lucio. *Il vedutismo a Napoli nella grafica dal XVII al XIX secolo.* Naples: Grimaldi, 1989.

Figlio, Karl. "Theories of Perception and the Physiology of the Mind in the Late Eighteenth Century." *History of Science* 13 (1975): 177–212.

Filangieri, Gaetano. *La scienza della legislazione* [1780–1789]. 3 vols. Florence: Le Monnier, 1864–1876.

Filippis, Vincenzo de. *De' terremoti della Calabria Ultra nel 1783 e 1789.* Catanzaro: Centro bibliografico calabrese, 1999.

Fiocca, Alessandra. "La geometria descrittiva in Italia, 1798–1838." *Bollettino di storia delle scienze matematiche* 12 (1992): 187–249.

Fiocca, Alessandra, Daniela Lamberini, and Cesare Maffioli. *Arte e scienza delle acque nel Rinascimento*. Venice: Marsilio, 2003.

Fiorentini, Erna. "Nuovi punti di vista: Giacinto Gigante e la camera lucida a Napoli." In *Pittura italiana dell'Ottocento*, edited by Martina Hansmann and Max Seidel, 535–57. Venice: Marsilio, 2005.

Fiorentino, Nicola. *Riflessioni sul Regno di Napoli, in cui si tratta degli studi, de' tribunali, delle arti, del commercio, de' tributi, dell'agricoltura, pastorizia, popolazione, e di altro*. Naples: De Bonis, 1794.

Fiorentino, Nicola. *Saggio sulle quantità infinitesime e sulle forze vive e morte*. Naples: n.d.

Flauti, Vincenzo. *Corso di analisi algebrica elementare*. Naples: Accademia di Marina, 1819.

Flauti, Vincenzo. *Corso di geometria elementare e sublime* [1810]. 4 vols. Naples: Gabinetto bibliografico e tipografico, 1826.

Flauti, Vincenzo. *Del metodo in matematiche, della maniera di ordinare gli elementi di queste scienze, e dell'insegnamento de' medesimi, con appendice che contiene una esposizione del corso di matematiche del professore Flauti*. Naples: Gabinetto bibliografico e tipografico, 1822.

Flauti, Vincenzo. *Elementi di geometria descrittiva*. Rome: Perego Salvioni, 1807.

Flauti, Vincenzo. *Elementi di geometria solida*. Naples: 1804.

Flauti, Vincenzo. "Elogio dell'abate Felice Giannattasio." *Rendiconto delle adunanze e de' lavori dell'Accademia Napolitana delle Scienze* 8 (1849): 340–52.

Flauti, Vincenzo. "Elogio di Giuseppe Scorza." *Rendiconto delle adunanze e de' lavori dell'Accademia delle scienze* 2 (1843): 191–200.

Flauti, Vincenzo. *Elogio storico di Nicola Fergola*. Naples: Gabinetto bibliografico e tipografico, 1824.

Flauti, Vincenzo. *Geometria di sito sul piano e nello spazio*. Naples: Società tipografica, 1815.

Flauti, Vincenzo. *Gli Elementi di geometria di Euclide, emendati in que' luoghi in cui una volta furono viziati da Teone, o da altri; e ne quali sono restituite alcune definizioni, e dimostrazioni dello stesso Euclide* [1810]. Naples: 1827.

Flauti, Vincenzo. "La vera misura del cilindroide wallisiano." In *ARAS* 4, 1839: 1–11.

[Flauti, Vincenzo]. *Produzioni relative al programma di tre quistioni geometriche proposto da un nostro professore*. Naples: 1840.

Flauti, Vincenzo. *Prospetto di un mezzo secolo di servizi scientifici resi dal Cav. V. Flauti, fino al 1849*. [Naples]: [1849].

Flauti, Vincenzo. "Soluzioni geometriche di alcuni principali problemi sulla piramide triangolare." In *ARAS* 1, 1819: 52–70.

Flauti, Vincenzo. *Tentativo di un progetto di riforma per la pubblica istruzione nel Regno di Napoli*. Naples: Accademia di Marina, 1820.

Fontana, Sandro. *La controrivoluzione cattolica in Italia, 1820–1830*. Brescia: Morcelliana, 1968.

Foscari, Giuseppe. "Ingegneri e territorio nel Regno di Napoli tra età francese e Restaurazione: Reclutamento, formazione e carriere." In *Avvocati medici ingegneri: Alle origini delle professioni modern*, edited by Maria Luisa Betri and Alessandro Pastore, 279–91. Bologna: Clueb, 1997.

Foucault, Michel. *Dits et écrits* [Texte #85]. Vol. 1. Paris: Gallimard, 2001.

Foucault, Michel. "Nietzsche, Genealogy, History." In *The Foucault Reader*, edited by Paul Rabinow, 76–100. New York: Pantheon Books, 1984.

Foucault, Michel. *Power/Knowledge: Selected Interviews and Other Writings 1972–1977.* New York: Pantheon, 1980.

Foucault, Michel. *The Archeology of Knowledge.* London: Tavistock, 1974.

Foucault, Michel. *The Order of Things: An Archaeology of the Human Sciences.* London: Routledge, 2002.

Foucault, Michel. "The Political Function of the Intellectual." *Radical Philosophy* 17 (1977): 126–33.

Foucher, Louis. *La philosophie catholique en France au XIXme siècle avant la Renaissance thomiste et dans son rapport avec elle, 1800–1880.* Paris: Vrin, 1955.

Fowler, David. *The Mathematics of Plato's Academy: A New Reconstruction.* Oxford: Clarendon Press, 1987.

Främgsmyr, Tore, John Heilbron, and Robin Rider, eds. *The Quantifying Spirit in the 18th Century.* Berkeley: University of California Press, 1990.

Francovich, Carlo. *Storia della massoneria in Italia dalle origini alla Rivoluzione francese.* Florence: La Nuova Italia, 1990.

Frasca, Elena. "Dalla Scozia all'Italia: Il dibattito sulla dottrina medica di John Brown tra rivoluzione e restaurazione." *Italianistica Debreceniensis* 21–22 (2016): 32–48.

Fraser, Craig. *Calculus and Analytical Mechanics in the Age of Enlightenment.* Aldershot: Variorum, 1997.

Fraser, Craig. "The Calculus as Algebraic Analysis: Some Observations on Mathematical Analysis in the 18th Century." *Archive for History of Exact Sciences* 39 (1989): 317–35.

Galanti, Giuseppe Maria. *Descrizione geografica e politica delle Sicilie.* 4 vols. Naples: Gabinetto letterario, 1786–1790.

Galasso, Giuseppe. *Italia democratica: Dai giacobini al Partito d'azione.* Florence: Le Monnier, 1986.

Galasso, Giuseppe. *La filosofia in soccorso de' governi: La cultura napoletana del Settecento.* Naples: Guida, 1989.

Galasso, Giuseppe. *L'altra Europa: Per un'antropologia storica del Mezzogiorno d'Italia.* Lecce: Argo, 1997.

Galasso, Giuseppe. "Scienze, filosofia e tradizione galileiana in Europa e nel Mezzogiorno d'Italia." In *Galileo e Napoli,* edited by Fabrizio Lomonaco and Maurizio Torrini, ix–lvi. Naples: Guida, 1987.

Galdi, Matteo. *Pensieri sull'istruzione pubblica, relativamente al Regno delle Due Sicilie.* Naples: Stamperia reale, 1809.

Galdi, Matteo. *Quadro politico delle rivoluzioni delle Provincie Unite e della Repubblica Batava e dello stato attuale del Regno d'Olanda.* Milan: Pirotta and Maspero, 1809.

Galdi, Matteo. *Saggio d'istruzione pubblica rivoluzionaria.* Milan: Stamperia de' Patriotti d'Italia, 1798.

Galdi, Vincenzo Ambrogio. *Instruzioni per gli amplissimi signori vice custodi dell'Arcadia Reale di Napoli.* Naples: 1800.

Gallo, Fernanda. "Gli hegeliani di Napoli e il Risorgimento: Bertrando Spaventa e Francesco De Sanctis a confronto (1848–1862)." *LEA—Lingue e letterature d'Oriente e d'Occidente* 6 (2017): 651–68.

Galluppi, Pasquale. *La filosofia di Vittorio Cousin.* 2 vols. Naples: Nuovo gabinetto letterario, 1831–1832.

Galluppi, Pasquale. *Lezioni di logica e metafisica* [1832–1834]. 2 vols. Milan: Borroni and Scotti, 1845.

Galluppi, Pasquale. *Saggio filosofico sulla critica della conoscenza o sia analisi distinta del pensiere umano con un esame delle più importanti questioni dell'ideologia, del kantismo, e della filosofia trascendentale.* 6 vols. Naples: Sangiacomo, 1819–1832.

Galluppi, Pasquale. *Sull'analisi e sulla sintesi* [1807]. Florence: Olschki, 1935.

Galluzzi, Massimo. "Geometria algebrica e logica tra Otto e Novecento." In *Storia d'Italia. Annali 3: Scienza e tecnica nella cultura e nella società dal Rinascimento ad oggi,* 1004–106. Turin: Einaudi, 1980.

Garrard, Graeme. *Counter-Enlightenments: From the Eighteenth Century to the Present.* London: Routledge, 2006.

Gatti, Serafino. *Elogio del cavaliere Giuseppe Saverio Poli.* Naples: Nobile, 1825.

Gatti, Serafino. *Lettera critica al sig. L. S. intorno ai libri apologetici della religione cristiana.* [Naples?]: 1821.

Gatto, Romano. "La discussione sul metodo e la sfida di Vincenzo Flauti ai matematici del Régno di Napoli." *Rendiconto dell'Accademia di Scienze Fisiche e Matematiche* 67 (2000): 181–233.

Gatto, Romano. "Tradizione e cartesianesimo nella matematica napoletana della prima metà del secolo XVIII." *Rendiconto dell'Accademia delle Scienze Fisiche e Matematiche* 73 (2006): 99–249.

Gavroglu, Kostas, Manolis Patiniotis, Faidra Papanelopoulou, Ana Simões, Ana Carneiro, Maria Paula Diogo, José Ramón Bertomeu Sánchez, Antonio García Belmar, and Agustí Nieto-Galan. "Science and Technology in the European Periphery: Some Historiographical Reflections." *History of Science* 46 (2008): 153–75.

Geertz, Clifford. "Deep Play: Notes on the Balinese Cockfight." In Clifford Geertz, *The Interpretation of Cultures,* 412–53. New York: Basic Books, 1973.

Geertz, Clifford. "Common Sense as a Cultural System." *Antioch Review* 33 (1975): 5–26.

Genovesi, Antonio. *Della diocesina, o sia della filosofia del giusto e dell'onesto* [1766]. Milan: Società tipografica dei classici italiani, 1835.

Genovesi, Antonio. "Discorso sopra il vero fine delle lettere e delle scienze" [1753]. In Antonio Genovesi, *Scritti,* 40–87. Turin: Einaudi, 1977.

Genovesi, Antonio. *Dissertatio physico-historica de rerum origine et constitutione* [1745]. Florence: Giunti, 2001.

Genovesi, Antonio. *Elementa metaphysicae mathematicum in morem adornata.* 4 vols. Naples: Gessari, 1743–1752.

Genovesi, Antonio. *Elementa Physicae Experimentalis usui Tironum aptatae auctore Antonio Genuensi P.P., accedunt nonnullae dissertationes Physico-Mathematicae conscriptae a Nicolao Fergola.* Naples: Terres, 1779.

Genovesi, Antonio. *Lettere accademiche su la questione se sieno più felici gl'ignoranti che gli scienziati* [1764]. Venice: Savioni, 1791.

Genovesi, Antonio. *Lezioni di commercio, o sia d'economia civile* [1765–1767]. Napoli: Istituto Italiano per gli Studi Filosofici, 2005.

Genovesi, Antonio. *Logica per gli giovanetti* [1766]. Bassano: Remondini, 1779.

Genovesi, Antonio. *Universae christianae theologiae elementa dogmatica, historica, critica.* 2 vols. Venice: Pasquali, 1771.

Gentile, Giovanni. *Storia della filosofia italiana dal Genovesi al Galluppi.* 2 vols. Milan: Treves, 1930.

Gerdil, Giacinto Sigismondo. "De l'infini absolu consideré dans la grandeur." *Mélanges de philosophie et de mathématique de la Société Royale de Turin* 2 (1760–1761): 1–45.

Gerdil, Giacinto Sigismondo. *Opere edite e inedite*. 7 vols. Naples: Diogene, 1853–1856.

Gerdil, Giacinto Sigismondo. *Précis d'un cours d'instruction sur l'origine, les droits, et les devoirs de l'autorité souveraine dans l'exercise des principales branches de l'administration*. Turin: Imprimerie de l'Academie R. des Sciences, 1799.

Gergonne, Joseph-Diaz. "Géométrie de la règle: Solution et construction, par la géométrie analitique, de deux problèmes dépendant de la géométrie de la règle." *Annales de mathématiques pures et appliquées* 7 (1817): 325–34.

Giampaolo, Paolo Nicola. *Dialoghi sulla religione, che comprendono una distinta ed ordinate apologia del cristianesimo* [1815]. 4 vols. Naples: Stamperia della Pietà de' Turchini, 1828.

Giampaolo, Paolo Nicola. *Elogio del commendatore Giuseppe Saverio Poli*. Naples: Giordano, 1825.

Giampaolo, Paolo Nicola. *Lezioni di metafisica*. Naples: Campo, 1803.

Giampaolo, Paolo Nicola. *Lezioni e catechismo di agricoltura*. 3 vols. Naples: Chianese, 1808.

Giannantonio, Pompeo, ed. *Alfonso M. De Liguori e la società civile del suo tempo*. 2 vols. Florence: Olschki, 1990.

Giannetti, Anna. "L'ingegnere moderno nell'amministrazione borbonica: La polemica sul Corpo di Ponti e Strade." In *Mezzogiorno preunitario: Economia, società e istituzioni*, edited by Angelo Massafra, 935–44. Bari: Dedalo, 1988.

Giarrizzo, Giuseppe. "Alla ricerca del giacobinismo italiano." In *L'eredità dell'Ottantanove e l'Italia*, edited by Renzo Zorzi, 227–35. Florence: Olschki, 1992.

Giarrizzo, Giuseppe. *Massoneria e illuminismo nell'Europa del Settecento*. Venice: Marsilio, 1994.

Giarrizzo, Giuseppe. "Massoneria e Risorgimento." *Hiram* 2 (1999): 43–46.

Giddens, Anthony. *The Consequences of Modernity*. Stanford, CA: Stanford University Press, 1990.

Gillispie, Charles. *Science and Polity in France, 1: The End of the Old Regime* [1980]; *2: The Revolutionary and Napoleonic Years*. Princeton, NJ: Princeton University Press, 2004.

Gillispie, Charles. "The Encyclopédie and the Jacobin Philosophy of Science: A Study in Ideas and Consequences." In *Critical Problems in the History of Science*, edited by Marshall Clagett, 255–89. Madison: University of Wisconsin Press, 1969.

Gingras, Yves. "What Did Mathematics Do to Physics?" *History of Science* 39 (2001): 383–416.

Ginzburg, Carlo, and Carlo Poni. "Il nome e il come: Mercato storiografico e scambio diseguale." *Quaderni storici* 40 (1979): 181–90.

Gioia, Melchiorre. *Filosofia della statistica*. 4 vols. Milan: Editori degli annali universali delle scienze e dell'industria, 1829–1830.

Giordano, Annibale. "Considerazioni sintetiche sopra un celebre problema piano e risoluzione di alquanti problemi affini." *Memorie di matematica e fisica della Società italiana delle scienze, detta dei XL* 4 (1786): 4–17.

Giordano, Annibale, and Carlo Lauberg. *Principi analitici delle matematiche*. 2 vols. Naples: Giaccio, 1792.

Giorgi Bertola, Aurelio de. *Della filosofia della storia*. Pavia: Bolzani, 1787.

Giuntella, Vittorio Emanuele. "L'esperienza rivoluzionaria." In *Nuove questioni di storia del Risorgimento e dell'Unità d'Italia*, vol. 1, 311–44. Milan: Marzorati, 1961.

Giustiniani, Lorenzo. *Dizionario geografico-ragionato del Regno di Napoli*. 13 vols. Naples, Manfredi [later de Bonis], 1797–1816.

Godechot, Jacques. *Les commissaires aux armées sous le Directoire: Contribution à l'étude des rapports entre les pouvoirs civils et militaires*. 2 vols. Paris: Presses Universitaires de France, 1941.

Godechot, Jacques. "Saliceti ministre du royaume de Naples sous Joseph Bonaparte et Murat." In *Studi in memoria di Nino Cortese*, 255–72. Rome: Istituto per la storia del Risorgimento italiano, 1976.

Godechot, Jacques. *The Counter-Revolution: Doctrine and Action, 1789–1804*. Princeton, NJ: Princeton University Press, 1981.

Goffman, Erving. *Strategic Interaction*. Philadelphia: University of Pennsylvania Press, 1969.

Goffman, Erving. *The Presentation of Self in Everyday Life*. New York: Anchor Books, 1959.

Goldstein, Catherine, Jeremy Gray, and Jim Ritter, eds. *L'Europe mathématique: Histoires, mythes, identités / Mathematical Europe: History, Myth, Identity*. Paris: Éditions de la Maison des sciences de l'homme, 1996.

Goldstein, Jan. *Post-Revolutionary Self: Politics and Psyche in France, 1750–1850*. Cambridge, MA: Harvard University Press, 2008.

Golinski, Jan. *Making Natural Knowledge: Constructivism and the History of Science*. Chicago: University of Chicago Press, 1998.

Golinski, Jan. "Science in the Enlightenment, Revisited." *History of Science* 49 (2011): 217–31.

Gooday, Graeme. "Vague and Artificial: The Historically Elusive Distinction between Pure and Applied Science." *Isis* 103 (2012): 546–54.

Gorman, Michael John. *The Scientific Counter-Revolution: The Jesuits and the Invention of Modern Science*. London: Bloomsbury Academic, 2020.

Grabiner, Judith. *The Calculus as Algebra: J. L. Lagrange, 1736–1813*. New York: Garland, 1990.

Grabiner, Judith. *The Origins of Cauchy's Rigorous Calculus*. Cambridge, MA: MIT Press, 1981.

Gracq, Julien. *The Shape of a City*. New York: Turtle Point Press, 2005.

Gramsci, Antonio. *Quaderni del carcere*. 4 vols. Turin: Einaudi, 1975.

Grandin, Karl, Nina Wormbs, and Sven Widmalm, eds. *The Science-Industry Nexus: History, Policy, Implications*. New York: Science History Publications, 2004.

Granger, Gilles-Gaston. *La mathématique sociale du Marquis de Condorcet*. Paris: Odile Jacob, 1989.

Grattan-Guinness, Ivor. *Convolutions in French Mathematics, 1800–1843: From the Calculus and Mechanics to Mathematical Analysis and Mathematical Physics*. 3 vols. Basel: Springer, 1990.

Grattan-Guinness, Ivor. "The Emergence of Mathematical Analysis and Its Foundational Progress, 1780–1880." In *From the Calculus to Set Theory, 1630–1910: An Introductory History*, edited by Ivor Grattan-Guinness, 94–148. Princeton, NJ: Princeton University Press, 1980.

Gray, Jeremy. "Anxiety and Abstraction." *Science in Context* 17 (2004): 23–47.

Gray, Jeremy. *Plato's Ghost: The Modernist Transformation of Mathematics*. Princeton, NJ: Princeton University Press, 2008.

Greco, Franco, ed. *La pittura napoletana dell'Ottocento*. Naples: Pironti, 1993.

Grimaldi, Domenico. *Istruzioni sulla nuova manifattura dell'olio introdotta nel Regno di Napoli*. Naples: Orsino, 1777.

Grimaldi, Domenico. *Memoria per lo ristabilimento dell'industria olearia nelle Calabrie*. Naples: Porcelli, 1783.

Grimaldi, Domenico. *Piano di riforma per la pubblica economia delle provincie del Regno di Napoli e per l'agricoltura delle Due Sicilie*. Naples: Porcelli, 1780.

Grimaldi, Domenico. *Relazione umiliata al Re d'un disimpegno fatto nella Ulteriore Calabria con alcune osservazioni economiche relative a quella provincia*. Naples: 1785.

Grimaldi, Francesco Antonio. *Riflessioni sopra l'ineguaglianza tra gli uomini*. 3 vols. Naples: Mazzola-Vocola, 1779–80.

Grove, Richard. *Green Imperialism: Colonial Expansion, Tropical Island Edens, and the Origins of Environmentalism, 1600–1860*. Cambridge: Cambridge University Press, 1995.

[Guarini, Luigi]. *Catalogo de' cappellani maggiori del Regno di Napoli e de' confessori delle persone reali*. Naples: Coda, 1819.

Guasti, Niccolò. "Antonio Genovesi's *Diocesina*: Source of the Neapolitan Enlightenment." *History of European Ideas* 32 (2006): 385–405.

Guccione, Eugenio, ed. *Gioacchino Ventura e il pensiero politico d'ispirazione cristiana dell'Ottocento*. Florence: Olschki, 1991.

Guerci, Luciano. "Democrazia e costituzione democratica nelle *Effemeridi repubblicane* di Matteo Galdi." In *Storia e vita civile: Studi in memoria di Giuseppe Nuzzo*, edited by Eugenio di Rienzo and Aurelio Musi, 115–39. Naples: Edizioni Scientifiche Italiane, 2003.

Guerci, Luciano. "Incredulità e rigenerazione nella Lombardia del triennio repubblicano." *Rivista storica italiana* 109 (1997): 83–85.

Guerci, Luciano. *Istruire nelle verità repubblicane: La letteratura politica per il popolo nell'Italia in rivoluzione (1796–1799)*. Bologna: Mulino, 1999.

Guerci, Luciano. *Uno spettacolo non mai più veduto nel mondo: La Rivoluzione francese come unicità e rovesciamento negli scrittori controrivoluzionari italiani, 1789–1799*. Turin: Utet, 2008.

Guerlac, Henry. "Some Aspects of Science during the French Revolution." *Scientific Monthly* 80 (1955): 93–101.

Guerra, Corinna. *Lavoisier e Parthenope: Contributo ad una storia della chimica del Regno di Napoli*. Naples: Società Napoletana di Storia Patria, 2017.

Guicciardini, Niccolò, ed. *Anachronism in History of Mathematics: Essays on the Historical Interpretation of Historical Texts*. Cambridge: Cambridge University Press, 2021.

Gurr, John Edwin. *Principle of Sufficient Reason in Some Scholastic Systems, 1750–1900*. Milwaukee, WI: Marquette University Press, 1959.

Habermas, Jürgen. *The Theory of Communicative Action*. 2 vols. Boston, MA: Beacon Press, 1984 and 1987.

Hachette, Jean. "Solution algébrique d'un problème de géometrie à trois dimensions." In *ARAS* 3, 1832: 3–34.

Hahn, Roger. *The Anatomy of a Scientific Institution: The Paris Academy of Sciences, 1666–1803*. Berkeley: University of California Press, 1971.

Haller, Karl Ludwig von. *Ristaurazione della scienza politica, ovvero teoria dello stato naturale sociale opposta alla supposizione di uno stato civile fattizio*. 8 vols. Naples: Biblioteca Cattolica, 1826–1828.

Hampster-Monk, Ian. "Burke and the Religious Sources of Sceptical Conservatism." In *The Skeptical Tradition around 1800*, edited by Johan van der Zande and Richard Popkin, 235–60. Dordrecht: Kluwer, 1998.

Hankins, Thomas. *Science and the Enlightenment*. Cambridge: Cambridge University Press, 1985.

Hanna, Gila, Hans Jahnke, and Helmut Pulte, eds. *Explanation and Proof in Mathematics: Philosophical and Educational Perspectives*. New York: Springer, 2010.

Hanson, Paul. *The Jacobin Republic Under Fire: The Federalist Revolt in the French Revolution*. University Park: Pennsylvania State University Press, 2003.

Hayek, Friedrich. *The Counter-Revolution of Science: Studies on the Abuse of Reason*. Glencoe, IL: Free Press, 1952.

Hazard, Paul. *La Révolution française et les lettres italiennes, 1789–1815*. Paris: Hachette, 1910.

Head, Brian. *Ideology and Social Science: Destutt de Tracy and French Liberalism*. Dordrecht: Nijohff, 1985.

Helvétius, Claude Adrien. *Dello Spirito*. 3 vols. Milan: Raffaele Netti, 1797–1799.

Helmholtz, Hermann von. *On the Sensations of Tone as a Physiological Basis for the Theory of Music*. New York: Dover Publications, 1954.

Hemingway, Andrew. *Landscape Imagery and Urban Culture in Early Nineteenth Century Britain*. Cambridge: Cambridge University Press, 1992.

Henderson, Linda. "Editor's Introduction." *Science in Context* 17 (2004): 423–66.

Henderson, Linda. *The Fourth Dimension and Non-Euclidean Geometry in Modern Art*. Cambridge, MA: MIT Press, 2013.

Herf, Jeffrey. *Reactionary Modernism: Technology, Culture and Politics in Weimar and the Third Reich*. Cambridge: Cambridge University Press, 1984.

Hills, Helen. "Silver's Eye: Naples, Excess, and Spanish Colonialism." In *Nature and the Arts in Early Modern Naples*, edited by Frank Fehrenbach and Joris van Gastel, 81–103. Berlin: De Gruyter, 2020.

Hirschman, Albert. *The Rhetoric of Reaction: Perversity, Futility, Jeopardy*. Cambridge, MA: Harvard University Press, 1991.

Hobbes, Thomas. *Leviathan, or, the Matter, Forme, and Power of a Common-Wealth, Ecclesiasticall and Civil*. London: Crooke, 1651.

Hocedez, Edgar. *Histoire de la théologie au XIXe siècle*. 3 vols. Bruxelles: L'Edition Universelle, 1947–1952.

[Holbach, Paul Henri Thiry d']. *Il buon senso, ossia idee naturali opposte alle soprannaturali*. 2 vols. Milan: Raffaele Netti, [1797 or 1798].

[Holbach, Paul Henri Thiry d']. *L'inferno distrutto, e esame ragionato del dogma dell'eternità delle pene*. Milan: Raffaele Netti, 1798.

Holbach, Paul Henri Thiry d'. *Systeme de la nature*. 2 vols. London [Paris]: 1793.

Iacovelli, Gianni. "Brownismo e Brownisti a Napoli nel primo '800." *Medicina nei secoli* 1 (1989): 321–37.

Iacovelli, Gianni. *Gli acquedotti di Cotugno: Medici pugliesi a Napoli tra Illuminismo e Restaurazione*. Galatina: Congedo, 1988.

Iaccarino, Giuliana. *I sogni della storia: Giovan Leonardo Marugi e la "Analisi ragionata de' libri nuovi."* Galatina: Congedo, 2004.

Imbruglia, Girolamo. "Enlightenment in Eighteenth-Century Naples." In *Naples in the Eighteenth Century: The Birth and Death of a Nation State*, edited by Girolamo Imbruglia, 70–94. Cambridge: Cambridge University Press, 2000.

Imbruglia, Girolamo. "Illuminismo e politica in una inedita memoria di Matteo Galdi del 1814." In *Le scienze nel Regno di Napoli*, edited by Roberto Mazzola, 47–74. Rome: Aracne, 2009.

Ingold, Tim. "The Textility of Making." *Cambridge Journal of Economics* 34 (2010): 91–102.

Ingram, James. "Jacobinism." *Krisis: Journal for Contemporary Philosophy* 2 (2018): 90–93.

Jacob, Margaret. *The Radical Enlightenment: Pantheists, Freemasons and Republicans*. London: Allen and Unwin, 1981.

Jacyna, Leon Stephen. "Immanence or Transcendence: Theories of Life and Organization in Britain, 1790–1835." *Isis* 74 (1983): 311–29.

Jahnke, Hans. "Algebraic Analysis in Germany, 1780–1840: Some Mathematical and Philosophical Issues." *Historia Mathematica* 20 (1993): 265–84.

Jahnke, Hans. *Mathematik und Bildung in der Humboldtschen Reform*. Göttingen: Vandenhoeck and Ruprecht, 1990.

Jannelli, Cataldo. *Saggio sulla natura e necessità della scienza delle cose e delle storie umane*. Naples: Porcelli, 1817.

Jasanoff, Sheila. *States of Knowledge: The Co-Production of Science and the Social Order*. London: Routledge, 2004.

Jay, Martin. *Reason after Its Eclipse: On Late Critical Theory*. Madison: University of Wisconsin Press, 2016.

Jéhan, Louis-François. *Dictionnaire de philosophie catholique*. Vol. 2 : *Psychologie et logique*. Paris: Migne, 1861.

Jemolo, Arturo. *Chiesa e stato in Italia negli ultimi cento anni*. Turin: Einaudi, 1975.

Jerocades, Antonio. *Il Paolo, o sia l'umanità liberata*. Naples: Porcelli, 1783.

Jerocades, Antonio *La lira focense*. [Naples]: [after 1784].

Jonas, Raymond. *France and the Cult of the Sacred Heart: An Epic Tale for Modern Times*. Berkeley: University of California Press, 2000.

Jones, Matt. *Reckoning with Matter: Calculating Machines, Innovation, and Thinking about Thinking from Pascal to Babbage*. Chicago: University of Chicago Press, 2016.

Jones, Matt. *The Good Life in the Scientific Revolution: Descartes, Pascal, Leibniz and the Cultivation of Virtue*. Chicago: University of Chicago Press, 2006.

Jones-Imhotep, Edward. "The Unfailing Machine: Mechanical Arts, Sentimental Publics and the Guillotine in Revolutionary France." *History of the Human Sciences* 30 (2017): 11–31.

Katz, Victor. *A History of Mathematics*. New York: Harper Collins, 1993.

Kennedy, Emmett. *A Philosophe in the Age of Revolution: Destutt de Tracy and the Origins of "Ideology."* Philadelphia, PA: American Philosophical Society, 1978.

Kennedy, Emmett. "Ideology from Destutt de Tracy to Marx." *Journal of the History of Ideas* 40 (1979): 353–68.

Kennedy, Michael. *The Jacobin Club of Marseille, 1790–1794*. Ithaca, NY: Cornell University Press, 1973.

Kleiner, Israel. "Rigor and Proof in Mathematics: A Historical Perspective." *Mathematics Magazine* 64 (1991): 291–314.

Klinck, David. *The French Counterrevolutionary Theorist, Louis de Bonald (1754–1840)*. New York: Peter Lang, 1996.

Kline, Morris. *Mathematics: The Loss of Certainty*. Oxford: Oxford University Press, 1980.

Kline, Ronald. "Construing Technology as Applied Science: Public Rhetoric of Scientists and Engineers in the United States, 1880–1945." *Isis* 86 (1995): 194–221.

Klonk, Charlotte. *Science and Perception of Nature: British Landscape Art in the Late Eighteenth and Early Nineteenth Centuries*. New Haven, CT: Yale University Press, 1996.

Kluge, Eike-Henner. "Abstraction: A Contemporary Outlook." *Thomist* 3 (1976): 337–65.

Kolnai, Aurel. "Notes sur l'utopie réactionnaire." *Cité libre* 13 (1955): 9–20.

Koyré, Alexandre. *Études d'histoire de la pensée philosophique*. Paris: Gallimard, 1971.

Knorr, Wilbur. *The Ancient Tradition of Geometric Problems*. New York: Dover, 1993.

Knorr Cetina, Karin. *Epistemic Cultures: How Sciences Make Knowledge*. Cambridge, MA: Harvard University Press, 1999.

Kusch, Martin. *Foucault's Strata and Fields: An Investigation into Archeological and Genealogical Science Studies*. Dordrecht: Kluwer, 1991.

Kusch, Martin. "Objectivity and Historiography." *Isis* 100 (2009): 127–31.

Labrot, Gérard, and Antonio Delfino. *Collections of Painting in Naples, 1600–1780*. Munich: Saur, 1992.

Lachterman, David. "Mathematics and Nominalism in Vico's *Liber Metaphysicus*." In *Sachkommentar zu Giambattista Vicos Liber Metaphysicus*, edited by Stephan Otto and Hermut Viechtbauer, 47–85. München: Wilhelm Fink Verlag, 1985.

Lachterman, David. *The Ethics of Geometry: A Genealogy of Modernity*. London: Routledge, 1989.

Lachterman, David. "Vico, Doria e la geometria sintetica." *Bollettino del Centro di Studi Vichiani* 10 (1980): 10–35.

Lagrange, Joseph-Louis. *Oeuvres*. 14 vols. Edited by J. -A. Serret and Gaston Darboux. Paris: Gautier-Villars, 1867–1892.

Lamennais, Félicité Robert de. *Oeuvres complètes*. 12 vols. Paris: Daubrée et Cailleux, 1836–1837.

Lampland, Martha, and Susan Leigh Star, eds. *Standards and Their Stories: How Quantifying, Classifying, and Formalizing Practices Shape Everyday Life*. Ithaca, NY: Cornell University Press, 2009.

Landi, Guido. *Istituzioni di diritto pubblico del Regno delle Due Sicilie (1815–1860)*. 2 vols. Milan: Giuffrè, 1977.

Latour, Bruno. *Science in Action: How to Follow Scientists and Engineers through Society*. Cambridge, MA: Harvard University Press, 1985.

Latour, Bruno. "Sur la pratique des théoriciens." In *Savoirs théoriques et savoirs d'action*, edited by Jean-Marie Barbier, 131–46. Paris: Presses Universitaires de France, 1996.

Latour, Bruno. *We Have Never Been Modern*. Cambridge, MA: Harvard University Press, 1993.

Lauberg, Carlo. *Analisi chimico-fisica sulle proprietà de' quattro principali agenti della natura, seguita da un saggio sulle principali funzioni degli esseri organizzati*. Naples: Coda, 1788.

Lauberg, Carlo. *Memoria sull'unità dei principi della meccanica*. [Naples]: [between 1787 and 1789].

Lauberg, Carlo. *Riflessioni sulle operazioni dell'umano intendimento*. Naples: [between 1786 and 1789].

Lauberg, Carlo. "Sull'alto prezzo delle cose." *Il Monitore Italiano* 12 (February 11, 1798).

Lauro, Agostino. *Il giurisdizionalismo pregiannoniano nel Regno di Napoli. Problema e bibliografia (1563–1723)*. Rome: Edizioni di Storia e Letteratura, 1974.

Lavoisier, Antoine Laurent. *Trattato elementare di chimica*. 2 vols. Naples: Campo, 1791–1792.

Lawrence, Christopher. "Incommunicable Knowledge: Science, Technology and Clinical Art in Britain, 1850–1914." *Journal of Contemporary History* 20 (1985): 503–20.

Lebrun, François, and Roger Dupuy, eds. *Les résistances à la Révolution*. Paris: Imago, 1987.

[Lefranc, Jacques-François]. *Le voile levé pour les curieux, ou les secrets de la Révolution de France révéles à l'aide de la Franc-Maçonnerie*. [n.p.]: 1791.

Leggett, Don, and Charlotte Sleigh, eds. *Scientific Governance in Britain, 1914–79*. Manchester: Manchester University Press, 2016.

Lehner, Ulrich. *The Catholic Enlightenment: The Forgotten History of a Global Movement*. New York: Oxford University Press, 2016.

Leoni, Francesco. *Storia della controrivoluzione in Italia (1789–1759)*. Naples: Guida, 1975.

Leopardi, Monaldo. *Catechismo filosofico per uso delle scuole inferiori, proposto dai redattori della Voce della Ragione*. Pesaro: Nobili, 1832.

Leopardi, Monaldo. *Dialoghetti sulle materie correnti dell'anno 1831*. Pesaro: 1832.

Leso, Erasmo. *Lingua e rivoluzione: Ricerche sul vocabolario politico italiano del triennio rivoluzionario 1796–1799*. Venice: Istituto veneto di scienze, lettere ed arti, 1991.

Levi, Giovanni. "On Microhistory." In *New Perspectives on Historical Writing*, edited by Peter Burke, 93–113. Cambridge: Polity Press, 1991.

Lezioni ad uso delle scuole normali di Francia raccolte per mezzo dei stenografi, e rivedute dai professori. Translated by Carlo Lauberg. 2 vols. Milan: Raffaele Netti, 1798.

Liberatore, Raffaele. "Necrologia: Melchiorre Delfico." *Il progresso delle lettere, delle scienze, e delle arti* 11 (1835): 292–318.

L[iberatore], R[affaele]. "Sulla Scuola di Applicazione annessa al Corpo de' Ponti e Strade del Regno di Napoli." *Il progresso delle lettere, delle scienze, e delle arti* 10 (1835): 328–35.

Liguori, Alfonso Maria de. *Le glorie di Maria*. 2 vols. Bassano: Remondini, 1789.

Lilla, Mark. *The Shipwrecked Mind: On Political Reaction*. New York: New York Review Books, 2016.

Livingston, Eric. "Cultures of Proving." *Social Studies of Science* 29 (1999): 867–88.

Livingston, Eric. *The Ethnomethodological Foundations of Mathematics*. London: Routledge, 1985.

Livingstone, David. *Putting Science in Its Place*. Chicago: University of Chicago Press, 2003.

Lombardi Satriani, Luigi, ed. *Antonio Jerocades nella cultura del Settecento*. Reggio Calabria: Falzea, 1998.

Lombardo, Enzo. "Il primo trattato italiano di Statistica di Luca di Samuele Cagnazzi ed i suoi interessi demografici." In *Da osservazione sperimentale a spiegazione razionale: Per una storia della statistica in Italia*, edited by Carlo Corsini, 33–48. Pisa: Pacini, 1989.

Lomonaco, Fabrizio, and Maurizio Torrini, eds. *Galileo e Napoli*. Naples: Guida, 1984.

Lomonaco, Francesco. *Rapporto al cittadino Carnot* [1800]. In Vincenzo Cuoco, *Saggio storico sulla rivoluzione napoletana del 1799, seguito dal Rapporto al cittadino Carnot di Francesco Lomonaco*, edited by Fausto Nicolini, 285–353. Bari: Laterza, 1913.

Longano, Francesco. *Dell'uomo naturale*. Cosmopoli: 1778.

Lorenat, Jemma. "Figures Real, Imagined, and Missing in Poncelet, Plücker, and Gergonne." *Historia Mathematica* 42 (2014): 155–92.

Loria, Gino. *Nicola Fergola e la scuola che lo ebbe a duce*. Genoa: Tipografia Sordo-Muti, 1892.

Lo Sardo, Eugenio. *Il mondo nuovo e le virtù civili: L'epistolario di Gaetano Filangieri: 1772–1788*. Naples: Fridericiana, 1999.

Lovejoy, Arthur. *The Great Chain of Being. A Story of the History of An Idea*. Cambridge, MA: Harvard University Press, 1971.

Luca, Ferdinando de. *Memoria per rivendicare alla scuola italica tutta l'antica geometria, cioè l'analisi geometrica, le sezioni coniche e i luoghi geometrici* [1832]. Naples: Fibreno, 1845.

Luca, Ferdinando de. *Nuovo sistema di studi geometrici, analiticamente dedotti dallo svolgimento successivo di una sola equazione* [1845]. Naples: Fibreno, 1847.

Lucas, Colin. "Résistances populaires à la Révolution dans le Sud-Est." In *Mouvements populaires et conscience sociale (XVIe-XIXe siècles)*, edited by Jean Nicolas, 473–88. Paris: Maloine, 1985.

Luciano, Domenico. *Domenico Grimaldi e la Calabria nel '700*. Assisi: Carucci, 1974.

Lugli, Emanuele. *Unità di misura: Breve storia del metro in Italia*. Bologna: Mulino, 2014.

Lukács, Georg. *The Destruction of Reason*. London: Verso, 2021.

Maclaurin, Colin. *A Treatise of Fluxions*. 2 vols. Edinburgh: Ruddimans, 1742.

MacKenzie, Donald. *An Engine, Not a Camera: How Financial Models Shape Markets*. Cambridge, MA: MIT Press, 2006.

MacKenzie, Donald. *Mechanizing Proof: Computing, Risk, and Trust*. Cambridge, MA: MIT Press, 2001.

MacKenzie, Donald. "On Invoking 'Culture' in the Analysis of Behavior in Financial Markets." In *Cultures without Culturalism: The Making of Scientific Knowledge*, edited by Karine Chemla and Evelyn Fox Keller, 29–48. Durham, NC: Duke University Press, 2017.

MacKenzie, Donald. "Slaying the Kraken: The Sociohistory of a Mathematical Proof." *Social Studies of Science* 29 (1999): 7–60.

MacKenzie, Donald. *Statistics in Britain, 1865–1930: The Social Construction of Scientific Knowledge*. Edinburgh: Edinburgh University Press, 1981.

Macry, Paolo. "Ceto mercantile e azienda agricola nel Regno di Napoli: il contratto alla voce nel XVIII secolo." *Quaderni Storici* 21 (1972): 851–909.

Maffioli, Cesare. *Out of Galileo: The Science of Waters, 1628–1718*. Rotterdam: Erasmus, 1994.

Magliari, Pietro. *Elogio istorico di Domenico Cotugno*. Naples: Stamperia francese, 1823.

Mahoney, Michael. "Another Look at Greek Geometrical Analysis." *Archive for the History of Exact Sciences* 5 (1968): 318–48.

Mahoney, Michael. *The Mathematical Career of Pierre de Fermat*. Princeton, NJ: Princeton University Press, 1973.

Maistre, Joseph de. *Oeuvres complètes*. 14 vols. Lyon: Vitte, 1884–1886.

Maiuri, Antonio. *Delle opere pubbliche nel Regno di Napoli e degl'ingegneri preposti a costituirle*. Naples: Stamperia e cartiera del Fibreno, 1836.

Mako, Paul. *Compendiaria metaphysicae institutio* [. . .] [1761]. Venice: Baseggio, 1797.

Mako, Paul. *Compendio di logica ad uso degli studenti di filosofia* [1760]. Venice: Andreola, 1819.

Mancosu, Paolo. *Philosophy of Mathematics and Mathematical Practice in the Seventeenth Century*. Oxford: Oxford University Press, 1996.

Mancosu, Jørgensen, and Pedersen Mancosu, eds. *Visualization, Explanation and Reasoning Styles in Mathematics*. Dordrecht: Springer, 2005.

Mandarini, Enrico. *I codici manoscritti della Biblioteca oratoriana di Napoli*. Naples: Festa, 1897.

Mandelbrote, Scott, and Helmut Pulte, eds. *The Reception of Isaac Newton in Europe*. 3 vols. London: Bloomsbury, 2018.

Mannheim, Karl. "Conservative Thought." In Mannheim, *Essays on Sociology and Social Psychology*, 74–164. London: Routledge, 1953.

Mannheim, Karl. *Ideology and Utopia*. London: Routledge, 1991.

Marchi, Tommaso. *Elogio funebre di Nicola Fergola*. Naples: Trani, 1824.

Marcialis, Maria Teresa. "Legge di natura e calcolo della ragione nell'ultimo Genovesi." *Materiali per una storia della cultura giuridica* 24 (1994): 315–40.

[Maréchal, Sylvain]. *Culte et loix d'une société d'hommes sans dieu*. Milan: Raffaele Netti, [1798].

Maresca, Luca. *Memoria in cui il metodo analitico degli antichi si applica alla risoluzione di vari difficili problemi, e di quelli specialmente che diconsi delle tazioni*. Naples: Stamperia reale, 1825.

Marino, Giuseppe. *La formazione dello spirito borghese in Italia*. Florence: La Nuova Italia, 1974.

Martino, Nicola di (de). *Elementa statices*. Naples: Mosca, 1727.

Martino, Pietro di (de). *Philosophiae naturalis institutionum libri tres*. Naples: Mosca, 1738.

Martirano, Maurizio. "Cuoco e la scienza." In *Le scienze nel Regno di Napoli*, edited by Roberto Mazzola, 29–46. Rome: Aracne, 2009.

Martorelli, Luisa. "Aspects of Collecting in Nineteenth-Century Naples: The Portfolios (*Cartiere*) of Giacinto Gigante." In *Ottocento: Romanticism and Revolution in 19th-Century Italian Painting*, edited by Roberta Olson, 59–64. New York: American Federation of the Arts, 1993.

Martorelli, Luisa. *Giacinto Gigante e la Scuola di Posillipo*. Naples: Electa, 1993.

Martuscelli, Stefania. *La popolazione del Mezzogiorno nella statistica di re Murat*. Naples: Guida, 1979.

Marugi, Giovanni Leonardo. *Stato attuale delle scienze*. 2 vols. Naples: 1792.

Marzucco, Giuseppe. *Riflessioni intorno alla quadratura del cerchio e delle curve, ove per comodo della gioventù si spiegano brevemente ancora li principi del calcolo differenziale ed integrale*. Naples: Azzolino, 1767.

Masnovo, Amato. *Il neotomismo in Italia: Origini e prime vicende*. Milan: Vita e Pensiero, 1923.

Massafra, Angelo, ed. *Il Mezzogiorno preunitario: Economia, società e istituzioni*. Bari: Dedalo, 1998.

Massafra, Angelo. "Una stagione di studi sulla feudalità nel Regno di Napoli." In *Fra storia e storiografia: Studi in onore di Pasquale Villani*, edited by Paolo Macry and Angelo Massafra, 103–29. Bologna: Mulino, 1994.

Mastellone, Salvo. "Il dibattito sulla democrazia nel triennio giacobino italiano (1796–1799)." In *Il modello politico giacobino e le rivoluzioni*, edited by Massimo Salvadori and Nicola Tranfaglia, 154–61. Florence: La Nuova Italia, 1984.

Matarazzo, Pasquale. "La stampa periodica a Napoli tra Decennio francese e Restaurazione. La *Biblioteca analitica* (1810–1823)." In *Le riviste a Napoli dal XVIII secolo al primo Novecento*, edited by Antonio Garzya, 287–302. Naples: Accademia Pontaniana, 2009.

Mathiez, Albert. *La Théophilantropie et le culte décadaire, 1796–1801*. Paris: Alcan, 1904.

Maturi, Walter. *Il principe di Canosa*. Florence: Le Monnier, 1944.

Mauer, Armand. "Thomists and Thomas Aquinas on the Foundations of Mathematics." *Review of Metaphysics* 47 (1993): 43–61.

Mazauric, Claude. *Jacobinisme et Révolution, autour du bicentenaire de Quatre-vingt-neuf*. Paris: Éditions sociales, 1984.

Mazzarella, Giuseppe. *Corso d'ideologia elementare*. Naples: Zambraja, 1826.

Mazzei, Raffaele. "Un calabrese del '700, patriota e scienziato: Vincenzo de Filippis." *Archivio Storico di Calabria e Lucania* 48 (1976): 161–99.

Mazzotti, Massimo. "Enlightened Mills: Mechanizing Olive Oil Production in Mediterranean Europe." *Technology and Culture* 45 (2004): 277–304.

Mazzotti, Massimo. "Newton in Italy." In *The Reception of Isaac Newton in Europe*, vol. 1, edited by Scott Mandelbrote and Helmut Pulte, 159–78. London: Bloomsbury, 2018.

McClellan, James. *The Applied Science Problem*. Jersey City, NJ: Jensen/Daniels Publishers, 2008.

McMahon, Darrin. *Enemies of the Enlightenment: The French Counter-Enlightenment and the Making of Modernity*. Oxford: Oxford University Press, 2002.

Mehrtens, Herbert. *Moderne-Sprache-Mathematik: Eine Geschichte des Streits um die Grundlagen der Disziplin und des Subjekts formaler Systeme*. Frankfurt am Main: Suhrkamp, 1990.

Mehrtens, Herbert, Henk Bos, and Ivo Schneider, eds. *Social History of Nineteenth Century Mathematics*. Boston, MA: Birkhäuser, 1981.

Melton, James van Horn. *The Rise of the Public in Enlightenment Europe*. Cambridge: Cambridge University Press, 2001.

Menozzi, Daniele. "Intorno alle origini del mito della cristianità." *Cristianesimo nella storia* 5 (1984): 523–62.

Menozzi, Daniele. *Sacro Cuore: Un culto tra devozione interior e restaurazione Cristiana della società*. Rome: Viella, 2001.

Menozzi, Daniele. "Tra riforma e restaurazione: Dalla crisi della società cristiana al mito della cristianità medievale, 1758–1848." In *Storia d'Italia, Supplemento 9: La chiesa e il potere politico dal medioevo all'età contemporanea*, 767–806. Turin: Einaudi, 1986.

Meyer, Jean. "La *Cristiada*: Peasant War and Religious War in Revolutionary Mexico, 1926–29." In *Religion and Rural Revolt*, edited by Janos Bak and Gerhard Benecke, 441–52. Dover, NH: Manchester University Press, 1984.

Miławicki, Marek. "The Dominicans in Saint Petersburg, 1816–1892: A Prosopographical Study." In *I Domenicani e la Russia*, edited by Viliam Doci and Hyacinthe Destivelle, 193–254. Rome: Angelicum University Press, 2019.

Montaudo, Aldo. *L'olio nel Regno di Napoli nel XVIII secolo: Commercio, annona e arrendamenti*. Naples: Edizioni Scientifiche Italiane, 2005.

Montone, Francesco. "Il *tópos* della *Campania felix* nella poesia latina." *Salternum* 24–25 (2010): 47–58.

Montucla, Jean-Étienne. *Histoire des mathématiques*. Vol. 1. Paris: Jombert, 1758.

Moravia, Sergio. *Il pensiero degli idéologues: Scienza e filosofia in Francia, 1780–1815*. Florence: La Nuova Italia, 1974.

Morrell, Jack. "Professors Robison and Playfair and the Theophobia Gallica: Natural Philosophy, Religion and Politics in Edinburgh, 1789–1815." *Notes Rec. R. Soc. Lond* 26 (1971): 43–63.

Morelli, Domenico, and Edoardo Dalbono. *La scuola napoletana di pittura nel secolo decimonono, ed altri scritti d'arte*. Edited by Benedetto Croce. Bari: Laterza, 1915.

Morelli di Gregorio, Nicola. "Cav. Giuseppe Saverio Poli." In *Biografia degli uomini illustri del Regno di Napoli, ornata de loro rispettivi ritratti, compilata da diversi letterati nazionali*, vol. 11. Naples: Gervasi, 1826 [unpaged].

Motooka, Wendy. *The Age of Reasons: Quixotism, Sentimentalism and Political Economy in Eighteenth-Century Britain*. London: Routledge, 1998.

Naddeo, Barbara Ann. "A Cosmopolitan in the Provinces: G. M. Galanti, Geography, and Enlightenment in Europe." *Modern Intellectual History* 10 (2013): 1–26.

Naddeo, Barbara Ann. *Vico and Naples: The Urban Origins of Modern Social Theory*. Ithaca, NY: Cornell University Press, 2011.

Napier, Francis. *Notes on Modern Painting at Naples*. London: Parker and Son, 1855.

Napoli Signorelli, Pietro. "Discorso istorico preliminare." In *ARAS* 1788: i–xcviii.

Neri, Carlo. *Giuseppe Ceva Grimaldi, marchese di Pietracatella. Cenni biografici*. Naples: Rinaldi e Sellitto, 1879.

Netz, Reviel. *The Shaping of Deduction in Greek Mathematics: A Study in Cognitive History*. Cambridge: Cambridge University Press, 1999.

Netz, Reviel. *The Transformation of Mathematics in the Early Mediterranean World: From Problems to Equations*. Cambridge: Cambridge University Press, 2004.

Nicolini, Nicola. *La spedizione punitiva del Latouche-Tréville (16 dicembre 1792) ed altri saggi sulla vita politica napoletana alla fine del secolo XVIII*. Florence: Le Monnier, 1939.

Nicolini, Nicola. *Luigi de Medici e il giacobinismo napoletano*. Florence: Le Monnier, 1935.

Niskanen, Kristi, and Michael Barany, eds. *Gender, Embodiment, and the History of the Scholarly Persona: Incarnations and Contestations*. Cham: Springer International Publishing, 2021.

Nouvelle biographie générale, depuis les temps les plus reculés jusqu'à nos jours, avec les renseignements bibliographiques et l'indication des sources à consulter. 46 vols. Paris: Firmin Didot, 1852–1866.

O'Connell, Joseph. "Metrology: The Creation of Universality by the Circulation of Particulars." *Social Studies of Science* 23 (1993): 129–73.

Odazi, Toriano. *Discorso pronunziato nella riapertura della cattedra di economia politica e commercio nella Regia Università degli studi di Napoli*. Naples: 1782.

Oldrini, Guido. *Gli hegeliani di Napoli: Augusto Vera e la corrente ortodossa*. Milan: Feltrinelli, 1964.

Oldrini, Guido. *Il primo hegelismo italiano*. Florence: Vallecchi, 1969.

Oldrini, Guido. *La cultura filosofica napoletana dell'Ottocento*. Bari: Laterza, 1973.

Olivier Poli, Gioacchino. *Cenno biografico sul cavalier commendatore Giuseppe Saverio Poli*. Naples: Marotta and Vanspandoch, 1825.

Onnis Rosa, Pia. *Filippo Buonarroti e altri studi*. Rome: Edizioni di storia e letteratura, 1971.

Opuscoli matematici della scuola del Sig. N. Fergola, parte già pubblicati e parte inediti. Vol. 1. Naples: Stamperia reale, 1811.

Opuscoli matematici e fisici di diversi autori. 2 vols. Milan: Giusti, 1832–1834.

Orlando, Pasquale. *Il tomismo a Napoli nel sec. XIX: La scuola del Sanseverino*. Vatican City: Pontificia Università Lateranense, 1968.

Orlov, Grigorii. *Mémoires historiques, politiques et littéraires sur le Royaume de Naples*. 5 vols. Paris: Chasseriau et Hécart, 1819–1821.

Ortolani, Sergio. *Giacinto Gigante e la pittura di paesaggio a Napoli e in Italia dal '600 all' '800*. Naples: Montanino, 1970.

Orwell, George. *Essays*. New York: Alfred Knopf, 2002.

Osborne, Michael. "Applied Natural History and Utilitarian Ideals: 'Jacobin Science' at the Muséum d'Histoire Naturelle, 1789–1870." In *Re-creating Authority in Revolutionary France*, edited by Bryant Ragan and Elizabeth Williams, 125–43. New Brunswick, NJ: Rutgers University Press, 1992.

Ostermann, Alexander, and Gerhard Wanner. *Geometry by Its History*. Berlin: Springer, 2014.

Ostuni, Nicola. *Iniziativa privata e ferrovie nel Regno delle Due Sicilie*. Naples: Giannini, 1980.

Ostuni, Nicola. *Le comunicazioni stradali nel Settecento meridionale*. Naples: Edizioni Scientifiche Italiane, 1991.

Ottonello, Franco. *Cultura filosofica nella stampa periodica dell'Italia meridionale della prima metà dell'Ottocento*. 5 vols. Milan: Marzorati, 1977–1989.

Ottonello, Franco. "Vincenzo De Ritis espositore e critico dell'ideologia del Tracy nella *Biblioteca analitica*." In Franco Ottonello, *Cultura filosofica*, vol. 1, 111–28. Milan: Marzorati, 1977.

Outram, Dorinda. *The Enlightenment*. Cambridge: Cambridge University Press, 1995.

Ozouf, Mona. "Federalism." In *A Critical Dictionary of the French Revolution*, edited by Mona Ozouf and François Furet, 54–64. Cambridge, MA: Harvard University Press, 1989.

Ozouf, Mona. *L'homme régénéré: Essai sur la Révolution française*. Paris: Gallimard, 1989.

Ozouf, Mona. "Public Opinion at the End of the Old Regime." *Journal of Modern History* 60 (1988): S1–S21.

Ozouf, Mona. "Public Spirit." In *A Critical Dictionary of the French Revolution*, edited by Mona Ozouf and François Furet, 771–80. Cambridge, MA: Harvard University Press, 1989.

Ozouf, Mona. "Regeneration." In *A Critical Dictionary of the French Revolution*, edited by Mona Ozouf and François Furet, 781–89. Cambridge, MA: Harvard University Press, 1989.

Pace, Antonio. *Benjamin Franklin and Italy*. Philadelphia, PA: American Philosophical Society, 1958.

Padula, Fortunato. *Raccolta di problemi di geometria risoluti con l'analisi algebrica*. Naples: Fibreno, 1838.

Pagano, Mario. *Considerazioni sul processo criminale*. Naples: Raimondi, 1787.

Pagano, Mario. "Discorso sull'origine e natura della poesia" [1791]. In Mario Pagano, *Opere*, vol. 4, 1–90. Lugano: Ruggia, 1832.

Pagano, Mario. *Saggi politici de' principi, progressi e decadenza delle società* [1783–85]. 2 vols. Lugano: Ruggia, 1831.

Palladino, Franco. *Metodi matematici e ordine politico: Lauberg, Giordano, Fergola, Colecchi. Il dibattito scientifico a Napoli tra Illuminismo, rivoluzione, e reazione.* Naples: Jovene, 1999.

Palmieri, Giuseppe. *Osservazioni su vari articoli riguardanti la pubblica economia.* Naples: Flauto, 1790.

Palmieri, Pasquale. *I taumaturghi della società: Santi e potere politico nel secolo dei Lumi.* Rome: Viella, 2010.

Palmisciano, Giuseppe. "Filosofia e intransigenza cattolica napoletana dopo l'Unità nella rivista *La Scienza e la Fede.*" In *Le filosofie del Risorgimento,* edited by Maurizio Martirano, 321–49. Milan: Mimesis, 2012.

Panvini, Pasquale. *Cenno biografico del Rev.mo padre D. Giuseppe Capocasale, insigne filosofo.* Naples: Gervasi, 1829.

Panza, Marco. "Classical Sources for the Concepts of Analysis and Synthesis." In *Analysis and Synthesis in Mathematics: History and Philosophy,* edited by Marco Panza and Michael Otte, 367–68. Dordrecht: Kluwer, 1997.

Panza, Marco, and Michael Otte, eds. *Analysis and Synthesis in Mathematics: History and Philosophy.* Dordrecht: Kluwer, 1997.

Parise, Marialuisa. "Francis Bacon ne *Il Galileo* di Francesco Colangelo." *Annali dell'Istituto Italiano per gli Studi Storici* 25 (2010): 373–96.

Parise, Marialuisa. "Lettere inedite a Francesco Colangelo nei manoscritti Ferrajoli 867 e 941 della Biblioteca Vaticana." *Giornale critico della filosofia italiana* 91 (2012): 44–60.

Parsons, Charles. "Infinity and Kant's Conception of the 'Possibility of Experience.'" In *Charles Parsons, Mathematics and Philosophy: Selected Essays,* 95–109. Ithaca, NY: Cornell University Press, 1983.

Pasnau, Robert. *After Certainty: A History of Our Epistemic Ideals and Illusions.* Oxford: Oxford University Press, 2017.

Patiniotis, Manolis. "Between the Local and the Global: History of Science in the European Periphery Meets Post-Colonial Studies." *Centaurus* 55 (2013): 361–84.

Patriarca, Silvana. *Numbers and Nationhood: Writing Statistics in Nineteenth-Century Italy.* Cambridge: Cambridge University Press, 1996.

Paty, Michel. "Rapports des mathématiques et de la physique chez d'Alembert." *Dix-huitième Siècle* 16 (1984): 69–79.

Payne, Christiana. *Toil and Plenty: Images of the Agricultural Landscape in England, 1780–1890.* New Haven, CT: Yale University Press, 1993.

Pavone, Mario. *Pittori napoletani del primo Settecento: Fonti e documenti.* Naples: Liguori, 1997.

Pedio, Tommaso. *Giacobini e sanfedisti nell'Italia meridionale,* 2 vols. Bari: Levante, 1974.

Pedio, Tommaso. *Massoni e giacobini nel Regno di Napoli: Emmanuele de Deo e la congiura del 1794.* Matera: Montemurro, 1976.

Pepe, Alfonso. *Il clero giacobino: Documenti inediti.* 2 vols. Naples: Procaccini, 1999.

Pepe, Luigi. "Sulla trattatistica del calcolo infinitesimale in Italia nel secolo XVIII." In *La storia delle matematiche in Italia,* edited by Oscar Grugnetti and Lucia Montaldo, 145–227. Cagliari: Università di Cagliari, 1984.

Perna, Maria Luisa. "L'universo comunicativo di Antonio Genovesi." In *Editoria e cultura Napoli nel XVIII secolo,* edited by Anna Maria Rao, 391–404. Naples: Liguori, 1998.

Perna, Maria Luisa. "Nota critica." In Antonio Genovesi, *Delle Lezioni di commercio o sia di Economia civile con Elementi del commercio*, 893–921. Naples: Istituto Italiano di Studi Filosofici, 2005.

Phillips, Christopher. "Robert Woodhouse and the Evolution of Cambridge Mathematics." *History of Science* 44 (2006): 69–93.

Piccone, Paul. *Italian Marxism*. Berkeley: University of California Press, 1983.

Pickering, Andrew, ed. *Science as Practice and Culture*. Chicago: University of Chicago Press, 1996.

Pickering, Andrew. *The Mangle of Practice: Time, Agency, and Science*. Chicago: University of Chicago Press, 1995.

Pickering, Andrew, and Adam Stephanides. "Constructing Quaternions: On the Analysis of Conceptual Practice." In *Science as Practice and Culture*, edited by Andrew Pickering, 139–67. Chicago: University of Chicago Press, 1996.

Picon, Antoine. *L'invention de l'ingénieur moderne: l'École des ponts et chaussées, 1747–1851*. Paris: Presses de l'École nationale des ponts et chaussées, 1992.

Picon, Antoine, and Michel Yvon. *L'ingénieur artiste: Dessins anciens de l'École des Ponts et chaussées*. Paris: Presses de l'École nationale des ponts et chaussées, 1989.

Piergili, Giuseppe. *Notizia della vita e degli scritti del Conte Monaldo Leopardi*. Florence: Sansoni, 1899.

Pigonati, Andrea. *La parte di strada degli Abruzzi da Castel di Sangro a Sulmona*. Naples, Morelli, 1783.

Pigonati, Andrea. *Le strade antiche e moderne del Regno di Napoli e riflessioni sopra li metodi di esecuzione e meccaniche*. Naples, Morelli, 1784.

Pii, Eluggero. *Antonio Genovesi: Dalla politica economica alla "politica civile."* Florence: Olschki, 1984.

[Piola, Gabrio]. *Lettere di Evasio ad Uranio intorno alle scienze matematiche*. Modena: Soliani, 1825.

Piola, Gabrio, "Saggio sulla metafisica dell'analisi pura." *Biblioteca italiana* 97 (1840): 315–41.

Piolanti, Antonio. *L'Accademia di Religione Cattolica: Profilo della sua storia e del suo tomismo*. Vatican City: Libreria Editrice Vaticana, 1977.

Placanica, Augusto. *Il filosofo e la catastrofe: Un terremoto del Settecento*. Turin: Einaudi, 1985.

Pliny. *Natural History, Volume II: Books 3–7*. Translated by H. Rackham. Loeb Classical Library 352. Cambridge, MA: Harvard University Press, 1942.

Pluquet, François André Adrien. *Esame del fatalismo, o sia esposizione e confutazione dei diversi sistemi di fatalismo che han diviso i filosofi sull'origine del mondo, sulla natura dell'anima, e sul principio delle azioni umane* [1757]. 3 vols. Naples: Giaccio, 1791.

Pocock, John Greville Agard. *Barbarism and Religion*. 6 vols. Cambridge: Cambridge University Press, 1999–2016.

Poli, Giuseppe Saverio. *Breve ragionamento intorno all'eccellenza dello studio della natura, ed a' sodi vantaggi che da quello si possono ritrarre*. Naples: Stamperia reale, 1780.

Poli, Giuseppe Saverio. *Elementi di fisica sperimentale* [1781]. 5 vols. Naples: Trani, 1822.

Poli, Giuseppe Saverio. *La formazione del tuono, della folgore, e di varie altre meteore, spiegata giusta le idee del signor Franklin*. Naples: Campo, 1772.

Poli, Giuseppe Saverio. *Ragionamento intorno allo studio della natura*. Naples: 1781.

Poli, Giuseppe Saverio. *Testacea utriusque siciliae eorumque historia et anatome tabulis aeneis illustrate*. 2 vols. Parma: [Bodoni], 1791–1795. A third volume was published, in two parts, in 1826–1827.

Poli, Giuseppe Saverio. *Viaggio celeste, poema astronomico*. 2 vols. Naples: Stamperia reale, 1805.

Poncelet, Jean Victor. "Philosophie mathématique: Réflexions sur l'usage de l'analyse algébrique dans la géométrie; suivies de la solution de quelques problèmes dépendant de la géométrie de la règle." *Annales de mathématiques pures et appliquées* 8 (1817): 141–55.

Popkin, Richard. "Scepticism and Anti-Scepticism in the Latter Part of the Enlightenment." In *Scepticism in the Enlightenment*, edited by Richard Popkin, Ezequiel de Olaso, and Giorgio Tonelli, 17–34. Dordrecht: Kluwer, 1997.

Popkin, Richard. *The History of Scepticism from Erasmus to Spinoza*. Berkeley: University of California Press, 1979.

Porter, Roy, and Mikulás Teich, eds. *The Enlightenment in National Context*. Cambridge: Cambridge University Press, 1981.

Porter, Theodore. *The Rise of Statistical Thinking, 1820–1900*. Princeton, NJ: Princeton University Press, 1986.

Porter, Theodore. *Trust in Numbers: The Pursuit of Objectivity in Science and Public Life*. Princeton, NJ: Princeton University Press, 1995.

Praz, Mario. *The Romantic Agony*. London: Humphrey Milford, 1933.

Presta, Giovanni. *Trattato degli ulivi, delle ulive, e della maniera di cavar l'olio, o si riguardi di primo scopo la massima perfezione, o si riguardi la massima possibile quantità del medesimo* [1794]. In Presta, *Opere*, vol. 2. Lecce: Edizioni del Grifo, 1988.

Pyenson, Lewis. *Neohumanism and the Persistence of Pure Mathematics in Wilhelmian Germany*. Philadelphia, PA: American Philosophical Society, 1983.

Quaini, Massimo. "Appunti per una archeologia del 'colpo d'occhio': Medici, soldati e pittori alle origini dell'osservazione sul terreno in Liguria." In *Studi di etnografia e dialettologia ligure in memoria di Hugo Plomteux*, edited by Lorenzo Coveri and Diego Moreno, 107–125. Genoa: SAGEP Editrice, 1983.

Rabouin, David. "Styles in Mathematical Practice." In *Cultures without Culturalism: The Making of Scientific Knowledge*, edited by Karine Chemla and Evelyn Fox Keller, 262–306. Durham, NC: Duke University Press, 2017.

Ragosta, Rosalba. *Napoli, città della seta: Produzione e mercato in età moderna*. Rome: Donzelli, 2009.

Rao, Anna Maria. "Eleonora de Fonseca Pimentel, le *Monitore Napoletano* et le problème de la participation politique." *Annales historiques de la Révolution francaise* 344 (2006): 179–91.

Rao, Anna Maria. "Enlightenment and Reform." In *Early Modern Italy, 1550–1796*, edited by John Marino, 229–52. Oxford: Oxford University Press, 2002.

Rao, Anna Maria. "Esercito e società a Napoli nelle riforme del secondo Settecento." *Studi Storici* 28 (1987): 623–77.

Rao, Anna Maria. *Esuli: L'emigrazione italiana in Francia, 1792–1802*. Naples: Guida, 1991.

Rao, Anna Maria. Introduction to Anna Maria Rao and Massimo Cattaneo, "L'Italia e la Rivoluzione Francese, 1789–1799." In *Bibliografia dell'età del Risorgimento, 1970–2001*, vol. 1, edited by Leo S. Olschki, 135–52. Florence: Olschki, 2003.

Rao, Anna Maria. *L'amaro della feudalità*. Naples: Guida, 1983.

Rao, Anna Maria. "La massoneria nel Regno di Napoli." In *Storia d'Italia: Annali*, vol. 21, *La Massoneria*, edited by G. M. Cazzaniga, 513–42. Turin: Einaudi, 2006.

Rao, Anna Maria. "La stampa francese a Napoli negli anni della Rivoluzione." *Mélanges de l'École française de Rome* 102 (1990): 469–520.

Rao, Anna Maria. "L'espace méditerranéen dans la pensée et les projets politiques des patriotes italiens: Matteo Galdi et la 'république du genre humain.'" In *Droit des gens et relations entre les peuples dans l'espace méditerranéen autour de la Révolution française*, edited by Marcel Dorigny and Rachida Tlili Sellaouti, 115–37. Paris: Société des études robespierristes, 2007.

Rao, Anna Maria. "Note sulla stampa periodica napoletana alla fine del '700." *Prospettive settanta* 10 (1988): 333–66.

Rao, Anna Maria. "Politica e scienza a Napoli tra Sette e Ottocento." In *Atti del bicentenario real museo mineralogico, 1801–2001*, edited by Maria Rosaria Ghiara and Carmela Petti, 16–35. Naples: Università degli Studi di Napoli, 2001.

Rao, Anna Maria. "Popular Societies in the Neapolitan Republic of 1799." *Journal of Modern Italian Studies* 4 (1999): 358–69.

Rao, Anna Maria. "Sociologia e politica del giacobinismo: Il caso napoletano." *Prospettive settanta* 2 (1979): 212–39.

Rao, Anna Maria. "The Feudal Question, Judicial Systems and the Enlightenment." In *Naples in the Eighteenth Century: The Birth and Death of a Nation State*, edited by Girolamo Imbruglia, 95–117. Cambridge: Cambridge University Press, 2000.

Rao, Anna Maria. "Unité et fédéralisme chez les jacobins italiens de 1794 à 1800." In *Les fédéralismes: Réalités et représentations, 1789–1874*, 381–90. Aix-en-Provence: Publications de l'Université de Provence, 1995.

Rao, Anna Maria, and Pasquale Villani. *Napoli 1799–1815: Dalla repubblica alla monarchia amministrativa*. Naples: Edizioni del sole, 1995.

Rascaglia, Mariolina. "Filosofia e scienza in età napoleonica: La lezione degli idéologues." In *Le scienze a Napoli tra Illuminismo e Restaurazione*, edited by Roberto Mazzola, 129–67. Rome: Aracne, 2011.

Ravera, Marco. *Introduzione al tradizionalismo francese*. Bari: Laterza, 1991.

Redondi, Pietro. "Cultura e scienza dall'illuminismo al positivismo." In *Storia d'Italia: Annali 3. Scienza e tecnica nella cultura e nella società dal Rinascimento ad oggi*, 677–811. Turin: Einaudi, 1980.

Reill, Peter. "Analogy, Comparison, and Active Living Forces: Late Enlightenment Responses to the Sceptical Critique of Causal Analysis." In *The Skeptical Tradition around 1800*, edited by Johan van der Zande and Richard Popkin, 203–12. Dordrecht: Kluwer, 1998.

Renda, Francesco. *La grande impresa: Domenico Caracciolo vicerè e primo ministro tra Palermo e Napoli*. Palermo: Sellerio, 2010.

Revel, Jacques. "L'histoire au ras du sol." In Giovanni Levi, *Le pouvoir au village*, i–xxxiii. Paris: Gallimard, 1989.

Rey, Roselyne. *Naissance et développement du vitalisme en France, de la deuxième moitié du 18e siècle à la fin du Premier Empire*. Oxford: Voltaire Foundation, 2000.

Richards, Joan. "Historical Mathematics in the French Eighteenth Century." *Isis* 97, no. 4 (2006): 700–13.

Richards, Joan. *Mathematical Visions: The Pursuit of Geometry in Victorian England*. New York: Academic Press, 1988.

Riet, Georges van. *L'epistémologie thomiste: Recherches sur le problème de la connaissance dans l'école thomiste contemporaine*. Louvain: Éditions de l'Institut Supérieur de Philosophie, 1946.

Riskin, Jessica. *Science in the Age of Sensibility: The Sentimental Empiricists of the French Enlightenment*. Chicago: University of Chicago Press, 2002.

[Ritis, Vincenzo de]. "Breve memoria filosofica sulla generazione e l'avanzamento delle scienze e delle arti." *Biblioteca analitica di scienze, letteratura e belle arti* 1 (1810): 1–35.

[Ritis, Vincenzo de]. "Programma." *Biblioteca analitica d'istruzione e di utilità pubblica* 1 (1812): 3–203.

Rivera, Carlo Afan de. *Circolari concernenti il servizio degl'ingegneri di acque e strade*. Naples: Stamperia reale, 1840.

Rivera, Carlo Afan de. *Considerazioni sui mezzi da restituire il valore proprio ai doni che la natura ha largamente conceduto al Regno delle Due Sicilie*. 2 vols. Naples: Fibreno, 1832.

Rivera, Carlo Afan de. *Considerazioni sul progetto di prosciugare il lago Fucino e di congiungere il mare Tirreno all'Adriatico per mezzo di un canale navigabile*. Naples: Reale tipografia della guerra, 1823.

Rivera, Carlo Afan de. *Del bonificamento del lago Salpi coordinato a quello della pianura della Capitanata: Delle opera eseguite e dei vantaggi ottenuti; Dell'applicazione del metodo stesso al bonificamento del bacino inferiore del Volturno*. Naples: Fibreno, 1845.

Rivera, Carlo Afan de. *Della restituzione del nostro sistema di misure, pesi e monete alla sua antica perfezione*. Naples: Fibreno, 1838.

Rivera, Carlo Afan de. *Tavole di riduzione dei pesi e delle misure delle Due Sicilie, in quelle statuite dalla legge de' 6 aprile del 1840*. Naples: Fibreno, 1840.

Roberts, Lissa. "The Senses in Philosophy and Science: Blindness and Insight." In *A Cultural History of the Senses in the Age of Enlightenment*, edited by Anne Vila, 109–32. London: Bloomsbury, 2014.

Robertson, John. "Enlightenment and Revolution: Naples 1799." *Transactions of the Royal Historical Society* 10 (2000): 17–44.

Robertson, John. *The Case for the Enlightenment: Scotland and Naples 1680–1760*. Cambridge: Cambridge University Press, 2005.

Robin, Corey. *The Reactionary Mind: Conservatism from Edmund Burke to Sarah Palin*. Oxford: Oxford University Press, 2011.

Robison, John. *Proofs of a Conspiracy against All the Religions and Governments of Europe, Carried On in the Secret Meetings of Free Masons, Illuminati, and Reading Societies*. Edinburgh: Creech, 1797.

Rohkrämer, Thomas. *Eine andere Moderne? Zivilisationskritik, Natur und Technik in Deutschland 1880–1933*. Paderborn: Ferdinand Schöningh, 1999.

Romagnoli, Giovanni. "La scoperta degli acquedotti del Cotugno e le prime reazioni del mondo scientifico." *Pagine di storia della medicina* 12 (1968): 78–88.

Romagnosi, Gian Domenico. "Questioni sull'ordinamento delle statistiche." *Annali universali di statistica* 14 (1827): 281–94; 15 (1828): 113–31; 16 (1828): 170–91; 17 (1828): 3–15; 25 (1830): 131–202.

Romano, Antonella. *La contre-réforme mathématique: Constitution et diffusion d'une culture mathématique jésuite à la Renaissance (1540–1640)*. Rome: École Française de Rome, 1999.

Romani, Roberto. "Un popolo da disciplinare: l'economia politica di Melchiorre Gioia come sapere amministrativo." In *Melchiorre Gioia (1767–1829): Politica, società, economia tra riforme e Restaurazione*, edited by Carlo Capra, 303–29. Piacenza: Cassa di Risparmio, 1990.

Romano, Paolo. "Un antagonista del Galluppi: Ottavio Colecchi." *Archivio storico per la Calabria e la Lucania* 13 (1943): 157–70.

Rosa, Mario. *Settecento religioso: Politica della ragione e religione del cuore*. Venice: Marsilio, 1999.

Rosenfeld, Sophia. *A Revolution in Language: The Problem of Signs in Late 18th-Century France.* Stanford, CA: Stanford University Press, 2001.

Rosenfeld, Sophia. *Common Sense: A Political History.* Cambridge, MA: Harvard University Press, 2011.

Rosenfeld, Sophia. "The Social Life of the Senses: A New Approach to 18th-Century Politics and Public Life." In *A Cultural History of the Senses in the Age of Enlightenment,* edited by Anna Vila, 21–40. London: Bloomsbury, 2014.

Rosental, Claude. *Weaving Self-Evidence: A Sociology of Logic.* Princeton, NJ: Princeton University Press, 2008.

Rossi, Giovanni Camillo. *La dottrina di Gesù Cristo sulla chiesa, sulla grazia e sulla sovranità, difesa contra gli attentati della teologia del tempo, sulle testimonianze specialmente delle chiese di Francia.* 2 vols. Naples: Verriento, 1794–1795.

Rowe, David. "New Trends and Old Images." In *Vita Mathematica: Historical Research and Integration with Teaching,* edited by Ronald Calinger, 3–16. Washington, DC: MAA, 1996.

Rowe, David. Review of Jeremy Gray, *Plato's Ghost. Bulletin of the American Mathematical Society* 50 (2013): 513–21.

Ruffini, Paolo. *Teoria generale delle equazioni, in cui si dimostra impossibile la soluzione algebraica delle equazioni generali di grado superior al quarto.* 2 vols. Bologna: Stamperia di S. Tommaso d'Aquino, 1799.

Ruffini, Paolo. *Della immaterialità dell'anima.* Modena: Soliani, 1806.

Ruffini, Paolo. *Opere matematiche.* 3 vols. Rome: Cremonese, 1953–1954.

Ruffini, Paolo. *Riflessioni critiche sopra il saggio filosofico intorno alle probabilità del sig. conte Laplace.* Modena: Società tipografica, 1821.

Russo, Giuseppe. *La scuola d'ingegneria di Napoli: 1811–1967.* Naples: Istituto Editoriale del Mezzogiorno, 1967.

Sabatini, Gaetano. *Ottavio Colecchi: Nuove notizie e nuovi documenti.* Rome: Tipografia delle Mantellate, 1929.

Sætnan, Ann Rudinow, Heidi Mork Lomell, and Svein Hammerm, eds. *The Mutual Construction of Statistics and Society.* London: Routledge, 2011.

Saggio di un corso di matematiche per uso della Real Scuola Politecnica e Militare. 12 vols. Naples: Sangiacomo, 1813–1815.

Saitta, Armando. *Alle origini del Risorgimento: I testi di un "celebre" concorso (1796).* 3 vols. Rome: Istituto storico italiano per l'età moderna e contemporanea, 1964.

Saladini, Girolamo. *Compendio d'analisi.* 2 vols. Bologna: S. Tommaso d'Aquino, 1775.

Salfi, Francesco Saverio. *Saggio di fenomeni antropologici relativi al tremuoto, ovvero riflessioni sopra alcune oppinioni pregiudiziali alla pubblica o privata felicità fatte per occasion de' tremuoti avvenuti nelle Calabrie l'anno 1783 e seguenti.* Naples: Flauto, 1787.

Salis Marschlins, Carlo Ulisse de. *Nel Regno di Napoli: Viaggi attraverso varie provincie nel 1789* [1793]. Lecce: Congedo, 1979.

Salvatorelli, Luigi. *Chiesa e stato dalla Rivoluzione francese ad oggi.* Florence: Le Monnier, 1955.

Salvatorelli, Luigi. *Il pensiero politico italiano dal 1700 al 1870.* Turin: Einaudi, 1959.

Salvatori, Massimo. *Il mito del buongoverno: La questione meridionale da Cavour a Gramsci.* Turin: Einaudi, 1963.

Salvemini, Biagio. *Economia politica e arretratezza meridionale nell'età del Risorgimento: Luca de Samuele Cagnazzi e la diffusione dello smithianesimo nel Regno di Napoli.* Lecce: Milella, 1981.

Salvemini, Biagio. *L'innovazione precaria: Spazio, mercati e società nel Mezzogiorno tra Sette e Ottocento*. Rome: Donzelli, 1995.

Sangro, Giuseppe. "Nuova soluzione del noto problema sul cilindroide wallisiano." *ARAS* 1, 1819: 83–96.

Sarnelli, Gennaro Maria. *L'anima desolata, confortata a patire cristianamente colla considerazione delle massime eterne, operetta utilissima per le persone tribolate che attendono all'esercizio dell'orazione, ed al cammino della perfezione: Aggiuntavi una lettera della B. Vittoria sopra l'amore della croce* [1740]. Bassano: Remondini, 1838.

Scarpato, Giovanni. "Le fatiche di un giornalista reazionario: Gioacchino Ventura e l'*Enciclopedia Ecclesiastica* (1821–1822)." *Studi Storici* 58 (2017): 605–43.

Schaffer, Simon. "Enlightened Automata." In *The Sciences in Enlightened Europe*, edited by William Clark, Jan Golinski, and Simon Schaffer, 126–65. Chicago: University of Chicago Press, 1999.

Schaffer, Simon. "Genius in Romantic Natural Philosophy." In *Romanticism and the Sciences*, edited by Andrew Cunningham and Nicholas Jardine, 82–98. Cambridge: Cambridge University Press, 1990.

Schaffer, Simon. "Machine Philosophy: Demonstration Devices in Georgian Mechanics." *Osiris* 9 (1994): 157–82.

Schaffer, Simon. "Newtonianism." In *Companion to the History of Modern Science*, edited by Robert Olby, Geoffrey Cantor, John Christie, and Jonathan Hodge, 610–26. London: Routledge, 1990.

Schaffer, Simon. "Newton on the Beach: The Information Order of *Principia Mathematica*." *History of Science* 47, no. 3 (September 2009):243–76.

Schettini, Glauco. "'La fucina dello spirito pubblico': L'organizzazione dei circoli costituzionali nella prima cisalpina (1797–1799)." *Società e storia* 150 (2015): 689–719.

Schettini, Glauco. "'Niente di più bello ha prodotto la rivoluzione': La Teofilantropia nell'Italia del triennio (1796–1799)." *Rivista di storia e letteratura religiosa* 50 (2014): 379–433.

Schettino, Edvige. "Franklinists in Naples in the 18th Century." In *Proceedings of the XX Congresso Nazionale della Società Italiana degli Storici della Fisica e dell'Astronomia*, 347–52. [Naples]: [CUEN], 2001.

Schettino, Edvige. "L'insegnamento della fisica sperimentale a Napoli nella seconda metà del Settecento." *Studi settecenteschi* 18 (1999): 367–76.

Schipa, Michelangelo. "Il secolo decimottavo." In Francesco Torraca et al., *Storia della Università di Napoli*, 433–66. Naples: Ricciardi, 1924.

Schipa, Michelangelo. *Un ministro napoletano del secolo XVIII (Domenico Caracciolo)*. Naples: Pierro, 1897.

Schmidt, James. "Inventing the Enlightenment: Anti-Jacobins, British Hegelians and the *Oxford English Dictionary*." *Journal of the History of Ideas* 64 (2003): 421–43.

Schneider, Jane, ed. *Italy's 'Southern Question': Orientalism in One Country*. New York: Berg, 1998.

Schroeder, Paul. *Transformation of European Politics, 1763–1848*. Oxford: Clarendon Press, 1994.

Scirocco, Alfonso. "I corpi rappresentativi nel Mezzogiorno dal *decennio* alla Restaurazione: Il personale dei consigli provinciali." *Quaderni Storici* 37 (1978): 102–25.

Scorza, Giuseppe. *Divinazione della geometria analitica degli antichi ovvero del metodo usato dalle greche scuole nella risoluzione de' problemi*. Naples: 1823.

Scorza, Giuseppe. *Euclide vendicato, ovvero gli* Elementi *di Euclide illustrati ed alla loro integrità ridotti*. Naples: Stamperia reale, 1828.

Scotti, Angelo Antonio. *Elogio storico del cavalier D. Domenico Cotugno*. Naples: Stamperia reale, 1823.

Scotti, Angelo Antonio. *Iosephi Xaverii Polii Elogium*. Naples: Stamperia reale, 1825.

Scotti Galletta, Bernardo. *Osservazioni critiche su la scuola sintetica napoletana*. Naples: Ariosto, 1843.

Semmola, Mariano. *Istituzioni di filosofia* [1797]. 2 vols. Naples: Sangiacomo, 1811.

Sereni, Emilio. *Storia del paesaggio agrario italiano*. Bari: Laterza, 1961.

Shank, J. B. *The Newton Wars and the Beginning of the French Enlightenment*. Chicago: University of Chicago Press, 2008.

Shapin, Steven. *A Social History of Truth: Civility and Science in Seventeenth-Century England*. Chicago: University of Chicago Press, 1994.

Shapin, Steven. "History of Science and Its Sociological Reconstructions." *History of Science* 20 (1982): 157–211.

Shapin, Steven. "Of Gods and Kings: Natural Philosophy and Politics in the Leibniz-Clarke Disputes." *Isis* 72 (1981): 187–215.

Shapin, Steven. *Scientific Life: A Moral History of a Late Modern Vocation*. Chicago: University of Chicago Press, 2008.

Shapin, Steven. "Social Uses of Science." In *The Ferment of Knowledge: Studies in the Historiography of Eighteenth-Century Science*, edited by G. S. Rousseau and Roy Porter, 93–139. Cambridge: Cambridge University Press, 1980.

Shaw, Matthew. *Time and the French Revolution: The Republican Calendar 1789—Year XIV*. Woodbridge: Boydell and Brewer, 2011.

Sheehan, Jonathan, and Dror Wahrman. *Invisible Hands: Self-Organization and the Eighteenth Century*. Chicago: University of Chicago Press, 2015.

Silvestrini, Maria Teresa. "Free Trade, Feudal Remnants and International Equilibrium in Gaetano Filangieri's *Science of Legislation*." *History of European Ideas* 32 (2006): 502–24.

Simioni, Attilio. *Le origini del Risorgimento politico nell'Italia meridionale*. 2 vols. Messina: Principato, 1925–1930.

Skinner, Gillian. *Sensibility and Economics in the Novel, 1740–1800: The Price of a Tear*. London: Macmillan, 1999.

Sofia, Francesca. *Scienza per l'amministrazione statistica e pubblici apparati tra età rivoluzionaria e restaurazione*. Roma: Carucci, 1988.

Solkin, David. *Richard Wilson: The Landscape of Reaction*. London: Tate Gallery, 1982.

Sombart, Werner. *Sozialismus und soziale Bewegung*. Jena: Fischer, 1919.

Soriga, Renato. *Le società segrete, l'emigrazione politica e i primi moti per l'indipendenza*. Modena: Società tipografica modenese, 1942.

Spagnoletti, Angelantonio. *Storia del Regno delle Due Sicilie*. Bologna: Il Mulino, 1997.

Spaventa, Silvio. *Dal 1848 al 1861: Lettere scritti documenti*. Naples: Morano, 1898.

Spedalieri, Nicola. *De' diritti dell'uomo, libri VI, nei quali si dimostra che la più sicura custodia de' medesimi nella società civile è la religione cristiana, e che però l'unico progetto utile alle presenti circostanze è di far rifiorire essa religione*. Assisi [Rome]: Salvioni, 1791.

Spini, Giorgio. Review of Delio Cantimori, *Giacobini italiani, vol. 1*. *Rassegna storica del Risorgimento* 43 (1956): 792–95.

Spinosa, Nicola. *La Collezione Angelo Astarita al Museo di Capodimonte: Giacinto Gigante e la scuola di Posillipo*. Introduction by Raffaello Causa. Naples: Stamperia napoletana, 1972.

Spinosa, Nicola. *Pittura napoletana del Settecento*. 2 vols. Naples: Electa, 1986–1987.

Spinosa, Nicola. *Vedute napoletane del Settecento.* Naples: Electa, 1989.

Stapelbroek, Koen. "Preserving the Neapolitan State: Antonio Genovesi and Ferdinando Galiani on Commercial Society and Planning Economic Growth." *History of European Ideas* 32 (2006): 406–29.

Starobinski, Jean. *Action and Reaction: The Life and Adventures of a Couple.* New York: Zone Books, 2003.

Staum, Martin. *Cabanis: Enlightenment and Medical Philosophy in the French Revolution.* Princeton, NJ: Princeton University Press, 1980.

Stella, Pietro. *Il prete piemontese dell'Ottocento tra la Rivoluzione francese e la Rivoluzione industriale.* Turin: Fondazione Agnelli, 1973.

Sternell, Zeev. *The Anti-Enlightenment Tradition as a Common Framework of Fascism and the Contemporary Far Right.* New Haven, CT: Yale University Press, 2010.

Stevens, Peter. *The Development of Biological Systematics: Antoine-Laurent de Jussieu, Nature, and the Natural System.* New York: Columbia University Press, 1994.

Stone, Harold Samuel. *Vico's Cultural History: The Production and Transmission of Ideas in Naples, 1685–1750.* Leiden: Brill, 1997.

Storchenau, Sigismund. *Institutiones logicae* [1769]. Naples: Marotta, 1816.

Storchenau, Sigismund. *Institutiones metaphysicae* [1771]. 4 vols. Naples: Marotta, 1816.

Sutherland, Donald. *The Chouans: The Social Origins of Popular Counter-Revolution in Upper Brittany, 1770–1796.* Oxford: Clarendon Press, 1982.

Taddei, Ferdinando, ed. *I consulti medici di Paolo Ruffini: Il carteggio Pasquali-Ruffini.* Modena: STEM Mucchi, 2016.

Tamburini, Pietro. *Lettere teologico-politiche sulla presente situazione delle cose ecclesiastiche.* 2 vols. Pavia: Comini, 1794.

Taparelli d'Azeglio, Luigi. "Di due filosofie." *Civiltà Cattolica* 1 (1853): 484.

Tataranni, Onofrio. *Catechismo nazionale pe'l cittadino.* Naples: 1799.

Tataranni, Onofrio. *Ragionamento sulle sovrane leggi della nascente popolazione di S. Leucio.* Naples: 1789.

Tataranni, Onofrio. *Saggio d'un filosofo politico amico dell'uomo* [. . .]. 5 vols. Naples: Di Bisogno, 1784–1788.

[Telesio, Luigi]. *Appendicetta all'elogio di Nicola Fergola.* Naples: 1836.

[Telesio, Luigi]. *Elogio di Niccolò Fergola scritto da un suo discepolo.* Naples: Trani, 1830.

Tëmkin, Ilya. "The Art and Science of *Testacea utriusque Siciliae,* by Giuseppe Saverio Poli." In *Atti del bicentenario del Museo Zoologico, 1813–2013,* edited by Maria Carmela del Re, Rosanna del Monte, and Maria Rosaria Ghiara, 147–68. Naples: CMSNF, 2015.

Terrall, Mary. *The Man who Flattened the Earth: Maupertuis and the Sciences in the Enlightenment.* Chicago: University of Chicago Press, 2002.

Tessitore, Fulvio. *Da Cuoco a De Sanctis: Studi sulla filosofia napoletana del primo Ottocento.* Naples: Edizioni Scientifiche Italiane, 1988.

Themelly, Mario, and Vito Lo Curto. *Gli scrittori cattolici dalla Restaurazione all'Unità.* Bari: Laterza, 1984.

Thévenot, Laurent. "La politique des statistiques: Les origines sociales des enquêtes de mobilité sociale." *Annales E.S.C.* 45 (1990): 1275–300.

Thoenes, Christof. "Das unendliche Leben dieser unvergleichlichen Stadt: Arte e natura nelle piante e vedute di Napoli." In *Nature and the Arts in Early Modern Naples,* edited by Frank Fehrenbach and Joris van Gastel, 45–58. Berlin: De Gruyter, 2020.

Timmermans, Stefan, and Steven Epstein. "A World of Standards but Not a Standard World: Toward a Sociology of Standards and Standardization." *Annual Review of Sociology* 36 (2010): 69–89.

Tocci, Giovanni. *Le comunità in età moderna: Problemi storiografici e prospettive di ricerca.* Rome: Carocci, 1997.

Tocci, Giovanni. *Le comunità negli stati italiani di antico regime.* Bologna: Clueb, 1989.

Tomasi di Lampedusa, Giuseppe. *The Leopard.* New York: Pantheon, 2007.

Tommasi, Donato. *Elogio storico del Cavaliere Gaetano Filangieri.* Naples: Raimondi, 1788.

Tommaso, Michele de, and Ascanio Orsi. *Catechismo su i dritti dell'uomo / Cathéchisme sur le droits de l'homme.* Fort-d'Ercule: Straforelli, [1794].

Tonelli, Giorgio. "The Law of Continuity in the Eighteenth Century." *Studies on Voltaire and the Eighteenth Century* 27 (1963): 1619–38.

Tonelli, Giorgio. "The Weakness of Reason in the Age of Enlightenment." In *Scepticism in the Enlightenment*, edited by Richard Popkin, Ezequiel de Olaso, and Giorgio Tonelli, 35–50. Dordrecht: Kluwer, 1997.

Torino, Marielva. "The Dismantling of Giuseppe Saverio Poli Collections and the *damnatio memoriae* of His Scientific Heir." In *Atti del XXXVI Convegno annuale della Società Italiana degli Storici della Fisica e dell'Astronomia*, edited by Salvatore Esposito, 81–86. Pavia: Pavia University Press, 2017.

Torraca, Francesco, et al. *Storia della Università di Napoli.* Naples: Ricciardi; 1924.

Tortarolo, Edoardo. *L'Illuminismo: Ragioni e dubbi della modernità.* Rome: Carocci, 2020.

Tortora, Alfonso. *L'eruzione vesuviana del 1631: Una storia d'età moderna.* Rome: Carocci, 2014.

Tortora, Alfonso, Domenico Cassano, and Sean Cocco, eds. *L'Europa moderna e l'antico Vesuvio.* Battipaglia: Laveglia and Carlone, 2017.

Toscano, Maria. "Conoscenza scientifica e pratica artistica: La presenza dell'arte nelle collezioni scientifiche e l'interesse degli artisti per la scienza." In *Nature and the Arts in Early Modern Naples*, edited by Frank Fehrenbach and Joris van Gastel, 223–38. Berlin: De Gruyter, 2020.

Toscano, Maria. "Giuseppe Saverio Poli as a Collector between Natural History and Antiquarianism." In *Atti del XXXVI Convegno annuale della Società Italiana degli Storici della Fisica e dell'Astronomia*, edited by Salvatore Esposito, 87–103. Pavia: Pavia University Press, 2017.

Toscano, Maria. "Il Museo Poliano e l'interesse per la zoologia a Napoli tra Sette e Ottocento." In *Atti del bicentenario del Museo Zoologico, 1813–2013*, edited by Maria Carmela del Re, Rosanna del Monte, and Maria Rosaria Ghiara, 169–185. Naples: CMSNF, 2015.

Traité des trois imposteurs. Milan: Raffaele Netti, [1797 or 1798]. After a 1796 *robespierriste* edition, published in Paris by Claude Mercier de Compiegne.

Trampus, Antonio, ed. *Diritti e costituzione: L'opera di Gaetano Filangieri e la sua fama europea.* Bologna: Il Mulino, 2005.

Trampus, Antonio. "La genesi e la circolazione della *Scienza della legislazione*. Saggio bibliografico." *Rivista storica italiana* 117 (2005): 309–59.

Trampus, Antonio. *Storia del costituzionalismo italiano nell'età dei Lumi.* Bari: Laterza, 2009.

Tresch, John. *The Romantic Machine: Utopian Science and Technology after Napoleon.* Chicago: University of Chicago Press, 2012.

Troisi, Tommaso. *Istituzioni metafisiche* [1796]. 3 vols. Naples: Di Napoli, 1829.

Trombetta, Vincenzo. *L'editoria a Napoli nel decennio francese: Produzione libraria e stampa periodica tra stato e imprenditoria privata (1806–1815).* Milan: Franco Angeli, 2011.

Trudi, Nicola. *Teoria de' determinanti e loro applicazioni.* Naples: Pellerano, 1862.

Tucci, Francesco. *Il problema del cerchio e de' tre punti risoluto con nuovo metodo analitico ed esteso alle rimanenti curve coniche.* Naples: Stamperia francese, 1818.

Tucci, Francesco. "Soluzione di alcuni problemi relativi alle curve coniche ed alle superficie generate dal rivolgimento di esse intorno a' loro assi primarii, eseguita coll'analisi degli antichi geometri." *Atti della Società Pontaniana* 3 (1819): 131–48.

Tucci, Francesco. "Soluzione di un problema creduto da Lagrange difficilissimo a trattarsi colla geometria." *Biblioteca analitica d'istruzione e utilità pubblica* 2 (1812): 321–26.

Uberti, Vincenzo degli. *Esercitazioni geometriche.* Naples: 1827.

Vaccarino, Giorgio. *I giacobini piemontesi (1794–1814).* 2 vols. Rome: Ministero per i beni ambientali e culturali, 1989.

Vaccarino, Giorgio. *I patrioti "anarchistes" e l'idea dell'unità italiana (1796–1799).* Turin: Einaudi, 1955.

Vaccarino, Giorgio. "L'inchiesta del 1799 sui giacobini in Piemonte." *Rivista storica italiana* 77 (1965): 27–73.

Valerio, Vladimiro. *Costruttori di immagini: Disegnatori, incisori e litografi nell'Officio Topografico di Napoli, 1781–1879.* Naples: Paparo Edizioni, 2002.

Valerio, Vladimiro. "Earth Sciences in the Correspondence of Ferdinando Visconti." *Annals of Geophysics* 52 (2009): 657–66.

Valerio, Vladimiro. "L'occhio mutevole: Militari e mappe tra rivoluzione e restaurazione." In *La cartografia europea tra primo Rinascimento e fine dell'Illuminismo,* edited by Diogo Ramada Curto, Angelo Cattaneo, and André Ferrand Almeida, 229–44. Florence: Olschki, 2003.

Valerio, Vladimiro. *Società, uomini, e istituzioni cartografiche nel Mezzogiorno d'Italia.* Florence: Istituto Geografico Militare, 1993.

Valsecchi, Antonino. *La religion vincitrice.* 2 vols. Genoa: Repetto, 1776.

Valson, Claude Alphonse. *La vie et les travaux du Baron Cauchy: Impression augmentée d'une introduction.* Paris: Blanchard, [1868] 1970.

Vecce, Carlo. "Maestra natura: La rappresentazione della natura nell'*Arcadia* di Sannazaro." In *Nature and the Arts in Early Modern Naples,* edited by Frank Fehrenbach and Joris van Gastel, 1–6. Berlin: De Gruyter, 2020.

Ventura, Gioacchino. *De methodo philosophandi.* Rome: Perego Salvioni, 1828.

Ventura, Gioacchino. *Elogi funebri e lettere necrologiche.* Genoa: Rossi, 1852.

Ventura, Gioacchino. *Opere.* 10 vols. Naples: Sarracino, 1856–1864.

Ventura, Gioacchino. *Osservazioni sulle opinioni filosofiche dei Sig. de Bonald, de Maistre, de La Mennais e Laurentie all'occasione di un articolo del giornale francese "Il Corrispondente."* Rome: Perego-Salvioni, 1829.

Venturi, Franco. "1764: Napoli nell'anno della fame." *Rivista storica italiana* 81 (1973): 394–472.

Venturi, Franco, ed. *Illuministi italiani.* Vol. 5: *Riformatori napoletani.* Milan: Ricciardi, 1962.

Venturi, Franco. *Italy and the Enlightenment: Studies in a Cosmopolitan Century.* New York: New York University Press, 1972.

Venturi, Franco. "La circolazione delle idee." *Rassegna storica del Risorgimento* 2–3 (1954): 203–22.

Venturi, Franco. "Napoli capitale nel pensiero dei riformatori illuministi." In *Storia di Napoli,* vol. 8, edited by Ernesto Pontieri, 3–73. Naples: Società Editrice Storia di Napoli, 1967–1978.

Venturi, Franco. *Settecento riformatore.* 5 vols. Turin: Einaudi, 1969–1990.

Venturi, Franco. *Utopia and Reform in the Enlightenment.* Cambridge: Cambridge University Press, 1971.

Verucci, Guido. "Chiesa e società nell'Italia della Restaurazione." *Rivista di storia della Chiesa in Italia* 30 (1976): 25–72.

Verucci, Guido. "Per una storia del cattolicesimo intransigente in Italia dal 1815 al 1848." *Rassegna storica Toscana* 4 (1958): 251–85.

Vila, Anne. *Enlightenment and Pathology: Sensibility in the Literature and Medicine of Eighteenth-Century France*. Baltimore, MD: Johns Hopkins University Press, 1998.

Vila, Anne. "Introduction: Powers, Pleasures and Perils of the Senses in the Enlightenment Era." In *A Cultural History of the Senses in the Age of Enlightenment*, edited by Anne Vila, 1–20. London: Bloomsbury, 2014.

Villari, Pasquale. *La pittura moderna in Italia e in Francia*. Florence: Pellas, 1869.

Villari, Rosario. "Antonio Genovesi e la ricerca delle forze motrici dello sviluppo sociale." *Studi storici* 11 (1970): 26–52.

Vincent-Buffault, Anne. *The History of Tears: Sensibility and Sentimentality in France*. New York: St. Martin's Press, 1991.

Viola, Paolo. "Les intellectuels napolitains face à la défaite de 1799." In *Les résistances à la Révolution*, edited by François Lebrun and Roger Dupuy, 321–27. Paris: Imago, 1987.

Visconti, Ferdinando. *Carteggio (1818–1847)*. Edited by Vladimiro Valerio. Florence: Olschki, 1995.

Vitale, Silvio. *Il principe di Canosa e l'epistola contro Pietro Colletta*. Naples: Berisio, 1969.

[Voltaire, François-Marie Arouet de]. *Istoria dello stabilimento del cristianesimo, di celeberrimo autore inglese*. Translated and edited by Flaminio Massa. [Milan]: Capelli, 1797.

Volterra, Vito. "Le matematiche in Italia nella seconda metà del secolo XIX." In Vito Volterra, *Saggi scientifici*. Bologna: Zanichelli, [1908] 1990.

Vovelle, Michel. *Les Jacobins: De Robespierre à Chevènement*. Paris: La Découverte, 2001.

Vovelle, Michel. *The Revolution against the Church: From Reason to the Supreme Being*. Columbus: Ohio State University Press, 1991.

Warwick, Anderson. "Asia as a Method in Science and Technology Studies." *East Asian Science, Technology and Society: An International Journal* 6 (2012): 445–51.

Warwick, Andrew. *Masters of Theory: Cambridge and the Rise of Mathematical Physics*. Chicago: University of Chicago Press, 2003.

Whiters, Charles. *Placing the Enlightenment: Thinking Geographically about the Age of Reason*. Chicago: University of Chicago Press, 2007.

Willey, Basil. *The Eighteenth Century Background: Studies on the Idea of Nature in the Thought of the Period*. London: Chatto and Windus, 1950.

Wittgenstein, Ludwig. *Philosophical Investigations*. Oxford: Macmillan, 1953.

Wittgenstein, Ludwig. *Remarks on the Foundations of Mathematics*. Oxford: Macmillan, 1956.

Wolfe, David. "Sydenham and Locke on the Limits of Anatomy." *Bulletin of the History of Medicine* 45 (1961):193–200.

Woolf, Harry, ed. *The Analytic Spirit: Essays in the History of Science in Honor of Henry Guerlac*. Ithaca, NY: Cornell University Press, 1981.

Zaghi, Carlo. "Il giacobinismo e il regime napoleonico in Italia." In *La storia: i grandi problemi dal medioevo all'età contemporanea*, vol. 5: *L'età moderna*, edited by Nicola Tranfaglia and Massimo Firpo, 735–93. Turin: Utet, 1986.

Zambelli, Paola. *La formazione filosofica di Antonio Genovesi*. Naples: Morano, 1972.

Zanoli, Paola, ed. *Giornale de' patrioti d'Italia*. 3 vols. Rome: Istituto storico per l'età moderna e contemporanea, 1988–1990.

Zazo, Alfredo. *Il giornalismo a Napoli nella prima metà del secolo XIX*. Naples: Procaccini, 1985.

Zazo, Alfredo. "La nomina del Galluppi a professore di Logica e Metafisica." *Logos* 8 (1925): 102–15.

Zazo, Alfredo. *L'istruzione pubblica e privata nel napoletano, 1767–1860*. Città di Castello: Il Solco, 1927.

Zazo, Alfredo. "L'ultimo periodo borbonico." In *Storia della Università di Napoli*. Bologna: Il Mulino, 1993.